Physiology of Salt Stress in Plants

Physiology of Salt Stress in Plants

Perception, Signalling, Omics and Tolerance Mechanism

Edited by

Pratibha Singh

Ranjan Plant Physiology and Biochemistry Laboratory, Department of Botany
University of Allahabad
Prayagraj, Uttar Pradesh, India

Madhulika Singh

Centre of Advanced Studies in Botany
Banaras Hindu University
Varanasi, Uttar Pradesh, India

Rajiv Kumar Singh

Horticultural Scientist
Krishi Vigyan Kendra, Sohaon
Ballia, Uttar Pradesh, India

Sheo Mohan Prasad

Ranjan Plant Physiology and Biochemistry Laboratory, Department of Botany
University of Allahabad
Prayagraj, Uttar Pradesh, India

WILEY

This edition first published 2022

Registered Office(s)
John Wiley & Sons, Inc., 111 River Street, Hoboken, NJ 07030, USA
John Wiley & Sons Ltd, The Atrium, Southern Gate, Chichester, West Sussex, PO19 8SQ, UK

Editorial Office
The Atrium, Southern Gate, Chichester, West Sussex, PO19 8SQ, UK

For details of our global editorial offices, customer services and more information about Wiley products, visit us at www.wiley.com.

Wiley also publishes its books in a variety of electronic formats and by print-on-demand. Some content that appears in standard print versions of this book may not be available in other formats.

Library of Congress Cataloging-in-Publication Data

Names: Singh, Pratibha (Plant physiology), editor.
Title: Physiology of salt stress in plants : perception, signalling, omics and tolerance mechanism / edited by Pratibha Singh, Ranjan Plant Physiology and Biochemistry Laboratory, Department of Botany, University of Allahabad, Prayagraj, Uttar Pradesh, INDIA, Madhulika Singh, Centre of Advanced Studies in Botany, Banaras Hindu University, Varanasi, Uttar Pradesh, INDIA, Rajiv Kumar Singh, Scientist Horticulture, Krishi Vigyan Kendra, Sohaon, Ballia 277504, Uttar Pradesh, INDIA, Sheo Mohan Prasad, Ranjan Plant Physiology and Biochemistry Laboratory, Department of Botany, University of Allahabad, Prayagraj, Uttar Pradesh, INDIA.
Description: First edition. | Hoboken, NJ : Wiley, 2022. | Includes bibliographical references and index.
Identifiers: LCCN 2021043242 (print) | LCCN 2021043243 (ebook) | ISBN 9781119700470 (cloth) | ISBN 9781119700487 (adobe pdf) | ISBN 9781119700494 (epub)
Subjects: LCSH: Crops–Effect of salt on. | Plants–Effect of salt on. | Crops–Effect of stress on. | Plants–Effect of stress on. | Soils, Salts in.
Classification: LCC QK753.S3 .P49 2022 (print) | LCC QK753.S3 (ebook) | DDC 581.7/88–dc23
LC record available at https://lccn.loc.gov/2021043242
LC ebook record available at https://lccn.loc.gov/2021043243

Cover Design by Wiley
Cover Images: © LutsenkoLarissa/Shutterstock, Oleksandr Yuchynskyi/ Shutterstock, JHK2303/Shutterstock

Set in 9.5/12.5pt STIXTwoText by Straive, Pondicherry, India
Printed and bound by CPI Group (UK) Ltd, Croydon, CR0 4YY

C9781119700470_210921

Contents

List of Contributors

Harish Chandra
Department of Botany and Microbiology
Gurukula Kangri Vishwavidayalaya
Haridwar, India

Atun Roy Choudhury
Department of Research and Development
Ramky Enviro Engineers Limited
Hyderabad, India

and

Department of Biological Sciences
Birla Institute of Technology and Science
Hyderabad, India

Anurag Dubey
Mycology and Plant pathology Laboratory,
School of Biological Sciences
Dr. Hari Singh Gour Central University
Sagar, India

Gunjan Dubey
Department of Botany
University of Rajasthan
Jaipur, Rajasthan, India

Abreeq Fatima
Ranjan Plant Physiology and Biochemistry
Laboratory, Department of Botany,
University of Allahabad
Prayagraj, India

Sankar Ganesh Palani
Department of Biological Sciences
Birla Institute of Technology and Science
Hyderabad, India

Gunjan Goyal
Department of Botany
University of Rajasthan
Jaipur,
Rajasthan, India

Ankush Gupta
Department of Biochemistry
Institute of Science
Banaras Hindu University
Varanasi, Uttar Pradesh
India

Arvind Gupta
Department of Biotechnology
Hemvati Nandan Bahuguna Garhwal
(Central) University
Srinagar
Uttarakhand, India

Ayushi Gupta
Institute of Environment and Sustainable
Development, Banaras Hindu University
Varanasi, India

Divya Gupta
Ranjan Plant Physiology and Biochemistry Laboratory, Department of Botany
University of Allahabad
Prayagraj, India

Shivam Gupta
Ganga River Ecology Research Laboratory, Environmental Sciences Division,
Centre of Advanced study in Botany, Institute of Sciences, Banaras Hindu University
Varanasi, India

Prateek Jain
Department of Biology
University of North Carolina at Chapel Hill
Chapel Hill, NC
USA

Durgesh Kumar Jaiswal
Institute of Environment and Sustainable Development, Banaras Hindu University
Varanasi, India

Lav Kumar Jaiswal
Department of Biochemistry
Institute of Science, Banaras Hindu University
Varanasi,
Uttar Pradesh, India

Aruba Khan
CSIR-National Botanical Research Institute
Lucknow, India

Ria Khare
CSIR-National Botanical Research Institute
Lucknow, India

and

Academy of Scientific and Innovative Research (AcSIR)
New Delhi, India

Abhay Kumar
Department of Biotechnology, ICAR-National Research Centre on Litchi
Muzaffarpur,
Bihar, India

Indrajeet Kumar
Department of Botany
Banaras Hindu University
Varanasi, India

Umesh Kumar
Department of Botany
Banaras Hindu University
Varanasi,
Uttar Pradesh, India

Pragati Kumari
Scientist Hostel-S-02, Chauras Campus,
Srinagar Garhwal, Uttarakhand
India

Neha Naaz
Ganga River Ecology Research Laboratory, Environmental Sciences Division
Centre of Advanced study in Botany, Institute of Sciences, Banaras Hindu University
Varanasi, India

Tanamyee Nayak
Department of Biochemistry
Institute of Science, Banaras Hindu University
Varanasi,
Uttar Pradesh, India

Jitendra Pandey
Ganga River Ecology Research Laboratory, Environmental Sciences Division
Centre of Advanced study in Botany, Institute of Sciences, Banaras Hindu University
Varanasi, India

Anuradha Patel
Ranjan Plant Physiology and Biochemistry Laboratory, Department of Botany
University of Allahabad
Prayagraj, India

Dilip Kumar Prajapati
Ganga River Ecology Research Laboratory, Environmental Sciences Division
Centre of Advanced study in Botany, Institute of Sciences, Banaras Hindu University
Varanasi, India

Sheo Mohan Prasad
Ranjan Plant Physiology and Biochemistry Laboratory, Department of Botany
University of Allahabad
Prayagraj, India

Gurpreet Sandhu
CSIR-National Botanical Research Institute
Lucknow, India

and

Academy of Scientific and Innovative Research (AcSIR)
New Delhi, India

Rajesh Kumar Sharma
Department of Botany
Banaras Hindu University
Varanasi, India

Garima Singh
Ranjan Plant Physiology and Biochemistry Laboratory, Department of Botany
University of Allahabad
Prayagraj, India

Madhulika Singh
Ganga River Ecology Research Laboratory, Environmental Sciences Division
Centre of Advanced study in Botany, Institute of Sciences, Banaras Hindu University
Varanasi, India

Neha Singh
Department of Research and Development
Ramky Enviro Engineers Limited
Hyderabad, India

Prabhakar Singh
Center of Advanced Study, Department of Botany
Institute of Science, Banaras Hindu University
Varanasi, India

Pratibha Singh
Ranjan Plant Physiology and Biochemistry Laboratory, Department of Botany
University of Allahabad
Prayagraj, India

Pratibha Singh
Department of Botany, School of Life Sciences
Mahatma Gandhi Central University
Motihari,
Bihar, India

Prince Kumar Singh
Department of Botany
Banaras Hindu University
Varanasi,
Uttar Pradesh, India

Rakesh Kumar Singh
Department of Biochemistry
Institute of Science
Banaras Hindu University
Varanasi, India

Shikha Singh
Ranjan Plant Physiology and Biochemistry
Laboratory, Department of Botany
University of Allahabad
Prayagraj, India

Sanjesh Tiwari
Ranjan Plant Physiology and
Biochemistry Laboratory
Department of Botany
University of Allahabad
Prayagraj, India

Santwana Tiwari
Ranjan Plant Physiology and
Biochemistry Laboratory
Department of Botany
University of Allahabad,
Prayagraj, India

Vivekanand Tiwari
Department of Fruit Tree Sciences,
Institute of Plant Sciences,
Agricultural Research Organization
(ARO)-Volcani Center
Rishon LeZion, Israel

Yashoda Nandan Tripathi
Center of Advance Study
Department of Botany
Institute of Science
Banaras Hindu University
Varanasi, Uttar Pradesh, India

Ram Sanmukh Upadhyay
Center of Advance Study
Department of Botany
Institute of Science
Banaras Hindu University
Varanasi, Uttar Pradesh, India

Nidhi Verma
Ranjan Plant Physiology and Biochemistry
Laboratory, Department of Botany
University of Allahabad
Prayagraj, India

Aruna Yadav
Department of Botany
University of Rajasthan
Jaipur,
Rajasthan, India

Poonam Yadav
Institute of Environment and
Sustainable Development,
Banaras Hindu University
Varanasi, India

Saurabh Yadav
Department of Biotechnology
Hemvati Nandan Bahuguna Garhwal
(Central) University
Srinagar,
Uttarakhand, India

Preface

This edited book entitled Physiology of Salt Stress in Plants: Perception, Signalling, Omics and Tolerance Mechanism is an important contribution to Plant Science containing information related to salt stress and its mitigation strategy by experimental techniques based on theoretical concepts. The salt-stress-related problems are rising in the soil and water due to natural and anthropogenic activities. Anthropogenic activities include repeated irrigation through canal system and heavy crop production practices which has led to enhanced salt level in crop/agricultural field, hence substantially declining the crop productivity. Therefore, study on salt toxicity is continued as an area of scientific interest in direction to understand their whole mechanism of its toxicity and their entry into crop plants.

In this book, the authors explain a number of approaches to ease the negative impact of salt stress in crop plants. These approaches include nutrients, antioxidants, osmolytes, phytohormones and extra cellular compounds, etc. They are endo as well as exogenous in nature. In this book, the adverse impact of salt ion toxicity on plants and implication of advance approaches in alleviating salt toxicity have briefly been reviewed. This work enables the scientific world to design strategies for reducing NaCl-mediated loss to crop by the application of different endo and exogenous substances in the farm soils. The governments and other organizations may design a holistic approach to reduce NaCl and other salt toxicity by different types of practices. Agriculturalists may be enlightened with several awareness programmers by the government and non-government actions wherein the content of this book may be used. It is widely useful for all post-graduate courses in the biological sciences. The idea of this work has a wide-ranging scientific and socio-economic utility.

All the editors thankfully acknowledge the contributions from all the scholars working across the Indian subcontinent and across the world. An authoritative book written by an individual that remains relevant over the coming years is rather cumbersome and instead requires the concerted effort of a team of expert scientists. All editors also gratefully acknowledge the team at John Wiley & Sons Limited which made possible the proposed book in its present form.

Editors
Pratibha Singh
Madhulika Singh
Rajiv Kumar Singh
Sheo Mohan Prasad

1

An Introduction to Salt Stress Perception and Toxicity Level: Worldwide Report at a Glance

Atun Roy Choudhury[1,2], Neha Singh[1], Ayushi Gupta[3], and Sankar Ganesh Palani[2]

[1] *Department of Research and Development, Ramky Enviro Engineers Limited, Hyderabad, India*
[2] *Department of Biological Sciences, Birla Institute of Technology and Science, Hyderabad, India*
[3] *Institute of Environment and Sustainable Development, Banaras Hindu University, Varanasi, India*

1.1 Soil Salinity: An Introduction

The threat of global warming is not limited only to the earthen atmosphere but slowly progressing toward the lithosphere. Attenuation of vegetative health due to the assimilation of brine substances is referred to as saline toxicity. The destruction caused solely due to the elevated level of sodium chloride (NaCl) is characterized as sodicity and is rather attainable to reclaim. Contrarily, the assorted salt stress enforces salinity, which makes land barren(Gul et al. 2015). Since the Green Revolution, the salinity footprint is engulfing the terrestrial territory quite rampantly. Presently, around one-tenth of the earth's crust accounting nearly 46 M ha has turned non-fertile (Hossain 2019). Excessive groundwater abstraction in the high zones may lead to premature desertification. The Indian Council of Agricultural Research (ICAR) predicted in a geospatial study that the coastal districts will be left with no aquifer water by 2050 without any technical interference (ICAR 2015). This further booms the inland intrusion of saline water. Figures are even more agitating, precisely in coastal regions. The rising sea levels often cause waterlogging in different parts of the world, precisely places located at lower elevations (EL-Raey et al. 1995). India, with a coastline length of approximately 7516 km, is presently under immense risk of temporary submergence. Saline water logging abnormally elevates the sodicity and thereby turns the lands nonproductive or unfertile.

The other reason that may trigger the soil sodicity is extreme groundwater abstraction. The negative aquifer pressure in the coastal regions causes brackish water intrusion and vertical rise by capillary action (Dillon et al. 2009). Whereas, the increase in soil salinity is a complex phenomenon. Studies showcased the discharge of industrial effluents into the water bodies successively raising the dissolved salt content resulting in increased salinity

Physiology of Salt stress in Plants: Perception, Signalling, Omics and Tolerance Mechanism, First Edition.
Edited by Pratibha Singh, Madhulika Singh, Rajiv Kumar Singh, and Sheo Mohan Prasad.

while utilized in irrigation. The other foremost reason for soil salinity involves drying out. Overutilization has already caused drying of a significant chunk of sweet water resources. In the absence of the desired water quality, farmers are moving toward alternate sources with high saline concentration, leading to the salinity of agricultural lands (Staniforth and Davies 2018).

The impact of salt stress is found to be most severe on agricultural crops. The primary issues involve the non-germination of seeds, reduced leaf surface area, retarded plant growth, strength, hampered yield, etc. Elevated soil salinity hampers the plants in various ways such as osmotic stress (OS), ionic toxicity, retarded cell division, reduced photosynthesis, to name a few. The inclusive impact of all the above factors boosts the mortality rate (Lauchli and Grattan 1970).

Immediate exposure to higher saline medium primarily increases the OS, causing reduced leaf surface area (i.e. due to repressed cell division and growth). Whereas, prolonged exposure imparts ionic stress leading to stomatal closure, immature senescence of mature leaves, chlorosis, necrosis, etc. The reduced biomass negatively affects photosynthesis and plant growth (Darko et al. 2019). In contrast, exposure to elevated sodicity, especially NaCl,affects the enzymatic system and augments cell swelling. The mutual impact leads to suppressed energy synthesis. Furthermore, excess exposure hinders all the growth-oriented processes like metabolism and protein synthesis (Acosta-Motos et al. 2017).

Therefore, prolonged exposure provoked the development of a defense mechanism in some species against salt stress and toxicity either by excluding through cells or by enhancing the salt tolerance. Additionally, synthetic species with transgenic properties are also synthesized by genetic engineering by altering the levels of gene expression (Carillo et al. 2011).

1.2 Salt Stress Perception and Current Scenario

Accumulation of excessive salt content in the soil causing direct and indirect adverse effects on flora and fauna is termed as salt stress (Shrivastava and Kumar 2015). The above situation can inhibit plant growth, and prolonged exposure may lead to a decrease. Higher saline level impacts the plants in various ways such as genotoxicity, alteration of metabolic processes, oxidative stress, water stress, ion toxicity, nutritional disorders, reduction of cell division and expansion, and membrane disorganization (Hasegawa et al. 2000; Munns 2002). The preliminary exposure to salt stress causes leaf surface area reduction. The immediate impacts include suppressed cell expansion and cell division and closure of stomata due to osmotic influence (Munns 2002; Flowers 2004). Furthermore, prolonged exposure imparts ionic stress leading to early senescence of mature leaves and thereby reducing the leaf surface area responsible for photosynthesis and plant growth.

The severity of salt stress is most predominant in the case of agricultural crops from a food security perspective; impacts include retarded seed germination, reduced biomass, and small yield. Higher abscisic acid (ABA) concentration results in the formation of specific genes through the plant defense mechanism which leads to counteracting its generation cause (Godoy et al. 1990; Lodeyro and Carrillo 2015). Generally, the acute level of salt toxicity causes instantaneous death in various species, whereas, in selected species, limited stress influences defense mechanisms mimicking halophytes. For instance, conversion of C_3 to CAM, amendment in epidermal bladder cell to withhold excessive NaCl enabling better survivability over the saline condition. Significant parts of the coastal irrigated areas face salination issues majorly due to the seawater intrusion. More than 45 M ha of cultivable land distributed among hundreds of countries covering more than 10% of the global land surface area have already been sacrificed due to saline irrigation. Additionally, approximately 1.5 M ha of fertile land becomes nonproductive every annum due to soil salinity (Munns and Tester 2008). Presently, about 1150 M ha of productive land are under induced stress, while80 M ha are only affected due to the anthropogenic activities (Rasool et al. 2013; Hossain 2019).

1.3 Types of Salt Stress

Based on the origin and root cause, there are two different categories of salinity, namely, primary and secondary. Primary salinity is a natural phenomenon and mostly occurs due to the former presence of salt lakes, slat clads, tidal swamp, etc., at a particular location. It is majorly a kind of sodicity. At the same time, secondary salinity is imposed due to man-made activities such as urbanization, saline irrigation, etc. (Shahid and Rahman 2011). Detailed reasons are delineated below.

Primary salinity:

i) Spreading from the saline artesian well.
ii) Capillary rise from saline groundwater.
iii) Seawater intrusion.
iv) Canopy formation due to the movement of fine sea sand by the sea breeze.
v) Waterlogging.

Secondary salinity:

i) Irrigation with impeded drainage
ii) Effluent discharge
iii) Excess fertilizer dosing
iv) Deforestation
v) Saline irrigation

Furthermore, based on the predominance of the type of anions present and the pH value, salt-affected soils are categorized as saline soil and sodic soil. Sodic soil typically comprises sodium carbonate and or bicarbonate ions with a pH value beyond 8.5, but contrarily, saline soil majorly incorporates chloride and sulphate ions with pH value below 8.5. Certain plant species manage to compensate the imparted stress through its metabolism and survive in the severe salt conditions known as halophytes. Remaining plant species are termed as glycophytes with a higher mortality rate overexposure to 10% or more concentration of saline water (Gorham 1995; Parida and Das 2005; Mane et al. 2011; Gupta and Huang 2014).

1.4 Origin of Problems

Primarily, hydro-geological activities contribute in escalating soil salinity and sodicity. Moreover, the soil is generated because of the weathering actions on intermediate and basic igneous rocks; sandstones already carry salt as a primary constituent. In the regions with moderate to low rainfall, a greater rate of evapotranspiration induces higher salinity and sodicity. Furthermore, coastal regions with tidal exposure may also develop salinity problems. A study conducted by Sultana et al. (2001) depicted that rice yield in coastal Asia gets often impaired due to the intrusion of saline Indian Ocean water. Inland precipitation also surprisingly elevates the soil sodicity. It is evidenced that rainwater can constitute up to a few milligrams of salt against each kilogram of a downpour with an electrical conductance (EC) value of 0.01 dS/m (Cucci et al. 2016; Corwin and Yemoto 2017; Hossain, 2019).

However, the deteriorating impacts of artificially induced salinity are more predominant. Over-irrigation or saline water irrigation is cited as one of the prima facie reason for human-induced salinity. Roughly, it is estimated that globally half of the irrigated lands are anyhow saltaffected. Other than irrigation, probable sources of inland salinity are the following:

i) Salt accumulation: Effluent and waste discharged into the surface water bodies from the industries and effluent treatment plants (ETPs) beyond absolute concentrations can accumulate and form salt films downstream to cause acute saline toxicity (Naidoo and Olaniran 2013).
ii) Reduction of greenbelt: Deforestation accelerates the salinization process by facilitating salt movement both through upper and lower soil layers. It further results in depleted annual precipitation and elevated soil temperature. Subsequent heating and cooling promote wear and tear, higher runoff, and substantial sedimentation to cause flooding and salt assimilation.
iii) Overdosing fertilizers: Post-Green Revolution, the usage of chemical fertilizers, herbicides, and pesticides has abruptly increased. Overdosing often results in underutilization and accumulation.
iv) Excessive grazing: Areas with scarce soil cover often suffer the root zone saline toxicity due to overgrazing. Surface waterlogging (i.e. either due to over-irrigation or riverbed sedimentation) in such areas can cause elevation of the water table and thereby facilitating salt migration from the deep aquifers.

1.5 Salt Toxicity Level: A Worldwide Report

Soil salinity and sodicity is a global issue faced by more than 115 nations with annual yield depletion of 7% or more (Yadav 2010). A total of 955 M ha of world surface area is either primarily or secondarily affected by salt pollution. Sodicity is predominant with impact over more than half of the land surface, e.g. Australia. Salinity issue dominates about 21% of comprehensive land footprint, especially arid regions of Asia and Pacific and areas with impeded drainage. Coming to India, the ambiguity of figures (salt-affected land) is quite a concern in the absence of liable evidence. The reported niche is found to be varying from

Table 1.1 Soil salinity/sodicity scenario in worst-affected partsof India.

State	Sodic soil (M ha)	Saline soil (M ha)	Total[a] (M ha)
Gujarat	0.54	1.7	2.24
Uttar Pradesh	1.35	0.02	1.37
Maharashtra	0.42	0.18	0.6
Rajasthan	0.18	0.195	0.375
Tamil Nadu	0.35	0.013	0.363

[a] 0.35 M ha: Threshold limit.
Source: Data from Mandal et al. (2018).

7 to 25 M ha (Rasool et al. 2013; Shahid et al. 2018; Isayenkov and Maathuis 2019). A location receiving low to moderate precipitation poses a tremendous threat to native agriculture.

The scenario is quite predominant in the southern region of India. These semiarid zones experience more than 300 sunny days per annum with high solar radiation, causing elevated evaporation rate and thereby moisture loss. The soil resource maps published by the National Bureau of Soil Survey and Land Use Planning, Nagpur (NBBS and LUP) were considered as the baseline data for the present study. The salinity of the soil was subcategorized into six basic classes depending on the EC of the saturated extract. The categories (based on severity) are as follows: very severe, severe, strong, moderately strong, moderate, and slight. The soil extracts portraying EC values between 200 and 400 mS/m were neglected for the above study. Furthermore, the sodicity of the soil was categorized into three major classes, namely, strong, moderate, and high. The above classification was done based on the presence of exchangeable sodium percentage (ESP), and the scale ranges from <5 to >15. The soil samples with sodicity <5 were considered as nonsodic, whereas the black soil samples with sodicity more than five were considered as alkali or sodic (Rasool et al. 2013; Shrivastava and Kumar 2015). The Rann of Kutch, Gujarat, an area with a footprint of 7500 sq. km mostly comprising salt marshes was marked as a separate entity by NBBS and LUP (Table 1.1).

As per the statistics published by the Food and Agriculture Organization (FAO) in 2007, about 770 000 sq. km of the global land surface area is already salt affected, and approximately 430 000 sq. km is under secondary salinity threat. The study further estimated that around one-third of the world's irrigated lands are either already affected by higher salinity or will be affected in the recent future.The global share distribution (in percentage) of saline soil across the countries is portrayed in Figure 1.1.

Multiple studies highlighted that abnormal abstraction of groundwater and waterlogging due to excessive irrigation often lead to desertification of fertile land. Therefore, urban local bodies (ULBs) and concerned government authorities should spread awareness among the farmers against salinization and over usage of water. Also, imposing water cess for misuse/overuse can be a productive measure to combat the typical tendency instantly, especially in the water-scarce areas.

The overall salinity profile of India was first estimated in 1966 under the accountability of the Ministry of Agriculture. The raising concern developed due to uncontrolled usage of

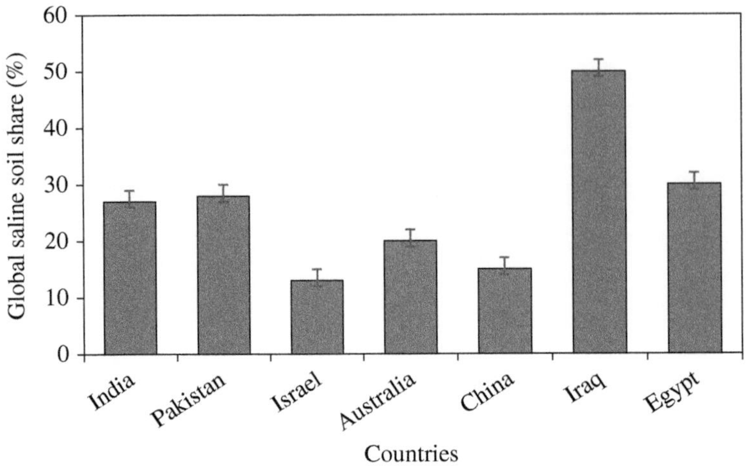

Figure 1.1 Worst-affected nations vs. salinity share. *Source:* Adopted from Rasool et al. (2013) and Mandal et al. (2018).

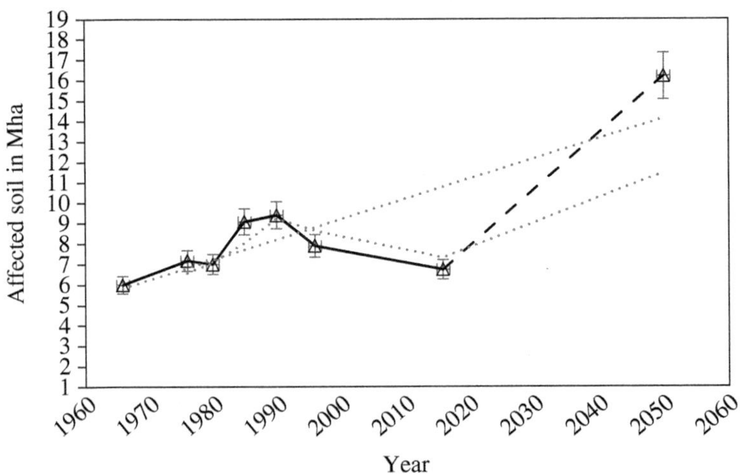

Figure 1.2 Salinity index and prediction for India since the Green Revolution. *Source:* Adopted from Sharma and Singh 2015.

the chemical fertilizer provoked the study. Furthermore, in the period of 1980–2015, a total saline footprint of approximately 2 Mha has been reclaimed under the effort of the ICAR. The retracted area presently yields around 16 MT of fodder grains per annum. The trajectory of the salinity footprint of the nation and foreseen values are portrayed in Figure 1.2.

The figure portrays the sudden escalation in salinity footprint (i.e. up to 16.2 Mha) between 2020 and 2050 as per the prediction made by the ICAR (ICAR 2015). The prediction is highly likely with no scientific intervention. Moreover, the usage of inferior irrigation water and overconsumption, leading to negative pressure, may further accelerate the salinization process.

1.6 Effect of Salt Stress on Flora and Fauna of the Ecosystem

Salt stress induces a diverse range of metabolic and growth-oriented detrimental changes in plants. Furthermore, protracted exposure can also inhibit crop yield. Primarily, saline exposure incurs OS, and it ultimately leads to ionic toxicities (Bano and Fatima 2009). Induced OS negatively impacts the root absorption capacity and accelerates the stomatal evaporation loss. Saline exposure elicits hyperosmotic pressure, which causes an adverse situation like the above. Initially, OS provokes several physicochemical amendments, which include membrane disruption and disfunctionality, disproportionate nutritive levels, retarded detoxification mechanism, and impaired photosynthesis rate (Munns and James 2003). Particularly, sodicity incites ionic stress by assimilating excess sodium and chloride ions into the plant tissue. Surplus accumulation of the above ions triggers ionic inequity leading to several growths related to detrimental changes. Elevated cell sodium ion concentration limits the required level of other essential plant nutrients such as potassium, thereby causing reduced yield and, ultimately, senescence (Ashraf 2004; Zhu 2007).

The inherent countermeasures also destructively impact plant biology. The generation of reactive oxygen species (ROS) can cause potential oxidation injury to the interstitial components like DNA structure, cell protein, and cell walls. The other sets of nondestructive defense mechanisms include elevation of photosynthetic rate, re-exercising ion and water relation in the vesicular system, etc.

The saline balance of the terrestrial water sources is globally affected due to continual anthropogenic activities, for example, new England Marshes (Williams 2002). Perennial inland sweet water sources are worst affected due to the waste charging throughout the trajectory. Whereas, the salinity of global waters is also sacrificed due to the melting of glaciers. Territorial brackish streams such as arid estuaries and salt-marshes are as well reasonably affected by the artificial agricultural discharge. But, this sudden change caters to a tremendous threat to the aquatic ecosystem. It incorporates abridged survivability, inhibited fertility, metabolic disorder, and retarded physiology (Velasco et al. 2019). Any varied salinity can mainly disrupt the osmotic balance between the surrounding hydrosphere and organismic, cellular fluid. The severity of impact varies between a minor metabolic malfunction and decease. To counteract the above, marine organisms develop an osmoregulation defense mechanism and eliminate hypo and hyperstress conditions. But, severe salinity or attenuation majorly conquers over the internal defense. Moreover, the consequences of successful defense are also majorly anonymous.

Additionally, investigations were also pursued to recognize the impact of ancillary factors such as temperature, but it seemed to be relatively insignificant. For instance, a study performed on zooplanktons reported metabolic issues upon exposure to altered salinity, but no apparent influence of temperature (Garreta-Lara et al. 2018). Basic metabolic disfunctionalities elicited due to OS are addressed by spontaneous excretion. But, stress ascertained due to bioaccumulation of metals is far more complex, so the defense is required. Moreover, each metal is highly specific in terms of threat enforced.

An advantageous fact is that the combined stress imposed by the salinity and metal toxicity gets tackled by a common mechanism in case of catadromous and anadromous fishes during the migration, in and out from the oceanic environment. For instance, a study reported the secretion of carbonic anhydrase in sheepshead minnow, an estuarine variety while exposed to the combined stress evolved due to amended salinity and copper toxicity (Velasco et al. 2019). Furthermore, OS plays a protagonist role against metal stress for invertebrates residing in saline waters by incurring the ion transport system. Lack of free radicals and ions in the freshwaters facilitate metal uptake in the inhabitants. Therefore, the vulnerability of the sweet water species against the metal toxication is indeed more severe (Halse et al. 1998; Velasco et al. 2019).

Dry farming also raises intricate contact interaction amid washed off pesticides and water salinity. These halogenated chemicals probably act as a neurotoxin and inhibit the counteract mechanisms of the nervous system. Remarkably, hypersalinity pivotally helps anadromous varieties such as Brown trout to subside the impacts of exposure efficaciously. The effective counter mechanism correlated with an interpretation of malfunctioning of the neural system upon instantaneous exposure to elevated salinity. Furthermore, collateral stress induced due to the combined effect of salinity and dehydration results in protraction of the osmoregulatory responses in some plants and marine bugs, thereby diminishing the moisture loss. It portrays the discordant individual stressors that abolish each other, while severe impacts were observed upon elementary exposure (Williams 1998; Kultz 2015; Cañedo-Argüelles et al. 2018).

1.7 Role in Sustainable Agriculture

The existence and survivability of the global population are mostly dependent on agriculture. It is estimated that about 99% or more consumable fodder sources are scattered across the lithosphere, whereas a hydrosphere contributes a negligible fraction of 0.5% or less. Thus, it is evident that a healthy and sufficient existence of earth crust is mandatory for the sustainable coexistence of the human being. Furthermore, soil erosion drastically impacts the agricultural yield. It is estimated that annually approximately 75 million tonnes of soil loss occurs only from the cultivable topographic regions worldwide. Other prominent effects of salinization include the erosion of the hilly terrains, which is probably less investigated (Aslam et al. 2017).

Saline soil mostly produces superficial seals due to two causes: (i) sodium pressure fragmentizes the soil structure and eliminates clay particles, resulting in clogging of interstitial voids and (ii) lean vegetative cover exposes the saline soil to precipitation compaction (Agassi et al. 1994, Singer and Lindquist 1998). Both the processes mainly decrease percolation and enhance surface runoff. Though the layer beneath gets safeguarded against vigorous erosion, the top layer gets severely imposed due to the disintegration caused by salinization (Agassi et al. 1994). Therefore, it is evident that soil salinity also can indirectly influence soil erosion up to a greater extent.

In this ever-raising context of fodder demand and versatile challenges, ensuring a hassle-free supply for the global population is a mammoth task. Amid eyeing for the alternate sources, existing challenges such as unavailability of the fertile land footprint, overconsumed natural resources, water and energy scarcity, and climate variance cannot be overlooked. Sustainability can only be achieved by compensating the need, not greed. Advanced issues need modern solutions, and indeed few are emerging as follows: reparation of sodicity with gypsum dosing, subsurface drainage of water-stagnant flood-planes, adaptation of agroforestry, and generating genetically engineered species and switching to them (ICAR 2015). The detailed pathway is delineated below:

i) The satellite-based remote-sensing approach with geographic information system (GIS) mapping and real-time ground truthing can provide an array of escalating salinity footprint (Singh et al. 2010).
ii) Gypsum-dosed alkali reparation techniques for soils affected with sodium toxicity.
iii) Reclamation of flooded wetlands through downward drainage– the method is quite useful in addressing multidimensional issues such as water stagnation and salinization.
iv) Chemical regeneration of saline soil with ameliorants is also practiced in some parts of the globe. The method is expensive and hence challenging to impose for more giant footprints.
v) Phytoremediation with salt-tolerant species is contrarily an inexpensive and eco-friendly mechanism.
vi) Multilayer agroforestry is a recent trend in the agricultural industry to mitigate rising demand. Anyhow, the method also assists in reclaiming saline soil by reducing the soil density and thereby causing an elevated percolation rate. Furthermore, the littered biomass improves soil fertility and yield (Kaur et al. 2000; Nosetto et al. 2007).
vii) Nonconventional techniques such as inland fishery have also gained limited popularity, majorly in the southern peninsula of the country. Flood-planes and wetlands near to the coastal regions are effectively serving as the source of alternate revenue generation.
viii) Microbialremediation: Desalination through microbial action is indeed rigorous. The inoculants are expensive and seek a suitable environment.

1.8 Unintended Effects of Salt-Containing Substance Application in Agricultural Land

Salinity intervenes with plant nutrition and growth by exerting osmotic and ionic stress. Higher salinity level in soil hinders water absorption ability, referred to as the osmotic effect. The utmost concern is when elevated concentration can deter biomass growth. OS in plants influences metabolic amendments similar to wilting and sometimes depicts genotype changes. Furthermore, factors such as ion toxicity and nutritional inequity ensure impeded plant growth. Thus, it is evident that the impact of salinity on vegetative growth is a time-variant. Therefore, a bi-phase kinetic model proposed by Munns et al. (1995) is considered as a benchmark for the present work. The primary phase is exceptionally speedy. OS resulting from internal water scarcity leads to growth retardation. Whereas, the secondary phase is relatively much slower and happens because of acute assimilation of salts in the shoot. But, still differentiating amid both the phases is a difficult task due to smooth transition array. High salinity downgrades the photosynthetic rate by reducing the availability of CO_2 caused by limiting diffusion and decreased concentration of pigments. For instance, salt assimilation in spinach entirely impedes photosynthesis by reducing the conductivity of CO_2 both in mesophyll and stomata. Also, by decreasing the chlorophyll concentration, salt stress can inhibit light absorption, thereby reducing photosynthesis. Furthermore, salinity causing reduced leaf expansion had reported an 80% reduction in growth rate in radish, while reduced conductance only retards body growth by up to 20% (Savci 2012).

Root zone salt assimilation activates OS and interrupts cell ion homeostasis by replacing the uptake of essential salts such as calcium and potassium nitrate with NaCl. Stem and leaf zone assimilation causes reduced photosynthetic rates, damaged chloroplasts, degraded metabolism, enzymatic malfunction, and other organelles. The impacts are predominant in adult leaves due to the longer accumulation tenure. Furthermore, nutrient deficiency and inequality occur in plants due to ionic substitution. Cations such as K^+, Ca^{2+} responsible for principal nutritive balance get replaced by Na^+ while NO^{3-} as a major anion gets substituted by Cl^-, leading to significant imbalance. In the case of higher soil sodium–calcium ratio, deficiency symptoms appear as a first sign. Though, plants such as tomato minimize the calcium absorbance to lower the rate of transpiration, and sodium competency plays a dormant factor there (Yadav et al. 2011; Zhang et al. 2018).

Primarily, a reduction in vegetative biomass, leaf surface area, and retarded plant growth is encountered chronologically in almost all the vegetative crops due to external salinity issues. The understanding of the interaction between plant root and salt-imposed stress is still clumsy. Conversely, root biomass found to be nearly unaffected when compared with upper ground organs, except cauliflower, broccoli, and tomato. Biomass reduction is prevalent in cauliflower and broccoli, whereas in tomato, root length and density reduction are observed. The signs of salinity exposure appear gradually in plants. The primary symptoms include the transformation of green leaves, wilting, and hindered growth. Furthermore, advanced symptoms such as chlorosis, leaf burning, scorching, necrosis, etc., start manifesting after two weeks and prolonged exposure. The visuals of the above issues negatively influence sellability and affect economics. Commercial varieties such as roots, fruits, and tubers are the worse affected. Also, rotting of blossom-end has been detected in tomato, eggplants, etc., due to saline irrigation (Maas 1993; Chandna et al. 2013).

Nonetheless, exposure to limited salinity also exerts some beneficial impacts, especially on vegetative crops. It improves the quality of the edible parts despite impinging certain visual defects. For instance, it reduces water content in fruits, enhances soluble solids and acid concentration in tomato, cucumber, and watermelon. Additionally, salinity can also improve the concentration of antioxidants and carotenoids in tomato and romaine lettuce. Studies also depicted that the beneficial nutritional properties (i.e. polyphenol concentration) of broccoli and spinach also flourish under a controlled saline environment with a dip in oxalic acid and nitrate ion content. All the prior mentioned effects are timedependent and only visible when subjected to the stress at the right moment (Thomas and Bohnert 1993; Chandna et al. 2013).

1.9 Role of Salt Toxicity in the Operation of Green Revolution

The third agricultural revolution colloquially termed as the Green Revolution was a path-breaking exercise that occurred between the early 1950s and late 1960s in the twentieth century. Impacts were majorly widespread in developing nations. Beside positive, the revolution also brought specific rampant detrimental impacts on both land and crop health (Pingali 2012). Several newly derived products and technologies such as synthetic fertilizers, herbicides, pesticides, and irrigation were introduced to amplify the crop yield. But, both in terms of dependency and severity, chemical fertilizers surpass the influence of introductions. The scarcity of scientific information deceived to excessive usage of the above without evaluating the requirement resulted in reduced soil buffer. Furthermore, to compensate for the disorder, an additional application of fertilizer becomes a mandate. Constituents such as sodium and potash salt present in abundance in those fertilizers influence pH imbalance and turn the soil alkaline (Shrivastava and Kumar 2015). The assimilation of salt makes the soil susceptible to degradation. It is accounted for globally approximately9 M ha of cultivable land transforms nonfertile every year due to enforced salinity. Moreover, salinity can also be induced due to flooding caused by over-irrigation and inadequate irrigation. Vegetable crops are more prone to vulnerability upon saline exposure resulting in depleted yield (Pingali 2012; Patel 2013; Sharma and Singh 2015; Eliazer-Nelson et al. 2019).

1.10 Reaching the Current Status and Conclusion

Subjectivity to the external stressors makes the plant vulnerable and critically impacts productivity. Salinity being a predominant stress factor accounts for a significant chunk of yield-oriented issues and up to further extent desertification. The usage of inferior quality of irrigation water and unlimited usage of chemical fertilizers often provoke saline intensification in arid and semiarid regions. Saline soil interrupts the unrestricted plant water uptake by hampering the osmotic balance. Sodic soil enforces the accumulation of sodium and chloride ions in the plant tissue. It leads to the scarcity of the essential minerals and inhibition of physiological growth due to retarded cell division rate. Plants exposed to sodicity often subject to ionic inequity and report an acute dearth of vital minerals such as potassium, magnesium, and calcium.

Moreover, excess sodium gets further transported to the xylem tissue of leaves and settle as a residue after evaporation (Baum et al. 2000). The accumulated salt mass disrupts the chloroplast structure and interferes with photosynthesis. As a preventive measure, the inherent water potential of the plant surge below that of the resident soil to withhold the negative turgor and uphold water balance. However, to mitigate a specific situation, plants internally enhance the solute concentration to uplift the osmotica (Carillo et al. 2011). Failing so may trigger the generation of ROS. Inequity between excessive synthesis and destitute scavenging often leaves insistent ROSs in the system. It oxidizes organics and destructs sensitive structures such as cell membranes, DNA and RNA structures, etc. (Gupta and Huang 2014).

Contrarily, the assimilation of metabolic solutes and other soil-driven solutes in the plant is a distinct scenario, while exposed to stressors (Thomas and Bohnert 1993). These solutes safeguard the vegetation against alien stresses by different mechanisms, including osmoregulation, ROS decontamination, shield the cell membrane against the rupture, and enhancement of enzymatic/protein balance (Hasegawa et al. 2000; Hasanuzzaman et al. 2014; Acosta-Motos et al. 2017). Furthermore, solutes such as osmoprotectants provide auxiliary protection to cellular components against desiccation. Solutes such as sucrose, glycine betaine, trehalose, and proline are appropriate examples of the above (Hanin et al. 2016).

While salinity is a violent threat steadily evolving across the globe, sustaining food security at this juncture is an enormous task before the agriculturists. Including genetic engineering, several approaches were embraced to produce salt-tolerant vegetative species mimicking halophytes. Unfortunately, none of the species could be commercialized due to certain limitations.

Reclamation of the deserted lands could be the other way around. Restoration of the sodic soil is more accepted. The conventional approach includes the application of gypsum or pyrite. At the same time, recent researches recommended the usage of biodegradable municipal waste and industrial sludge as a sustainable alternative. Further research needs to be carried out to comprehend the challenges oriented and develop this technique as the alternate disposal model, especially for a small town with no existing waste management mechanism (ICAR 2015).

Acknowledgments

The review work was assisted by Dr. DK Sharma, Emeritus Scientist and Ex. Director, CSSRI, Karnal. We also thank Dr. Prasenjit Mondal, Associate Professor, Chemical Engineering Department, IIT Roorkee, for extending his insights and expertise related to chemical desalination techniques. Authors would also like to acknowledge Dr. RD Tripathi, Chief Scientist, CSIR-NBRI, for his comments to escalate the technical quality of the manuscript. Ultimately, we would like to extend our gratitude to UGC for providing financial assistance for the literature study.

References

Acosta-Motos, J.R., Ortuno, M.F., Bernal-Vicente, A. et al. (2017). Plant responses to salt stress: adaptive mechanisms. *Agronomy* 7: 18.

Agassi, M., Bloem, D., and Ben-Hur, M. (1994). Effect of drop energy and soil and water chemistry on infiltration and erosion. *Water Resour. Res.* 30: 1187–1193.

Ashraf, M. (2004). Some important physiological selection criteria for salt tolerance in plants. *Flora* 199: 361–376.

Aslam, M., Ahmad, K., Akhtar, M.A., and Maqbool, M.A. (2017). Salinity stress in crop plants: effects of stress, tolerance mechanisms and breeding strategies for improvement. *J. Agric. Basic Sci.* 2: 70–85.

Bano, A. and Fatima, M. (2009). Salt tolerance in *Zea mays* (L.) following inoculation with *Rhizobium* and *Pseudomonas*. *Biol. Fertil. Soils* 45: 405–413.

Baum, S.F., Tran, P.N., and Silk, W.K. (2000). Effects of salinity on xylem structure and water use in growing leaves of sorghum. *New Phytol.* 146: 119–127.

Cañedo-Argüelles, M., Kefford, B., and Schäfer, R. (2018). Salt in freshwaters: causes, effects and prospects - introduction to the theme issue. *Philos. Trans. R. Soc. Lond. Ser. B Biol. Sci.* 374: 20180002.

Carillo, P., Grazia, M., Pontecorvo, G. et al. (2011). Salinity stress and salt tolerance. In: *Abiotic Stress in Plants - Mechanisms and Adaptations* (eds. A. Shanker and B. Venkateswarlu), 26–33. London, UK: IntechOpen Limited.

Chandna, R., Azooz, M.M., and Ahmad, P. (2013). Recent advances of metabolomics to reveal plant response during salt stress. In: *Salt Stress in Plants* (eds. P. Ahmad, M.M. Azooz and M.N.V. Prasad), 1–14. Basel, Switzerland: Springer.

Corwin, D.L. and Yemoto, K. (2017). Salinity: electrical conductivity and total dissolved solids. *Methods Soil Anal.* 2: 1–16.

Cucci, G., Lacolla, G., Mastro, M.A., and Caranfa, G. (2016). Leaching effect of rainfall on soil under four-year saline water irrigation. *Soil Water Res.* 11: 181–189.

Darko, E., Végh, B., Khalil, R. et al. (2019). Metabolic responses of wheat seedlings to osmotic stress induced by various osmolytes under iso-osmotic conditions. *PLoS One* 14: e0226151.

Staniforth, M.J. and Davies, P. (2018). Performance analysis and optimisation of a saline groundwater batch reverse osmosis desalination system for irrigation and education in the Jordan Valley. Master's thesis. Aston University, United Kingdom. doi:10.13140/RG.2.2.31624.16643.

Dillon, P., Kumar, A., Kookana, R. et al. (2009). Managed aquifer recharge - risks to groundwater dependent ecosystems - a review. *Water for a Healthy Country Flagship Report to Land and Water Australia.*

Eliazer-Nelson, A.R.L., Ravichandran, K., and Antony, U. (2019). The impact of the Green Revolution on indigenous crops of India. *J. Ethn. Food.* 6: 8.

El-Raey, M., Nasr, S., Frihy, O. et al. (1995). Potential impacts of accelerated sea-level rise on Alexandria Governorate, Egypt. *J. Coast. Res.* 11: 190–204.

Flowers, T.J. (2004). Improving crop salt tolerance. *J. Exp. Bot.* 55: 307–319.

Garreta-Lara, E., Campos, B., Barata, C. et al. (2018). Combined effects of salinity, temperature and hypoxia on Daphnia magna metabolism. *Sci. Total Environ.* 610–611: 602–612.

Godoy, G., Steadman, J.R., Dickman, M.B., and Dam, R. (1990). Use of mutants to demonstrate the role of oxalic acid in pathogenicity of Sclerotiniasclerotiorum on Phaseolus vulgaris. *Physiol. Mol. Plant Pathol.* 37: 179–191.

Gorham, J. (1995). Betaines in higher plants: biosynthesis and role in stress metabolism. In: *Amino Acids and Their Derivatives in Higher Plants* (ed. R.M. Wallsgrove), 171–203. Cambridge, UK: Cambridge University Press.

Gul, M., Wakeel, A., Saqib, M., and Wahid, A. (2015). Effect of NaCl-induced saline sodicity on the interpretation of soil potassium dynamics. *Arch. Agron. Soil Sci.* 62: 523–532.

Gupta, B. and Huang, B. (2014). Mechanism of salinity tolerance in plants: physiological, biochemical, and molecular characterization. *Int. J. Genomics* 2014: 1–18.

Halse, S.A., Shiel, R.J., and Williams, W.D. (1998). Aquatic invertebrates of Lake Gregory, northwestern Australia, in relation to salinity and ionic composition. *Hydrobiologia* 381: 15–29.

Hanin, M., Ebel, C., Ngom, M. et al. (2016). New insights on plant salt tolerance mechanisms and their potential use for breeding. *Front. Plant Sci.* 7: 1787.

Hasanuzzaman, M., Nahar, K., Alam, M.M. et al. (2014). Potential use of halophytes to remediate saline soils. *BioMed. Res. Int.* 2014: 1–12.

Hasegawa, P.M., Bressan, R.A., Zhu, J.K., and Bohnert, H.J. (2000). Plant cellular and molecular responses to high salinity. *Annu. Rev. Plant Physiol. Plant Mol. Biol.* 51: 463–499.

Hossain, M.S. (2019). Present scenario of global salt affected soils, its management and importance of salinity research. *Int. Res. J. Biol. Sci.* 1: 1–3.

Indian Council of Agricultural Research (ICAR) (2015). *Vision 2050*. New Delhi, India: M/s Royal Offset Printers.

Isayenkov, S.V. and Maathuis, F. (2019). Plant salinity stress; many unanswered questions remain. *Front. Plant Sci.* 10: 80.

Kaur, B., Gupta, S.R., and Singh, G. (2000). Soil carbon, microbial activity and nitrogen availability in agroforestry systems on moderately alkaline soils in northern India. *Appl. Soil Ecol.* 15: 283–294.

Kultz, D. (2015). Physiological mechanisms used by fish to cope with salinity stress. *J. Exp. Biol.* 218: 1907–1914.

Lauchli, A. and Grattan, S.R. (1970). Plant growth and development under salinity stress. In: *Advances in Molecular Breeding Toward Drought and Salt Tolerant Crops* (eds. M.A. Jenks, P.M. Hasegawa and S.M. Jain), 1–32. Dordrecht: Springer.

Lodeyro, A.F. and Carrillo, N. (2015). Salt stress in higher plants: mechanisms of toxicity and defensive responses. In: *Stress Responses in Plants* (eds. B. Tripathi and M. Müller), 1–33. Basel, Switzerland: Springer, Cham.

Maas, E.V. (1993). Plant growth response to salt stress. *Tasks Veg. Sci.* 1: 279–291.

Mandal, A.K., Singh, R., Joshi, P.K., and Sharma, D.K. (2018). Mapping and characterization of salt affected soils for reclamation and management: a case study from the trans-gangetic plain of India. In: *Sustainable Management of Land Resources an Indian Perspective* (eds. G.P. Obi Reddy, N.P. Patil and A. Chaturvedi), 191–201. Oxfordshire, UK: Taylor and Francis.

Mane, A.V., Karadge, B.A., and Samant, J.S. (2011). Salt stress induced alteration in growth characteristics of a grass *Pennisetumalopecuroides*. *J. Environ. Biol.* 32: 753–758.

Munns, R. (2002). Comparative physiology of salt and water stress. *Plant Cell Environ.* 25: 239–250.

Munns, R. and James, R.A. (2003). Screening methods for salinity tolerance: a case study with tetraploid wheat. *Plant Soil* 253: 201–218.

Munns, R. and Tester, M. (2008). Mechanisms of salinity tolerance. *Annu. Rev. Plant Biol.* 59: 651–681.

Munns, R., Schachtman, D., and Condon, A. (1995). The significance of a two-phase growth response to salinity in wheat and barley. *Aust. J. Plant Physiol.* 22: 561.

Naidoo, S. and Olaniran, A.O. (2013). Treated wastewater effluent as a source of microbial pollution of surface water resources. *Int. J. Environ. Res. Public Health* 11: 249–270.

Nosetto, M., Jobbágy, E., Tóth, T., and Di-Bella, C. (2007). The effects of tree establishment on water and salt dynamics in naturally salt-affected grasslands. *Oecologia* 152: 695–705.

Parida, A.K. and Das, A.B. (2005). Salt tolerance and salinity effects on plants: a review. *Ecotoxicol. Environ. Saf.* 60: 324–349.

Patel, R. (2013). The long green revolution. *J. Peasant Stud.* 40: 1–63.

Pingali, P.L. (2012). Green revolution: impacts, limits, and the path ahead. *Proc. Natl. Acad. Sci.* 109: 12302–12308.

Rasool, S., Hameed, A., Azooz, M.M. et al. (2013). Salt stress: causes, types and responses of plants. In: *Ecophysiology and Responses of Plants Under Salt Stress* (eds. P. Ahmad, M.M. Azooz and M. Prasad), 1–24. New York: Springer.

Savci, S. (2012). An agricultural pollutant: chemical fertilizer. *Int. J. Environ. Sci. Dev.* 3: 77–80.

Shahid, S.A. and Rahman, K.U. (2011). Soil salinity development, classification, assessment, and management in irrigated agriculture. In: *Handbook of Plant and Crop Stress* (ed. M. Pessarakli), 24–36. Oxfordshire: Taylor and Francis.

Shahid, S.A., Zaman, M., and Heng, L. (2018). Soil salinity: historical perspectives and a world overview of the problem. In: *Guideline for Salinity Assessment, Mitigation and Adaptation Using Nuclear and Related Techniques* (eds. S.A. Shahid, M. Zaman and L. Heng), 43–53. Basel, Switzerland: Springer, Cham.

Sharma, D.K. and Singh, A. (2015). Salinity research in India- achievements, challenges and future prospects. *Water Energy Int.* 58: 35–45.

Shrivastava, P. and Kumar, R. (2015). Soil salinity: a serious environmental issue and plant growth promoting bacteria as one of the tools for its alleviation. *Saudi J. Biol. Sci.* 22: 123–131.

Singer, M.A. and Lindquist, S. (1998). Multiple effects of trehalose on protein folding in vitro and in vivo. *Mol. Cell* 1: 639–648.

Singh, A., Krause, P., Panda, S.N., and Flugel, W.A. (2010). Rising water table: a threat to sustainable agriculture in an irrigated semi-arid region of Haryana, India. *Agric. Water Manag.* 97: 1443–1451.

Sultana, N., Ikeda, T., and Kashem, M.A. (2001). Effect of foliar spray of nutrient solutions on photosynthesis, dry matter accumulation and yield in seawater-stressed rice. *Environ. Exp. Bot.* 46: 129–140.

Thomas, J.C. and Bohnert, H.J. (1993). Salt stress perception and plant growth regulators in the halophyte *Mesembryanthemumcrystallinum. Plant Physiol.* 103: 1299–1204.

Velasco, J., Gutiérrez-Cánovas, C., Botella-Cruz, M. et al. (2019). Effects of salinity changes on aquatic organisms in a multiple stressor context. *Philos. Trans. R. Soc. B: Biol. Sci.* 374: 20180011.

Williams, W.D. (1998). Salinity as a determinant of the structure of biological communities in salt lakes. *Hydrobiologia* 381: 191–201.

Williams, W.D. (2002). Environmental threats to salt lakes and the likely status of inland saline ecosystems in 2025. *Environ. Conserv.* 29: 154–167.

Yadav, S.K. (2010). Heavy metals toxicity in plants: an overview on the role of glutathione and phytochelatins in heavy metal stress tolerance of plants. *S. Afr. J. Bot.* 76: 167–179.

Yadav, S., Irfan, M., Ahmad, A., and Hayat, S. (2011). Causes of salinity and plant manifestations to salt stress. *J. Environ. Biol.* 32: 667–685.

Zhang, H., Xu, N., Wu, X. et al. (2018). Effects of four types of sodium salt stress on plant growth and photosynthetic apparatus in sorghum leaves. *J. Plant Interact.* 13: 506–513.

Zhu, J.K. (2007). Plant salt stress. In: *Encyclopedia of Life Sciences*. Chichester: Wiley doi:10.1002/9780470015902.a0001300.pub2.

2

Effects of Salt Stress on Physiology of Crop Plants: At Cellular Level

Vivekanand Tiwari[1], Abhay Kumar[2], and Pratibha Singh[3]

[1] *Department of Fruit Tree Sciences, Institute of Plant Sciences, Agricultural Research Organization (ARO) - Volcani Institute, Rishon LeZion, Israel*
[2] *Department of Biotechnology, ICAR-National Research Centre on Litchi, Muzaffarpur, Bihar, India*
[3] *Department of Botany, School of Life Sciences, Mahatma Gandhi Central University, Motihari, Bihar, India*

2.1 Soil Salinity and Plants

Salt stress to plants occurs due to the accumulation of soluble salt in the plant rhizosphere beyond a threshold level, which can disturb the plants' optimal metabolic homeostasis. This accumulation of the salt ions may happen either due to natural means such as the weathering of rocks, oceanic salt carried by the rain and wind, flooding of the seawater and leaching of saline water from the sea to the underground water resources of the coastal area, or by uneven irrigations and excess use of chemical fertilizers (Munns and Tester 2008). Natural weathering of parental rocks releases chloride salts of sodium, magnesium, and calcium, of which the most soluble and maximum proportion is sodium chloride (NaCl) (Szabolcs 1989). In general, the soil is defined as saline if its measured electrical conductivity (EC) is equal or higher than 4 dS/m, which is equivalent to 40 mM of NaCl concentration (Munns and Tester 2008). Since most cultivated crops are sensitive to salt stress, at soil salinity higher than 4 dS/m, the reduction in crop productivity due to salt stress accounts 50–80% (Zörb et al. 2019). The land area across the globe affected by the salinity is more than 800 million hectares and facing the problem of moderate to extreme salinity (Munns and Tester 2008). A total of 230 million hectares agriculture lands have a proper irrigation system and are the source of maximum crop productivity. Surprisingly, the salinity analysis of these irrigated agriculture lands revealed that approximately 20% (45 million hectares) is affected by salt stress (Munns and Tester 2008).

The increasing soil salinity affects the plants negatively for their growth, survival, and productivity. In saline soil, accumulation of salt ions triggers osmotic stress to the plant's root cells, which are then being absorbed by the plants to adjust the osmotic balance. The excess entry of the salt ions to the root cells creates the ionic imbalance at the cellular level.

Physiology of Salt stress in Plants: Perception, Signalling, Omics and Tolerance Mechanism, First Edition.
Edited by Pratibha Singh, Madhulika Singh, Rajiv Kumar Singh, and Sheo Mohan Prasad.

However, the halophytic plant species have adapted to thrive even under these adverse conditions of high salinity and can complete their life cycle under the extreme saline soil conditions. The halophytes have a better ability to tolerate the salt stress than a glycophyte (plants sensitive to salt stress) and can survive and grow well in the saline soil with soil salinity equivalent or higher than 200 mM NaCl (Flowers and Colmer 2008).

The effect of salt stress on plant physiology and its productivity depends on the level of soil salinity and how long plants get the stress. Immediately after the exposure to the salt stress, plants induce the signaling cascades to adjust their metabolic pathways. The plant cells exhibit biphasic response under salt stress dealing with osmotic and ionic stress, which overlap at some points. Earlier researchers assumed that the osmotic stress signaling initiates immediately after the salt stress. In contrast, the signaling cascade and response to the ionic imbalance initiate later due to the slow accumulation of sodium ions (Na^+) in shoot tissues beyond a threshold level and corresponding inhibition of the photosynthesis (Zörb et al. 2019). Intriguingly, the new findings showed the root growth response specific to the Na^+ accumulation and rapid signaling cascade mediated by reactive oxygen species (ROS) or calcium ion (Ca^{2+}), specific to the salt ionic stress (Choi et al. 2014; Galvan-Ampudia et al. 2013; van Zelm et al. 2020). Apart from the ROS and Ca^{2+} signals, the phytohormones viz. absiscic acid (ABA), jasmonic acid (JA), salicylic acid (SA), gibberelic acid (GA), and ethylene play crucial role in signal transduction and regulation of expression and function of several proteins during salt stress (reviewed in Zhao et al. 2020). These signals are perceived at the organelle level or at the level of the nucleus and responded by the plant cell in terms of stress-responsive gene expression, different degrees of mRNA stability, and varied way of translational or post-translational regulation to change protein abundance and the activity. These responses depend not only on the extent and duration of the stress but also on the plants' genetic nature. The halophytes are evolutionary adapted to survive in the salt stress with unique genetic makeup, morphological, physiological, and anatomical adaptation (Munns and Tester 2008; van Zelm et al. 2020; Zhao et al. 2020). They are adapted to sequester the excess salt ions in the root or shoot vacuoles and secretion of excess salt through different kinds of salt glands and epidermal bladder cells [EBCs; (Zhao et al. 2020)]. However, salt-stress tolerance is a complex trait regulated by several genes and pathways; engineering the crops using a single gene is inefficient. Moreover, the pyramiding of several genes is time-consuming and seems less realistic to improve the salt-tolerance capacity of conventional crops. Cultivation of halophytes for food, forage, renewable energy, and phytoremediation emerged as an alternative and economic strategy in the salt-affected areas (Panta et al. 2014). Thus, in summary, to understand better the effect of salt stress on plants, a comprehensive approach is required to understand the cellular ion transport system in different tissues, major phytohormone, or osmotic stress-specific signaling pathways not only in the model plant *Arabidopsis thaliana* but also in the halophytic plant species (van Zelm et al. 2020) in order to understand the advantageous differences in the halophytes.

2.2 Crop Loss Due to Salt Toxicity – An Estimation Worldwide

The soil salinization is one of the three soil degradation processes that pose a threat to human health and crop productivity by affecting more than one billion hectares land across the globe (Ondrasek et al. 2011). The negative effect of salt stress on crop productivity indirectly affects the economy dependent on the agricultural produce, resulting in the loss of billion dollars annually. The economic loss caused by the salt stress can include two components: first, the loss of crop productivity (presented in t/ha) and thus, the loss of income generated from the agricultural production and, second, the cost spent for the restoration of degraded land. Estimating the global loss due to soil salinization can be heterogeneous among the different countries or geographical regions because of factors such as labor costs, the market price of the agricultural produce, fertilizers, seeds, and other operational costs affecting the total input and income differentially. There is the possibility that in some regions of the world (developed countries), even the moderate salinization of the soil could result in higher economic loss due to higher operational and labor costs. In many developing countries, most of the poor farmers depend on agriculture for their livelihood and loss of crop productivity due to salt stress affects their livelihood. In Asia, the Maldives is a low-lying country, always on risk of submergence due to increasing sealevel and salt deposition. The intrusion of seawater on its land area due to Tsunami and deposition of toxic salts caused degradation of more than 70% of the agricultural land (FAO 2005; Ondrasek et al. 2011). The salt deposition destroyed more than 3 70 000 fruit trees, with an estimated loss of AU $ 6.5 million, which affected around 15 000 farmers economically. In a previous report by Qadir et al. (2014), the total estimated economic loss globally was more than the 27 billion US dollars per year. The loss of productivity among the crop also varied depending upon their genetic makeup, for example, moderate salt stress of 8–10 dS/m results in the yield losses of 15%, 28%, and 55% in cotton, wheat, and corn cultivars, respectively (Satir and Berberoglu 2016; Zörb et al. 2019), showing that cotton performs better at moderate salt stress; however, at the higher salt stress, the cotton also became susceptible, and the yield loss at the 18 dS/m resulted in a 55% loss in cotton productivity (Satir and Berberoglu 2016). The yield loss estimation by salt stress in comparison with the healthy growth conditions in some of the major crops of Indian subcontinent revealed the loss of yield by 45, 39, 63, and 48%, in rice, wheat, cotton, and sugarcane, respectively (Qadir et al. 2014; Tripathi 2009), which again suggest the variation in yield losses could be the result of combined effects of many factors such as the cultivars used for the cultivation, environmental condition of that specific area, and extent and time of exposure to the saline soil conditions.

2.3 Effect of Salt Stress on Target and Nontarget Plants and Microorganisms

The adoption of agricultural practices leads to the colonization of humans. For agriculture, domestication of crop plants began a million years ago, and after that, the crops were selected for numerous beneficial traits. The selection continued along with the development of civilization and till the recent. During the selection process, the plant species with the higher yield and less fruit dehiscent property became target species for the selection and agriculture. In comparison, several nontarget plant species with better stress tolerance but less yield were ignored or not prioritized. Their further improvement for the better yield through conventional breeding approaches narrowed down the plant genetic diversity. Approximately, 30 plant species fulfill 90% of the total food demand of the world human population, but they are highly sensitive to salt stress (Zörb et al. 2019). The horticultural crops, vegetables, or fruits are a nutritious source of minerals, carotenoids, vitamins, and fiber in the human diet and are more sensitive to salt stress than the cereals. The salt-stress threshold level for the metabolic homeostasis in most horticultural crops is ≤2.5 dS/m (SNAPP et al. 1991). Among the cereal crops, the effect of salinity varies depending on the cultivars' salt-tolerance ability. Rice and barley are both the targeted cereal crops used in agriculture for food production but affected differently by salt stress. The rice is the most sensitive to salt stress, whereas the barley is the most tolerant of salt stress among all the cereal crops used for agriculture (Munns and Tester 2008). The salt tolerance ability among the different targeted dicot crop species varies greatly. For example, some legumes are very sensitive to the salt-stress conditions among the legume species. In contrast, the *Medicago* is tolerant of the salt stress up to the level of salinity equivalent to the seawater (Munns and Tester 2008).

Recent advances in the field of salinity stress tolerance and research identified close relative of the crop species, which were never targeted for agriculture but are halophytes and survive the salt stress more efficiently. The halophyte *Eutrema salsugineum* belongs to the same family as the horticultural crop cabbage or the oilseed crop mustard, and the model plant *A. thaliana*. Under 100 mM of NaCl salt stress which is deleterious for cabbage (Pavlovic et al. 2019) and *A. thaliana*, the *E. salsugineum* not only survived but completed its life cycle with a negligible effect on their growth (Kant et al. 2006). The close relative of the wheat, the wheatgrass (*Thinopyrum ponticum*, syn. *Agropyronelongatum*) is one of the most salt-tolerant monocotyledonous plants (Munns and Tester 2008), which can complete its life cycle at the soil salinity equivalent to the seawater salinity. Few halophytic nontarget plant species grow like weeds in the saltmarshes and have their ability to grow and complete the life cycle in the highly saline soil (Flowers and Colmer 2008; van Zelm et al. 2020).

The soil microbial community is an indispensable component of the soil nutrient cycling and supports crop plants growth and immunity. The microbial community helps plants absorb nutrients by participating in chemical modification of mineral, which becomes suitable for the plants' easy absorption, increasing the soil porosity and organic matter, and soil aggregation (Xu et al. 2016; Zhang et al. 2019). Importance of rhizospheric microbial communities in plant nutrient availability and health suggests that any impact of the soil microbial community diversity or structure would directly affect the crop productivity in those soils. The incident of salt stress in soil has exhibited a reduction in the rhizosphere respiration, enzyme activity, soil microbial community size, and their growth rate, which ultimately affects the biogeochemical cycle negatively (Tripathi et al. 2006; Zhang et al. 2019). The increasing soil salinity has negatively impacted the bacterial community of *Planctomyces* and *Archangium*, associated with the soil organic matter input and stabilization (Zhang et al. 2019). The fungal mycorrhizal community helps plant better absorption of the mineral from the soil, and the activity of these mycorrhiza communities affects the soil nutrient availability (Babu and Reddy 2011). Under salt-stress conditions, the arbuscular mycorrhizae *Glomus* was suggested to enhance the phosphorus and potassium uptakes by the plants (Porcel et al. 2012), but the increasing salt stress reduces the *Glomus* availability in the soil (Zhang et al. 2019). The fungus *Hydropisphaera,* which is involved in degrading the lignin in the salt marsh, brings a negative impact on the plants' growth and survival during the salt-stress conditions. In salt stress, the *Hydropisphaera* got dominated by the rhizosphere. Degradation of the lignin by *Hydropisphaera* reduces the soil organic matter and soil fertility (Zhang et al. 2019) posing a threat to the survival of the plants under the salt-stress conditions. The salt stress to the horticultural crop tomato showed the increased incidence of growth of the root-rot disease-causing fungal pathogen *Phytophthora parasitica* (SNAPP et al. 1991), showing the importance of understanding the effect of salt stress on the nontargeted organism in parallel. There are several beneficial halotolerant plant growth-promoting rhizobacteria (PGPR) that assist plants in their survival under the salt-stress conditions. The PGPRs and their functional role are described in detail by Kumari et al. (2019), and studying these PGPRs will help us improve the salt-stress tolerance in plants up to some extent.

2.4 Effect of Salt Stress on Physiology of Crop Plants

Accumulation of salt ions in the plant rhizosphere causes dual stress to the plants, the osmotic imbalance between the soil and root cells, and cellular ionic imbalance due to passive or active intake of salt ions. The salinity-induced osmotic stress suppresses the water uptake by the plants immediately, which then triggers various physiological and metabolic adjustments in the plants. Osmotic stress inhibits the expansion of the root and shoot cells within minutes due to reduced turgor pressure (Fricke et al. 2004; Munns et al. 2000) and the stress signal transduced rapidly at a speed of the sound through different signaling cascades from the root to the shoot tissue for the closure of stomata to minimize water loss, which seizes shoot metabolism (Christmann et al. 2013; Shabala et al. 2016). The mechanosensitive ion channels of the leaf guard cells sense the drop in xylem pressure and participate in stomatal closure, reducing the CO_2 exchange and photosynthetic assimilation (Shabala et al. 2016). For osmotic adjustment, plants start the intake of the salt ions from the soil, which then enters different root and shoot cells creating the ionic stress to plants (Shabala and Lew 2002).

The NaCl dominates the saline soil, and, therefore, the Na^+ is the most critical ion, which is assumed to pose toxic effects on the plant cell physiology and metabolism. For their survival and optimal physiological activity, the ionic homeostasis is essential in the plants, and different ion channels and transporters in root or shoot tissues are involved in this process. The Na^+ enters into the root cells through the glutamate receptor-like (GLRs) channels, cyclic nucleotide-gated (CNGCs) nonselective cation channels, high-affinity K^+ transporters (HKT2, HAK5), Shaker-type K^+ channels (AKT1), the low-affinity cation transporter (LCT1), and aquaporins (PIP2;1) (Zhao et al. 2020). The entry of salt ions to the root cells and the osmotic stress spikes the Ca^{2+} in the root cells activating the salt overly sensitive (SOS) signaling pathway for the extrusion of the excess Na^+ out of the cell (Hong et al. 2019; Shi et al. 2002). The function of active Na^+ extrusion from root cells to the soil is performed by the plasma membrane Na^+/H^+ antiporter named salt overly sensitive 1 (SOS1) antiporter (Shi et al. 2002).

However, during salt stress, a higher concentration of salt ions in soil and continuous intrusion of salt ions to the root cells, the extrusion of Na^+ by SOS1 is not sufficient to maintain the cellular homeostasis of the Na^+. Thus, the plants sequester the excess Na^+

into the vacuole by the vacuolar Na^+/H^+ antiporter (NHX1), loaded to the xylem vessel to distribute it all over the plants by nonselective cation channel (NCCS), SOS1, high-affinity K^+ family transporter 2 (HKT2) or cation-chloride cotransporter (CCC), or retrieved to the phloem from leaf cells by HKT1 (Zhao et al. 2020). The general limit of external salt concentration for glycophytes is 0.5% of seawater (≈86 mM NaCl) above which the stress-coping mechanism fails, and severe yield loss in plants are reported (Bose et al. 2017). We discuss the effect of the salt stress on different compartments of the plant cell and their physiology or metabolic activities in the following sections.

2.4.1 Effect of Salt Stress on Chlorophyll Biosynthesis, Chloroplast Functioning, and Photophosphorylation

In addition to photosynthesis, the chloroplast is the hub of several other metabolic activities: the biosynthesis of chlorophyll, carotenoid, ABA, vitamins, amino acids, and fatty acids and lipids (Zhao et al. 2020). Therefore, any environmental stress affects the metabolic pathways that operate in the chloroplast. Minutes after the salt stress, the stomata are closed, suppressing the exchange of the CO_2 resulting in photosynthesis inhibition. In the light reaction of photosynthesis, the electron generated from the photo-oxidation of water passes through the electron transport chain (ETC) on the thylakoid membrane to the electron acceptor proteins and molecules in the stroma. The transfer of electrons through the ETC generates a proton gradient across the thylakoid membrane, creating a proton motive force (PMF). This PMF drives the function of the chloroplast ATP synthase for the synthesis of ATP from the ADP, a process defined as the photophosphorylation (Sudhir and Murthy 2004). The plants use the ATP produced by the photophosphorylation in their metabolic pathways that operates in the chloroplast.

2.4.1.1 Chlorophyll Biosynthesis in Salt Stress

The chlorophyll biosynthesis in plants occurs in the plastid involving more than 17 enzymes encoded by the nucleus (Pattanayak and Tripathy 2002). Various environmental stresses, including salt stress, downregulate the chlorophyll biosynthesis. Exposure of young and etiolated rice seedlings with salt stress showed that the decrease in the chlorophyll biosynthesis was the result of the reduced transcript expression and protein abundance of biosynthetic enzymes 5-aminolevulinic acid (ALA) dehydratase, porphobilinogen deaminase, coproporphyrinogen III oxidase, protoporphyrinogen IX oxidase, Mg-protoporphyrin IX chelatase, and protochlorophyllideoxidoreductase (Turan and Tripathy 2015). Although ALA synthesis decreased, the gene/protein expression of glutamyl-tRNA reductase increased, suggesting that it may play a role in acclimation to salt stress. In mature Sunflower plants, the salt-stress-induced downregulation of chlorophyll biosynthesis in leaves showed a correlation with the reduced level of glutamate and 5-ALA accumulation (Santos 2004). Salinity downregulates the expression of chlorophyllide-*a*-oxygenase (CAO), which is involved in Chl b synthesis. The reduced Chl b synthesis contributes to an increased Chl a/b ratio observed in salt stress (Pattanayak and Tripathy 2002). The adaptation of plants to downregulate the chlorophyll biosynthesis is essential for protecting the chloroplast from the photo-oxidative damage by the ROS generated by light and triplet chlorophylls in the absence of photosynthesis (Turan and Tripathy 2015). Thus, the downregulation of chlorophyll biosynthesis might be part of the stress avoidance and protection mechanism in plants.

2.4.1.2 Salt Stress Affects Chloroplast Function

Salt stress affects chloroplast more severely than another organelle, as the ROS increases rapidly due to chlorophyll triplet and oxygen molecules produced due to the photolysis of water, leading to its destruction. The salt stress leads to increased Na^+ and Cl^- concentrations and decreased K^+ concentration inside chloroplasts resulting in the swelling of thylakoid and unstacking of grana, thereby altering the photosynthetic structure (Santos 2004). The regulation of K^+ and Cl^- across the thylakoid membrane is important in regulating the pH difference, which regulates the photoprotective nonphotochemical quenching mechanism and optimal photosynthesis (Finazzi et al. 2015). Any sort of disorder in this mechanism due to insufficient K^+ during salt stress in the chloroplast may hamper its photoprotective mechanism. Accumulation of anions in the chloroplast affects both the acceptor and donor sides of photosystem II (PSII) by binding to a specific site on PSII (Jajoo et al. 1994, 2005). When the anions permeate the membrane during the salt stress, the influx of protons to balance the resultant charge causes structural reorganization in thylakoid membranes resulting in energy redistribution. All these salt stress components ultimately upset the ionic balance within chloroplasts, resulting in poor photosynthetic performance and reduced growth and yield. A shift in anion balance has shifted the solar energy distribution from PSII to PSI, affecting the function of the water-splitting complex and electron transport to and from the plastocyanin (Singh-Rawal et al. 2011). At the same time, soluble cations also cause alteration in the excitation energy transfer in the photosynthetic system at physiological temperatures by affecting several primary processes in thylakoids: initial energy distribution or "spillover" from PSII to PSI and rate constant of nonphotochemical quenching (Wong et al. 1980).

During salt stress, the maximum quantum yield of PSII (Fv/Fm), photochemical quenching coefficient, and electron transport rate decreases, whereas nonphotochemical quenching increases (Yang et al. 2020), possibly due to unstacking of the grana. The reduced photochemistry and carbon fixation lead to an increase in ROS production which are potent in damaging chloroplast and its membrane (Hernández et al. 1995). To cope with this situation, the level of enzymatic (catalase, ascorbate peroxidase, and glutathione reductase) and nonenzymatic (ascorbate, total carotenoids, phenolics, and flavonoids) antioxidants increases (Taïbi et al. 2016). The ROS production, especially H_2O_2, inhibits the synthesis of D1 protein (Murata et al. 2007). Salt stress modifies the Q_B plastoquinone binding site of D1 at the acceptor side. It stabilizes oxygen-evolving complex (OEC) in S2 state at the donor side by salt-induced migration of the PsbO subunit of PSII to the lumen (Sasi et al. 2018). This change in the PSII and OEC slows the water-splitting process, which may be part of the plant's defense strategy. The enzyme plastoquinol terminal oxidase (PTOX) in halophyte *E. salsugineum* chloroplast minimizes the damage by diverting electrons from plastoquinol to oxygen and producing water molecules (Stepien and Johnson 2009).

Salt has an essential role in maintaining enzyme activity. If the salt concentration is below optimum, the charged amino acid side chains of the enzyme will attract each other, thus denaturing it. At the same time, if the salt concentration is too high, regular interaction of charged groups will be blocked, new interactions will occur, and again the enzyme will denature. Some of the enzymes involved in the Calvin–Benson–Bassham cycle share their functional activity with the chloroplast glycolytic pathway. Therefore, one of the direct effects of ionic stress is the denaturation of enzymes involved in the Calvin–Benson–Bassham cycle, leading to the transduction of signals for the upregulation or downregulation of several genes encoding enzymes involved in photosynthesis. We will discuss the effect of salt stress on their activity in the following glycolytic section.

2.4.1.3 Photophosphorylation in Salt Stress

The PSI is relatively less sensitive to salt stress than PSII and participates in the cyclic ETC in algae during the salt stress. The PSI could play a vital role in salt tolerance by increasing cyclic ETC generating ATP by photophosphorylation while avoiding the build-up of toxic reducing species (Bose et al. 2017). The excess ATP generated through cyclic ETC around PSI has been suggested to prevent Na^+ overaccumulation in the chloroplasts of soybean (He et al. 2015). Two chloroplasts envelope antiporters *CHX23* and *NHD1* help plants to maintain the Na^+ homeostasis between the chloroplast and the cytosol (Song et al. 2004). Light-induced thylakoid swelling in salt-stressed plants also facilitates the diffusion of the plastocyanin between cytochrome *b6f* complex and a PSI reaction center, enhancing the overall electron transfer rate (Kirchhoff et al. 2011) by cyclic ETC and producing ATP by photophosphorylation to avoid oxidative damage of chloroplast. In salt stress, the increased accumulation of subunits of soybean NDH complex was found to be involved in cyclic ETC (He et al. 2015) and a gene encoding the protein essential for the assembly of ATP synthase in sorghum (Sui et al. 2015) suggested the importance of cyclic ETC and photophosphorylation during the salt stress.

2.4.2 Glycolysis, Kreb'sCycle Enzymes, Oxidative Phosphorylation, and Other Mitochondrial Functioning

The sugar molecule fixed by photosynthesis is the carbon source of plants used for biosynthesis of structural components to support growth or in the respiration for supplying energy for maintenance of metabolic activity. The regulatory mechanisms involved in allocating the carbon to either growth or the respiratory pathway are defined as the carbon balance of plants (Lambers et al. 2008). Mitochondria are the energy house of the cells producing chemical energy in the form of ATP by oxidation of sugar molecules and supplying energy for metabolic activities. Under optimum growth conditions, energy production from a sugar molecule involves glycolysis, followed by Kreb's (tricarboxylicacid, TCA) cycle and the mitochondrial electron transport chain (mtETC) coupled with oxidative phosphorylation in mitochondria (Fernie et al. 2004). In salt-stress conditions, the demand for ATP production increased suddenly for ion transporters to maintain the ionic homeostasis, detoxification of ROS, and synthesis of osmolytes to maintain the cellular osmotic balance (Che-Othman et al. 2017). The breakdown of glucose by glycolysis in plants operates at the cytoplasm and in the plastids (Plaxton 1996), where some of the isoforms of the chloroplastic glycolytic pathway participate in the Calvin–Benson–Bassham cycle (Dumont and Rivoal 2019).

2.4.2.1 Glycolytic Pathway in Salt Stress

The enzyme fructose-1,6-bisphosphatase (FBPase) catalyzing the hydrolysis of fructose-1,6-bisphosphate to fructose-6-phosphate and fructose-1,6-bisphosphatase aldolase (FBP aldolase) catalyzes the breakdown of fructose-1,6-bisphosphate into glyceraldehyde 3-phosphate and dihydroxyacetone phosphate. The FBPase and FBP aldolase expression increased during salt stress (Kim et al. 2005). However, salt-stress treatment of rice seedlings exhibited increased levels of enzymes involved in ethanolic fermentation and glycolate metabolism (Abbasi and Komatsu 2004), and salt-stress-treated soybean and leaves of grass pea also showed a similar response (Chattopadhyay et al. 2011). These findings suggest that salt stress may also induce anaerobic metabolism in plants to fulfill the increased energy demand.

2.4.2.2 TCA Cycle in Salt Stress

The glycolytic breakdown product pyruvate enters the mitochondria from the cytoplasm and serves as the TCA cycle substrate. Another possible source of pyruvate is the synthesis of pyruvate in the mitochondrial matrix with the help of malic enzyme (Che-Othman et al. 2017). The TCA cycle provides carbon skeleton essential for the biosynthesis of amino acids, fatty acids, nucleic acids, isoprenoids, and secondary metabolites, and reductants to be used in the mtETC (Plaxton 1996; Sweetlove et al. 2010). The TCA cycle is regulated at various steps of the cycle depending upon the environmental conditions, developmental age, and plant species. Only half of the TCA cycle proteins have shown an increased abundance in salt stress (Che-Othman et al. 2017). The protein abundance of pyruvate dehydrogenase complex (PDC) subunits and the enzyme succinyl coenzyme A synthase enhanced in salt-sensitive plants upon exposure to salt stress. However, the increased abundance of PDC subunits was not consistent in all the plant species (Che-Othman et al. 2017). In contrast, the abundance of isocitrate dehydrogenase decreases in salt-sensitive plants under salt stress.

Apart from being a significant step of respiration, the TCA cycle also contributes to the biosynthesis of amino acids. This role of the TCA cycle links the carbon to nitrogen metabolism. Although the TCA cycle is more efficient in energy production, plants require to maintain the balance between the carbon and nitrogen metabolism even under the increased energy demand under stress. Under salinity stress, the TCA network channelizes the carbon source to the malate/pyruvate pathway (Kazachkova et al. 2013), and the γ-aminobutyric acid (GABA) shunt (Renault et al. 2010; Zhao et al. 2020) which provides the metabolic flexibility to plants during the stress. In parallel to the increased energy demand, the salt stress exerts pressure on the nitrogen metabolism for the synthesis of polyamines or other nitrogen-containing osmolytes in plants. The extraction of oxaloacetate for the nitrogen metabolic pathways disturbs the cyclic continuation of the TCA cycle. At this condition, plants activate the alternate malate/pyruvate pathway, where the malic enzyme converts the excess malate into pyruvate replenish and restart the TCA cycle (Che-Othman et al. 2017). The protein abundance and activity of the ME increased in salt stress in rice and wheat, respectively (Lima et al. 2012). GABA is an amino acid that accumulates in plants during the abiotic stresses and involves in carbon metabolism, pH regulation, nitrogen storage, and functions as osmoticum (Kinnersley and Turano 2010). The GABA shunt pathway involves four enzymes, bypassing the activity of 2-oxoglutarate dehydrogenase and succinyl-CoA synthase of the TCA cycle (Che-Othman et al. 2017). The increase in cytosolic Ca^{2+} concentration and lower pH, the changes occur during salt stress, activates the enzymes of GABA shunt pathway, whereas the accumulation of sufficient NADH and ATP deactivates the GABA shunt pathway (Busch et al. 2000). The activity of the last enzyme of the GABA shunt, succinate semialdehyde dehydrogenase, synthesizes succinate and thus probably helps to minimize the ROS accumulation in stress and proper functioning of the mtETC (Bao et al. 2015).

2.4.2.3 Salt Stress and Oxidative Phosphorylation

Oxidative phosphorylation is the process of ATP production on the mitochondrial membrane by mitochondrial ATP synthase using the electrochemical gradient generated by the mtETC involving protein complexes arranged on the inner membrane of mitochondria. The non-photosynthetic tissues, like root cells, depend on the mitochondrial oxidative phosphorylation for their energy demand. However, the root cells have limited access to an oxygen supply, have minimal ability to increase the ATP production during stress, and increasing salt ions further suppresses the activity of the complex I and complex II of the mtETC (Jacoby et al. 2011). The negative membrane potential of the mitochondrial membrane assisted the intake of Na^+ to the mitochondria (Che-Othman et al. 2017), of which excess accumulation inhibits the mtETC by denaturing the proteins of the mtETC complexes or by disassembly of the complexes (Flowers 1972). At the moderate or lower salt stress in glycophytes, the ATP synthase activity increased without affecting different subunits of the ATP synthase. However, the ATP synthase subunits decreased significantly at higher salt stresses (Kosova et al. 2013). Due to presence of alternative mtETC pathways like NAD(P)H-dehydrogenases, and alternative oxidase (AOX), the ETC is more complex and flexible in mitochondria (Jacoby et al. 2011). Suppression or inhibition of the mtETC negatively impacts the oxidative phosphorylation and ATP production. Moreover, the altered mtETC facilitates the electron's leakage to the free oxygen molecule and produces an excess amount of the ROS. However, the ionic component of the salt stress is more deleterious to the mtETC than the osmotic, affecting the NADH dehydrogenase (complex I) and succinate dehydrogenase (complex II) activity (Hamilton and Heckathorn 2001). Two competing respiratory chains in plant mitochondria help plants maintain the mtETC up to some extent in moderate salt stress. When salt stress inhibits the cytochrome-mediated respiratory chain activity, the AOX-mediated chain remains unaffected (Jacoby et al. 2011) and provides metabolic adjustment to the mitochondria at low or moderate salt stress.

2.4.3 Peroxisome Functioning

Peroxisomes play a crucial role in plant physiology. It is the organelle where β-oxidation of fatty acids, part of the photorespiratory pathway, glyoxylate cycle, biosynthesis of phytohormones: auxin, JA and SA, and ROS metabolism occurs (Fahy et al. 2017). The peroxisome is the primary site of ROS generation, detoxification, and also involved in energy-producing metabolism. Hypothetically, it is plausible that in salt stress, the peroxisomes' biogenesis and functional activities might be increased to cope with increased photorespiratory pathway, increased energy demand, and increased ROS level. In several previous studies, the increased number of peroxisomes was reported in plants subjected to salt stress (Mitsuya et al. 2010). A peroxisomal membrane protein family PEX11 is known to regulate the size and number of peroxisome in plants, animals, and yeast (Mitsuya et al. 2010). During salt stress in plants, the expression of proteins from the PEX11 families is upregulated (Fahy et al. 2017; Mitsuya et al. 2010). However, there is no evidence found so far, which supports the hypothesis of increased functional activity of the peroxisome and to understand that a detailed study is still missing.

2.5 Halophytes and Their Physiology

Halophytes are the plant species that can survive and complete their lifecycle in the soil with salinity equivalent or higher than 200 mM of NaCl (Flowers and Colmer 2008). The plasticity of halophytes to tolerate saline soil conditions is attributed to genetic, morphological, anatomical, and physiological adaptations (Flowers et al. 1977). Decades of research accumulated knowledge about the halophytes suggest that they evolved to maintain efficient cellular ion homeostasis, better osmotic adjustments, and secretion of excessive salt by the salt gland or by the epidermal bladder cells (EBC). How halophytes enable them to tolerate the salt stress and differed from the glycophytesis, summarized in Figure 2.1.

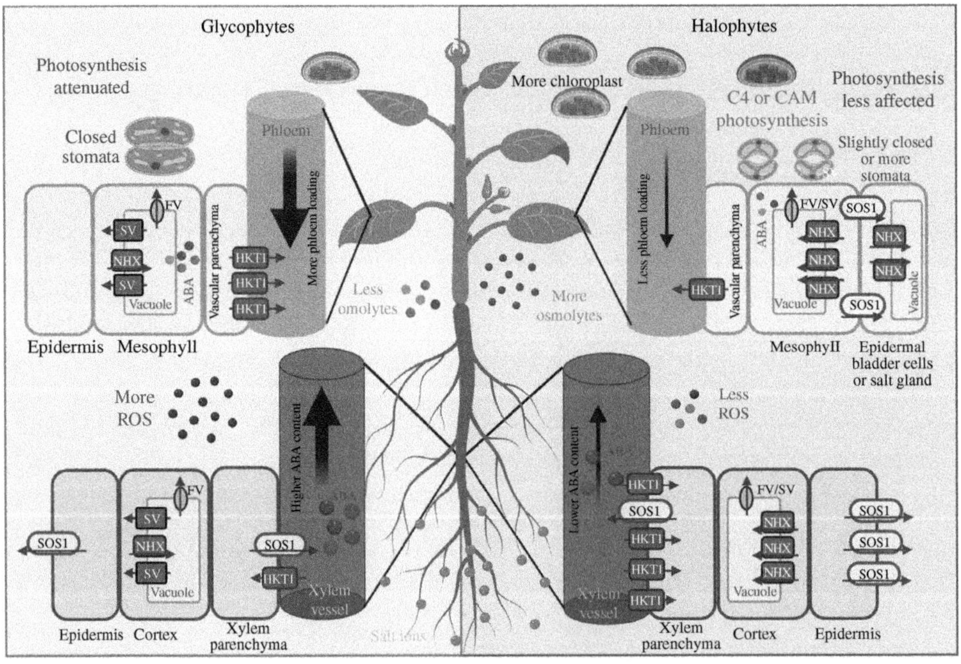

Figure 2.1 Schematic illustration shows comparison of ion homeostasis and physiological changes in glycophytes and halophytes during salt stress. The illustration shows only some of the important differences between a glycophyte and a halophyte to provide simplicity of the scheme. The arrows represent the direction of Na^+ movement during salt stress, whereas the thickness of the arrows indicates the amount of Na^+ movement. The number of transporters on the membranes represents the amount of Na^+ being transported to the direction of the arrow. ABA, absiscic acid; CAM, Crassulacean acid metabolism; FV, fast-activating Na^+ channel; HKT1, high- affinity K^+-1 transporter; NHX, vacuolar Na^+/H^+ antiporter protein family; SOS1, salt overly sensitive 1 (plasma membrane Na^+/H^+ antiporter); SV, slow-activating Na^+ channel. Scheme was created using BioRender.com. *Source:* Modified from (Bose et al. 2017; Zhao et al. 2020).

2.5.1 Ion Homeostasis in a Halophyte

The halophytes maintain better cellular ionic homeostasis by efficient Na^+ exclusion and vacuolar sequestration, xylem loading and retrieval of the Na^+, minimized recirculation of Na^+ through the phloem tissue, and secretion of salts through the salt gland and EBCs (Zhao et al. 2020). The constitutively expressing plasma membrane Na^+/H^+ antiporter SOS1 in halophytes performs the task of Na^+ extrusion out of the cell more efficiently than the glycophytes (Shi et al. 2002). The sequestration of the Na^+ and Cl^- in the vacuole of root and leaf cells of halophytes are facilitated by constitutively expressing Na^+/H^+ antiporter (NHX), vacuolar H^+-inorganic pyrophosphatase (V-PPase), and vacuolar H^+-ATPase (V-ATPase) (Jha et al. 2011). However, to avoid the leakage of the ions through Na^+ permeable channels, slow- (SV) and fast- (FV) activating ion channels, halophytes minimize their activity (Shabala et al. 2020). This efficient sequestration avoids the toxicity of ions on cytoplasmic physiology and also requires less organic osmolytes to adjust the osmotic balance between the cytoplasm and vacuole. The cytoplasm contributes only 10% of the cell volume, and thus halophytes spend relatively very less energy for cytoplasmic osmotic adjustment in comparison with glycophytes in osmolytes biosynthesis (Zhao et al. 2020).

The excess Na^+ in the root cortical or parenchyma cell is then loaded to the xylem vessels by NCCS, SOS1, HKT2, or CCC (Ishikawa et al. 2018; Shi et al. 2002) for dilution of the salt ions in the root cells. However, to avoid the ionic imbalance in the photosynthetically active shoot tissues, plants attempt to retrieve back the Na^+ from the xylem vessel to the root cells for extrusion. The transporters of HKT1 family retrieve the Na^+ from xylem vessel to the root cells and from leaves to the phloem (Munns et al. 2012; van Zelm et al. 2020). Interestingly, the model plant species *A. thaliana* contains only one gene encoding for the HKT1 in their genome. Whereas, its close halophytic relative *E. salsugineum* has five genes encoding proteins belonging to the HKT1 family (Wu et al. 2012) suggesting the better Na^+ retrieval strategy in the halophytes. However, the anatomical structure of the root very unlikely allows the unloading of the Na^+ coming from the shoot tissue through the phloem which remained circulated in the phloem and ultimately creates damage to the young growing tissues and meristematic region (Zhao et al. 2020). The additional checkpoint in halophytes minimizes the damage of the young meristematic tissues by reducing the Na^+ retrieval in the phloem tissue. A comparative analysis revealed that barley allows only 10% of the shoot Na^+ retrieval to the phloem, whereas in a salinity-sensitive lupin species, the retrieval rate was 50% of the shoot Na^+ concentration (Jeschke et al. 1992). Here then, the question arises, if halophytes are not recirculating their excess Na^+ in the shoot through the phloem, then how can they establish the ionic homeostasis in the shoot tissue? The answer to this question emerged from the study on the halophytes, which revealed the development of salt gland or bladder in approximately 50 species of the halophytes (Zhao et al. 2020), playing a role in sequestering the Na^+ and Cl^- away from the metabolically active cells and secreting them when accumulates in access. These unique structural developments in halophytes with yet unknown mechanisms showed structural and functional variation among themselves. The exo-recretohalophytes have the salt gland on the leaves' surface, while the endo-recretohalophytes collect salt in the vacuole of specialized EBCs (Dassanayake and Larkin 2017).

2.5.2 Osmotic Adjustment

During salt stress, plants have two options for their osmotic adjustments: de-novo synthesis of organic osmolyte or the uptake of inorganic ions from the soil. Energetically, the former option is costly, whereas the latter option is economic by an order of magnitude to the plants (Munns et al. 2020; Shabala and Shabala 2011), but depends on the plants' ability to establish an ionic homeostasis during salt stress. At the onset of salt stress, the relative concentration of Na^+ in the soil is much higher than the K^+ concentration. Thus, the plants are unable to accumulate the most favorable inorganic ion (K^+) as an osmolyte. Contrary to the glycophytes, the halophytes adapted to use the Na^+ as the cheap osmoticum to maintain cellular turgor pressure, cell elongation, and stomatal operation (Zhao et al. 2020). The halophytes achieved this ability by efficiently sequestering the Na^+ and Cl^- to the vacuole and organic osmolyte only for the cytosol, which contributes only 10% of the cell volume and thus energetically cheaper to synthesize than the osmolyte for a whole cell (Zhao et al. 2020). Succulence is another necessary morphological adaptation of some of the halophytes from the Chenopodioideae, and Salicornioideae order (Flowers et al. 2015), however, the detailed mechanism of succulence development are not understood yet (Qi et al. 2009). The succulent cells in the halophytes provide them the ability to store the excess Na^+ and K^+ in those cells, higher H^+-ATPase activity, and nonenzymatic antioxidant activity in this tissue (Zeng et al. 2018; Zhao et al. 2020). Moreover, these succulent cells retain a constitutively lower number of open SV vacuolar channels and suppress the activity of FV in the vacuole.

2.5.3 Physiological and Metabolic Adaptation of Halophytes

Osmotic shock to the plants in salt stress drops the xylem pressure. The guard cells perceive the drop in xylem pressure and different salinity-induced signaling cascades, to close the stomata with a purpose to minimize water loss. This closure of the stomata or reduction in the stomatal conductance poses a penalty on the plants by reduced CO_2 assimilation and decreased growth rate of the plants. The stomatal density and the stomatal aperture size regulate the gaseous exchange through the stomata. The halophytes adapted themselves efficiently to control the stomatal aperture and the stomatal density in salt stress with a minimal negative impact on photosynthesis (Zhao et al. 2020). The level of ABA content in the xylem sap and leaves were maintained much lower in halophytes in comparison with the glycophytes (Hedrich and Shabala 2018). More strikingly, the halophyte guard cells have a different sensitivity to the ROS produced during the salt stress, and the Na^+ regulates the opening and closing of the stomatal aperture in halophytes contrary to the K^+ in the glycophytes (Chiang et al. 2016). This adaptation provides advantages to the halophytes over glycophytes in maintaining the better gas exchange and the efficient photosynthetic ability to the halophytes in saline soil condition. Halophytes change the architecture of the photosynthetic complexes by omitting the salt-stress-sensitive component from the complexes and avoid the ROS-induced photoinhibition of PSII and PSI (Pagliano et al. 2009; Trotta et al. 2012). For maintaining ROS homeostasis during photosynthesis, halophytes use Na^+ accumulated in the chloroplast, for the exchange of ascorbate, pyruvate, and phosphate to the chloroplast through different Na^+ transporters on the chloroplast membrane

(Bose et al. 2017). The activity of the enzymes involved in the Calvin–Benson–Bassham cycle could be highly sensitive to the change of the ionic balance and availability in the chloroplast during the salt stress. This example comes from the enzyme FBPase of rice whose *in vitro* activity was suppressed at very low concentration of the NaCl, whereas the same enzyme from a close halophytic relative of rice, *Porteresia coarctata*, showed very less inhibition on its activity. The sequence analysis of *P. coarctata* from rice and *P. coarctata* exhibited a mutation of a few amino acids in PcFBPase, which resulted in reduced sensitivity of the enzyme to the salt stress (Ghosh et al. 2001).

2.6 Halophytes in Agriculture and Land Management

Approximately, 90% of the cultivated crops are sensitive to salt stress (Zörb et al. 2019), and thus for the irrigation of these crops, freshwater is used extensively. An estimation revealed that 70% of all available freshwater is used for irrigation at the global level. In comparison, in developing countries, the statistics reach 90% (Panta et al. 2016). Thus, the use of crop species with better water-use efficiency may provide an alternative to this adversary and save freshwater resources. However, with the increasing trend of the soil salinization either due to the natural cause or due to the anthropogenic activity, this approach seems to be less realistic. In the past decade, several research articles were published to provide information about the improved salt-stress tolerance in different crops under the lab conditions. However, due to the social concern and strict regulatory processes, only a few have been tested in the field conditions. The fact is that transferring only one salt-stress responsive gene in a crop is not sufficient to improve the salt-stress tolerance in the crop plants under the field conditions. Instead, it may be possible by pyramiding of several regulatory and stress-tolerance genes, which may require several years of research and economic costs. The better alternative strategy would be to use the halophytic plant species in the land degraded by salinity and use of the saline water for the irrigation of these plants. There is no limitation of the saline water coming either from the sea or as the industrial wastewater. This strategy provides no competition for the available freshwater, and the wasteland will be recycled. Fortunately, in recent years, several halophytic plants are being cultivated for the food, fodder, fuel production purposes, and landscaping purpose (Panta et al. 2014, 2016). *Salicornia* is a halophyte from the Amaranthaceae family, which grows in the temperate and subtropical coastal areas. The plants from this genus are being used to produce the cooking oil, biomass with high fiber contents, and fresh vegetable products (Glenn et al. 1991). However, in India, only one species represent the genus *Salicornia* named as *Salicornia brachiata* Roxb. (Chaudhary et al. 2018). In the coastal salt marshes of Gujarat, India, the *S. brachiata* is a dominant halophyte. It serves as a model plant for the study and understanding of the salt tolerance mechanisms in the plant. The *S. brachiata* serves as the source of several novel abiotic stress-responsive genes (Mishra and Tanna 2017; Singh et al. 2016) which are shown to improve the salinity tolerance in the glycophytes through transgenic approach and novel stress-responsive alternative promoters for the genetic engineering of the crops (Tiwari et al. 2019). Traditionally, the *S. brachiata* at their early developmental stages are used to prepare vegetable salad and pickles, and the seed oil is rich with the polyunsaturated fatty acids (Chaudhary et al. 2018). The *S. brachiata* can accumulate salt up to 40% of the dry biomass (Chaudhary et al. 2018), suggesting *S. brachiata* as a tool for the phytoremediation of the land degraded by the salinity. The halophytic shrub *Atriplex lentiformis* has been cultivated as forages, revegetation of the salinity-degraded land, and rangeland enhancement programs across the globe (Jordan et al. 2009).

Halophytes to be consumed as food, are getting popularity, and the best example is the halophyte quinoa (*Chenopodium quinoa*), which can tolerate the 40 dS/m of salinity. The quinoa is gluten-free and rich in vitamins and minerals, which helped it gain popularity even in countries that are not affected by the salinity (Panta et al. 2014). The halophyte *Atriplex triangularis* can be grown in the soil salinity of 30 g/l with the yield potential of 21.2 t/ha fresh weight (Gallagher 1985). The taste of leaves from the *A. triangularis* is similar to the spinach. Therefore, it is a component of the human food in European countries like the Netherlands, Belgium, and Portugal (Panta et al. 2014). The use of halophytes grown in the wasteland or salinity-affected land as forage may reduce the load on agricultural land and freshwater resources for cultivation of forage. However, the candidate halophyte selection depends on their biomass productivity, nutrient quantity, and quality, which should not affect livestock productivity. The halophytes saltbush (*Atriplex* spp.) and blue bush (*Maireana* spp.) are being cultivated in salt-affected coastal areas and used as a forage crop for feeding the animals. Several halophytes were grown successfully using the diluted seawater irrigation practices and feeding the cattle of the poultry and meat industry without affecting the quality and productivity of the meat (Panta et al. 2014).

Production of halophytes as the oilseed provides another dimension of the use of the salinity-degraded land and gain alternative agriculture practices in the salt-stress conditions. Several species from the genus *Salicornia*, *Suaeda*, and other halophytes showed no accumulation of the salt in the seed, no matter what saline conditions they were grown (Panta et al. 2014). The *Salicornia bigelovii* showed the higher yield potential of the seeds, even at the soil salinity equivalent to 70 g/l NaCl concentration (Glenn et al. 1999). When the *S. bigelovii* growing area was irrigated with the seawater having salt concentration equivalent to 40 g/l, it was able to produce seed at a rate of 2 t/ha (Glenn et al. 1999). This yield potential is equivalent to the oilseed production from the conventional crop in the nonsaline conditions (Panta et al. 2014). It must be noted that *S. bigelovii* not only has the higher yield potential in the salt-affected areas but its seed is also rich in the oil (30%) with similar properties to the oil derived from the safflower (Glenn et al. 1991; Zerai et al. 2010) and protein (35%) content (Panta et al. 2014). These findings provide hope to the farmers across the globe whose agricultural land is degraded due to the salinity and can help us minimize demand load on the glycophytic crop to increase their productivity to feed the growing population. The higher oil yield potential of the halophyte seed opens the opportunity to use the oil produced from the halophytes for biodiesel production without competing with oilseed crops used for food. In a trial at the Mexico coast for oilseed productivity, the *S. bigelovii* showed higher seed productivity in seawater irrigated fields than the other oil seeds, such as the mustard, safflower, and sunflower grown under nonsaline conditions (Glenn et al. 2009; Panta et al. 2014). The perennial halophytic *Tamarix* species can grow and survive in drought and flooding conditions and is salinity tolerant (Panta et al. 2014). The plant has very high yield potential ranging from 19.5 to 52 t/ha/year and heat value relative to standard coal at the ratio of 1/0.7 (Panta et al. 2014). Higher heat value and biomass yield potential make it an important halophytic candidate for the thermal electricity generation and lignocellulosic second generation biofuel production. The cooperation of Boeing, Etihad Airways, and UOP Honeywell with the Masdar Institute of Science and Technology to grow *Salicornia* species at the Abu Dhabi for production of aviation industry-grade biofuel can exemplify the importance of biofuel production using the halophytes (Panta et al. 2014).

The unique ability of halophytes to accumulate salt ions in the root and shoot vacuole with very low leakage provides them the unique ability to extract salts from the soil. It thus may act in the desalinization of degraded land. Some halophytes such as *S. brachiata* (40% of dry weight) and *Atriplex* sp. (39% of the dry weight) serve as the salt accumulator (Barrett-Lennard 2002) and therefore can be extensively used for the phytoremediation. Panta et al. (2014) have provided a table of the halophytes from the previously published report, which describes different species from the genus *Suaeda, Atriplex, Tectocornia, Sesuvium, Anthrocnemum, Excoecarcia, Ipomoea, Batis,* and *Salicornia*, which were used in past or have a high potential in coming future for the desalinization and removal of the heavy metals from the degraded soil.

2.7 Conclusion and Future Perspectives

Salt stress affects plants' survival and productivity. Research on understanding the mechanism of salinity stress gave us significant insight into the plants' response to salt stress and how halophytes respond differently than the glycophytes. However, the complete mechanistic understanding of the salt tolerance mechanism in the halophytes at the developmental and intracellular molecular levels is lacking. To meet the food, feed, fiber, and fuel demand of the growing human population and improve the abiotic stress resilience of the crop, there are two ways to pursue in parallel; one, adopting the halophyte for food, forage, and fuel generation by growing them in the salt-affected land, and, second, understanding the salt-stress tolerance mechanism in the halophytes and then engineering the genome of glycophytic crop plants for better salt-stress tolerance. Halophytes grown by seawater irrigation have shown their significance by producing a higher yield of oil and protein-rich seeds than a conventional oilseed crop, which suggested that the salt-degraded land can be used for growing halophytes for edible oil or biofuel. Halophytes were also used to restore the wasteland by desalinization and phytoremediation of the heavy metals, used as forage for the cattle and vegetable salads in different parts of the world. The halophyte *C. quinoa* seeds are rich in minerals and vitamins and gained popularity as mainstream food because of its gluten-free nature. For improving the salt-stress tolerance in the crop plants, the current requirement is to understand salt-stress sensory system at the root, root to shoot signaling, and intracellular sensory and responses at the cell organelles such as chloroplasts, mitochondria, or peroxisomes in halophytes. In halophytes, yet very little is known about the development and functioning mechanism of the EBC or different types of the salt glands. Understanding these mechanisms and engineering, the crop using the genome-editing approach may improve the salt-stress tolerance of the crops within a shorter time frame.

References

Abbasi, F.M. and Komatsu, S. (2004). A proteomic approach to analyze salt-responsive proteins in rice leaf sheath. *Proteomics* 4: 2072–2081.

Babu, G.A. and Reddy, S.M. (2011). Diversity of arbuscular mycorrhizal fungi associated with plants growing in fly ash pond and their potential role in ecological restoration. *Curr. Microbiol.* 63: 273–280.

Bao, H., Chen, X., Lv, S. et al. (2015). Virus-induced gene silencing reveals control of reactive oxygen species accumulation and salt tolerance in tomato by γ-aminobutyric acid metabolic pathway. *Plant Cell Environ.* 38: 600–613.

Barrett-Lennard, E.G. (2002). Restoration of saline land through revegetation. *Agric. Water Manag.* 53: 213–226.

Bose, J., Munns, R., Shabala, S. et al. (2017). Chloroplast function and ion regulation in plants growing on saline soils: lessons from halophytes. *J. Exp. Bot.* 68: 3129–3143.

Busch, K., Piehler, J., and Fromm, H. (2000). Plant succinic semialdehyde dehydrogenase: dissection of nucleotide binding by surface plasmon resonance and fluorescence spectroscopy. *Biochemistry* 39: 10110–10117.

Chattopadhyay, A., Subba, P., Pandey, A. et al. (2011). Analysis of the grasspea proteome and identification of stress-responsive proteins upon exposure to high salinity, low temperature, and abscisic acid treatment. *Phytochemistry* 72: 1293–1307.

Chaudhary, D.R., Rathore, A.P., and Jha, B. (2018). Aboveground, belowground biomass and nutrients pool in *Salicornia brachiata* at coastal area of India: interactive effects of soil characteristics. *Ecol. Res.* 33: 1207–1218.

Che-Othman, M.H., Millar, A.H., and Taylor, N.L. (2017). Connecting salt stress signalling pathways with salinity-induced changes in mitochondrial metabolic processes in C3 plants. *Plant Cell Environ.* 40: 2875–2905.

Chiang, C.P., Yim, W.C., Sun, Y.H. et al. (2016). Identification of ice plant (*Mesembryanthemum crystallinum* L.) microRNAs using RNA-seq and their putative roles in high salinity responses in seedlings. *Front. Plant Sci.* 7: 1143.

Choi, W.G., Toyota, M., Kim, S.H. et al. (2014). Salt stress-induced Ca^{2+} waves are associated with rapid, long-distance root-to-shoot signaling in plants. *Proc. Natl. Acad. Sci. U. S. A.* 111: 6497–6502.

Christmann, A., Grill, E., and Huang, J. (2013). Hydraulic signals in long-distance signaling. *Curr. Opin. Plant Biol.* 16: 203–300.

Dassanayake, M. and Larkin, J.C. (2017). Making plants break a sweat: the structure, function, and evolution of plant salt glands. *Front. Plant Sci.* 8: 1–20.

Dumont, S. and Rivoal, J. (2019). Consequences of oxidative stress on plant glycolytic and respiratory metabolism. *Front. Plant Sci.* 10: 1–16.

Fahy, D., Sanad, M.N.M.E., Duscha, K. et al. (2017). Impact of salt stress, cell death, and autophagy on peroxisomes: Quantitative and morphological analyses using small fluorescent probe N-BODIPY. *Sci. Rep.* 7: 1–17.

FAO (2005). Salt-Affected Soils from Sea Water Intrusion: Strategies for Rehabilitation and Management. Report of the regional workshop on salt-affected soils from sea water intrusion: Strategies for rehabilitation and management, RAP PUBLICATION 2005/11. Bangkok, Thailand 6 pp.

Fernie, A.R., Carrari, F., and Sweetlove, L.J. (2004). Respiratory metabolism: glycolysis, the TCA cycle and mitochondrial electron transport. *Curr. Opin. Plant Biol.* 7: 254–261.

Finazzi, G., Petroutsos, D., Tomizioli, M. et al. (2015). Ions channels/transporters and chloroplast regulation. *Cell Calcium* 58: 86–97.

Flowers, T.J. (1972). Salt tolerance in *Suaeda maritima* (L.) dum: the effect of sodium chloride on growth, respiration, and soluble enzymes in a comparative study with Pisum sativum L. *J. Exp. Bot.* 23: 310–321.

Flowers, T.J. and Colmer, T.D. (2008). Salinity tolerance in halophytes. *New Phytol.* 179: 945–963.

Flowers, T.J., Troke, P.F., and Yeo, A.R. (1977). The mechanism of salt tolerance in halophytes. *Annu. Rev. Plant Physiol.* 28: 89–121.

Flowers, T.J., Munns, R., and Colmer, T.D. (2015). Sodium chloride toxicity and the cellular basis of salt tolerance in halophytes. *Ann. Bot.* 115: 419–431.

Fricke, W., Akhiyarova, G., Veselov, D., and Kudoyarova, G. (2004). Rapid and tissue-specific changes in ABA and in growth rate in response to salinity in barley leaves. *J. Exp. Bot.* 55: 1115–1123.

Gallagher, J.L. (1985). Halophytic crops for cultivation at seawater salinity. *Plant Soil* 89: 323–326.

Galvan-Ampudia, C.S., Julkowska, M.M., Darwish, E. et al. (2013). Halotropism is a response of plant roots to avoid a saline environment. *Curr. Biol.* 23: 2044–2050.

Ghosh, S., Bagchi, S., and LahiriMajumder, A. (2001). Chloroplast fructose-1,6-bisphosphatase from Oryza differs in salt tolerance property from the Porteresia enzyme and is protected by osmolytes. *Plant Sci.* 160: 1171–1181.

Glenn, E.P., O'Leary, J.W., Watson, M.C. et al. (1991). *Salicornia bigelovii* Torr.: an oilseed halophyte for seawater irrigation. *Science* 251: 1065–1067.

Glenn, E.P., Brown, J.J., and Blumwald, E. (1999). Salt tolerance and crop potential of halophytes. *CRC Crit. Rev. Plant Sci.* 18: 227–255.

Glenn, E.P., Mckeon, C., Gerhart, V. et al. (2009). Deficit irrigation of a landscape halophyte for reuse of saline waste water in a desert city. *Landsc Urban Plan* 89: 57–64.

Hamilton, E.W. and Heckathorn, S.A. (2001). Mitochondrial adaptations to NaCl. Complex I is protected by anti-oxidants and small heat shock proteins, whereas complex II is protected by proline and betaine. *Plant Physiol.* 126: 1266–1274.

He, Y., Fu, J., Yu, C. et al. (2015). Increasing cyclic electron flow is related to Na^+ sequestration into vacuoles for salt tolerance in soybean. *J. Exp. Bot.* 66: 6877–6889.

Hedrich, R. and Shabala, S. (2018). Stomata in a saline world. *Curr. Opin. Plant Biol.* 46: 87–95.

Hernández, J.A., Olmos, E., Corpas, F.J. et al. (1995). Salt-induced oxidative stress in chloroplasts of pea plants. *Plant Sci.* 105: 151–167.

Hong, M., Li, N., Li, J. et al. (2019). Adenosine monophosphate-activated protein kinase signaling regulates lipid metabolism in response to salinity stress in the red-eared slider turtle *Trachemys scripta elegans*. *Front. Physiol.* 10: 1–11.

Ishikawa, T., Cuin, T.A., Bazihizina, N., and Shabala, S. (2018). Xylem ion loading and its implications for plant abiotic stress tolerance. *Adv. Bot. Res.* 87: 267–301.

Jacoby, R.P., Taylor, N.L., and Millar, A.H. (2011). The role of mitochondrial respiration in salinity tolerance. *Trends Plant Sci.* 16: 614–623.

Jajoo, A., Dube, A., and Bharti, S. (1994). Mg^{2+}-induced lipid phase transition in thylakoid membranes is reversed by anions. *Biochem. Biophys. Res. Commun.* 202: 1724–1730.

Jajoo, A., Bharti, S., and Kawamori, A. (2005). Interactions of chloride and formate at the donor and the acceptor side of photosystem II. *J. Bioenerg. Biomembr.* 37: 49–54.

Jeschke, W.D., Wolf, O., and Hartung, W. (1992). Effect of NaCI salinity on flows and partitioning of C, N, and mineral ions in whole plants of white lupin, *Lupinus albus* L. *J. Exp. Bot.* 43: 777–788.

Jha, A., Joshi, M., Yadav, N.S. et al. (2011). Cloning and characterization of the Salicornia brachiata Na^+/H^+ antiporter gene *SbNHX1* and its expression by abiotic stress. *Mol. Biol. Rep.* 38: 1965–1973.

Jordan, F.L., Yoklic, M., Morino, K. et al. (2009). Consumptive water use and stomatal conductance of *Atriplex lentiformis* irrigated with industrial brine in a desert irrigation district. *Agric. For. Meteorol.* 149: 899–912.

Kant, S., Kant, P., Raveh, E., and Barak, S. (2006). Evidence that differential gene expression between the halophyte, *Thellungiella halophila*, and *Arabidopsis thaliana* is responsible for higher levels of the compatible osmolyteproline and tight control of Na+ uptake in *T. halophila*. *Plant Cell Environ.* 29: 1220–1234.

Kazachkova, Y., Batushansky, A., Cisneros, A. et al. (2013). Growth platform-dependent and -independent phenotypic and metabolic responses of Arabidopsis and its halophytic relative, *Eutrema salsugineum*, to salt stress. *Plant Physiol.* 162: 1583–1598.

Kim, D.W., Rakwal, R., Agrawal, G.K. et al. (2005). A hydroponic rice seedling culture model system for investigating proteome of salt stress in rice leaf. *Electrophoresis* 26: 4521–4539.

Kinnersley, A.M. and Turano, F.J. (2010). Gamma Aminobutyric Acid (GABA) and plant responses to stress. *CRC Crit. Rev. Plant Sci.* 2689: 37–41.

Kirchhoff, H., Hall, C., Wood, M. et al. (2011). Dynamic control of protein diffusion within the granal thylakoid lumen. *Proc. Natl. Acad. Sci. U. S. A.* 108: 20248–20253.

Kosova, K., Prasil, I.T., and Vitamvas, P. (2013). Protein contribution to plant salinity response and tolerance acquisition. *Int. J. Mol. Sci.* 14: 6757–6789.

Kumari, N., Malik, K., Rani, B. et al. (2019). Insights in the physiological, biochemical and molecular basis of salt stress tolerance in plants. In: *Microorganisms in Saline Environments: Strategies and Functions* (eds. B. Giri and A. Verma), 353–374. Switzerland: Springer Nature.

Lambers, H., Chapin, F.S., and Pons, T.L. (2008). Respiration. In: *Plant Physiological Ecology*, 2e, 101–150. New York: Springer.

Mishra, A. and Tanna, B. (2017). Halophytes: potential resources for salt stress tolerance genes and promoters. *Front. Plant Sci.* 8: 1–10.

Mitsuya, S., El-Shami, M., Sparkes, I.A. et al. (2010). Salt stress causes peroxisome proliferation, but inducing peroxisome proliferation does not improve NaCI tolerance in *Arabidopsis thaliana*. *PLoS One* 5: e9408.

Munns, R. and Tester, M. (2008). Mechanisms of salinity tolerance. *Annu. Rev. Plant Biol.* 59: 651–681.

Munns, R., Guo, J., Passioura, J.B., and Cramer, G.R. (2000). Leaf water status controls day-time but not daily rates of leaf expansion in salt-treated barley. *Aust. J. Plant Physiol.* 27: 949–957.

Munns, R., James, R.A., Xu, B. et al. (2012). Wheat grain yield on saline soils is improved by an ancestral Na^+ transporter gene. *Nat. Biotechnol.* 30: 360–364.

Munns, R., Day, D.A., Fricke, W. et al. (2020). Energy costs of salt tolerance in crop plants. *New Phytol.* 225: 1072–1090.

Murata, N., Takahashi, S., Nishiyama, Y., and Allakhverdiev, S.I. (2007). Photoinhibition of photosystem II under environmental stress. *Biochem.Biophys. Acta. - Bioenerg.* 1767: 414–421.

Ondrasek, G., Rengel, Z., and Veres, S. (2011). Soil salinisation and salt stress in crop production. In: *Abiotic Stress in Plants - Mechanisms and Adaptations* (ed. A. Shanker), 171–190. Croatia: InTech.

Pagliano, C., La Rocca, N., Andreucci, F. et al. (2009). The extreme halophyte *Salicornia veneta* is depleted of the extrinsic PsbQ and PsbP proteins of the oxygen-evolving complex without loss of functional activity. *Ann. Bot.* 103: 505–515.

Panta, S., Flowers, T., Lane, P. et al. (2014). Halophyte agriculture: success stories. *Environ. Exp. Bot.* 107: 71–83.

Panta, S., Flowers, T., Doyle, R. et al. (2016). Growth responses of *Atriplex lentiformis* and *Medicago arborea* in three soil types treated with saline water irrigation. *Environ. Exp. Bot.* 128: 39–50.

Pattanayak, G.K. and Tripathy, B.C. (2002). Catalytic function of a novel protein protochlorophyllide oxidoreductase C of *Arabidopsis thaliana*. *Biochem. Biophys. Res. Commun.* 291: 921–924.

Pavlovic, I., Mlinaric, S., Tarkowska, D. et al. (2019). Early *Brassica* crops responses to salinity stress: a comparative analysis between chinese cabbage, white cabbage, and kale. *Front. Plant Sci.* 10: 1–16.

Plaxton, W.C. (1996). The organization and regulation of plant glycolysis. *Annu. Rev. Plant Physiol. Plant Mol. Biol.* 47: 185–214.

Porcel, R., Aroca, R., and Ruiz-Lozano, J.M. (2012). Salinity stress alleviation using arbuscular mycorrhizal fungi. A review. *Agron. Sustain. Dev.* 32: 181–200.

Qadir, M., Quillérou, E., Nangia, V. et al. (2014). Economics of salt-induced land degradation and restoration. *Nat. Resour. Forum* 38: 282–295.

Qi, C.H., Chen, M., Song, J., and Wang, B.S. (2009). Increase in aquaporin activity is involved in leaf succulence of the euhalophyte *Suaeda salsa*, under salinity. *Plant Sci.* 176: 200–205.

Renault, H., Roussel, V., El Amrani, A. et al. (2010). The *Arabidopsis pop2-1* mutant reveals the involvement of GABA transaminase in salt stress tolerance. *BMC Plant Biol.* 10: 1–16.

Santos, C.V. (2004). Regulation of chlorophyll biosynthesis and degradation by salt stress in sunflower leaves. *Sci. Hortic. (Amsterdam)* 103: 93–99.

Sasi, S., Venkatesh, J., Daneshi, R.F., and Gururani, M.A. (2018). Photosystem II extrinsic proteins and their putative role in abiotic stress tolerance in higher plants. *Plan. Theory* 7: 100.

Satir, O. and Berberoglu, S. (2016). Crop yield prediction under soil salinity using satellite derived vegetation indices. *F. Crop Res.* 192: 134–143.

Shabala, S.N. and Lew, R.R. (2002). Turgor regulation in osmotically stressed arabidopsis epidermal root cells. Direct support for the role of inorganic ion uptake as revealed by concurrent flux and cell turgor measurements. *Plant Physiol.* 129: 290–299.

Shabala, S. and Shabala, L. (2011). Ion transport and osmotic adjustment in plants and bacteria. *Biomol. Concepts* 5: 407–419.

Shabala, L., Zhang, J., Pottosin, I. et al. (2016). Cell-type-specific H^+-ATPase activity in root tissues enables K^+ retention and mediates acclimation of barley (*Hordeum vulgare*) to salinity stress. *Plant Physiol.* 172: 2445–2458.

Shabala, S., Chen, G., Chen, Z.H., and Pottosin, I. (2020). The energy cost of the tonoplast futile sodium leak. *New Phytol. 225* (3): 1105–1110.

Shi, H., Quintero, F.J., Pardo, J.M., and Zhu, J.K. (2002). The putative plasma membrane Na^+/H^+ antiporter SOS1 controls long-distance Na^+ transport in plants. *Plant Cell* 14: 456–477.

Singh, D., Yadav, N.S., Tiwari, V. et al. (2016). A SNARE-like superfamily protein SbSLSP from the halophyte Salicornia brachiata confers salt and drought tolerance by maintaining membrane stability, K^+/Na^+ ratio, and antioxidant machinery. *Front. Plant Sci.* 7: 737.

Singh-Rawal, P., Zsiros, O., Bharti, S. et al. (2011). Mechanism of action of anions on the electron transport chain in thylakoid membranes of higher plants. *J. Bioenerg. Biomembr.* 43: 195–202.

Snapp, S.S., Shennan, C., and Van Bruggen, A.H.C. (1991). Effects of salinity on severity of infection by *Phytophthora parasitica* Dast., ion concentrations and growth of tomato, *Lycopersicon esculentum* Mill. *New Phytol.* 119: 275–284.

Song, C.P., Guo, Y., Qiu, Q. et al. (2004). A probable $Na^+(K^+)/H^+$ exchanger on the chloroplast envelope functions in pH homeostasis and chloroplast development in *Arabidopsis thaliana*. *Proc. Natl. Acad. Sci. U. S. A.* 101: 10211–10216.

de Souza Lima, M.D., Lopes, N.F., Zimmer, P.D. et al. (2012). Enzyme expression in indica and japonica rice cultivars under saline stress. *Acta. Sci. Biol. Sci.* 34: 473–481.

Stepien, P. and Johnson, G.N. (2009). Contrasting responses of photosynthesis to salt stress in the glycophyte Arabidopsis and the halophyte Thellungiella: Role of the plastid terminal oxidase as an alternative electron sink. *Plant Physiol.* 149: 1154–1165.

Sudhir, P. and Murthy, S.D.S. (2004). Effects of salt stress on basic processes of photosynthesis. *Photosynthetica* 42: 481–486.

Sui, N., Yang, Z., Liu, M., and Wang, B. (2015). Identification and transcriptomic profiling of genes involved in increasing sugar content during salt stress in sweet sorghum leaves. *BMC Genomics* 16: 1–18.

Sweetlove, L.J., Beard, K.F.M., Nunes-Nesi, A. et al. (2010). Not just a circle: flux modes in the plant TCA cycle. *Trends Plant Sci.* 15: 462–470.

Szabolcs, I. (1989). *Salt-Affected Soils*. CRC Press, Inc.

Taïbi, K., Taïbi, F., Abderrahim, A. et al. (2016). Effect of salt stress on growth, chlorophyll content, lipid peroxidation and antioxidant defence systems in *Phaseolus vulgaris* L. *South African. Aust. J. Bot.* 105: 306–312.

Tiwari, V., Patel, M.K., Chaturvedi, A.K. et al. (2019). Cloning and functional characterization of the Na^+/H^+ antiporter (NHX1) gene promoter from an extreme halophyte *Salicornia brachiata*. *Gene* 683: 233–242.

Tripathi, R.S. (2009). *Alkali Land Reclamation*. New Delhi: Mittal Publications.

Tripathi, S., Kumari, S., Chakraborty, A. et al. (2006). Microbial biomass and its activities in salt-affected coastal soils. *Biol. Fertil. Soils* 42: 273–277.

Trotta, A., Redondo-Gomez, S., Pagliano, C. et al. (2012). Chloroplast ultrastructure and thylakoid polypeptide composition are affected by different salt concentrations in the halophytic plant *Arthrocnemum macrostachyum*. *J. Plant Physiol.* 169: 111–116.

Turan, S. and Tripathy, B.C. (2015). Salt-stress induced modulation of chlorophyll biosynthesis during de-etiolation of rice seedlings. *Physiol. Plant.* 153: 477–491.

Wong, D., Jee, G., and Merkelo, H. (1980). Effects of bulk pH and of monovalent and divalent cations on chlorophyll a fluorescence and electron transport in pea thylakoids. *Biochem. Biophys. Acta. - Bioenerg.* 592: 546–558.

Wu, H.J., Zhang, Z., Wang, J.Y. et al. (2012). Insights into salt tolerance from the genome of *Thellungiella salsuginea*. *Proc. Natl. Acad. Sci. U. S. A.* 109: 12219–12224.

Xu, J., Feng, Y., Wang, Y. et al. (2016). The foliar spray of *Rhodopseudomonas palustris* grown under Stevia residue extract promotes plant growth via changing soil microbial community. *J. Soils Sediments* 16: 916–923.

Yang, Z., Li, J.L., Liu, L.N. et al. (2020). Photosynthetic regulation under salt stress and salt-tolerance mechanism of sweet Sorghum. *Front. Plant Sci.* 10: 1–12.

van Zelm, E., Zhang, Y., and Testerink, C. (2020). Salt tolerance mechanisms of plants. *Annu. Rev. Plant Biol.* 71: 403–433.

Zeng, F., Shabala, S., Maksimović, J.D. et al. (2018). Revealing mechanisms of salinity tissue tolerance in succulent halophytes: a case study for *Carpobrotus rossi*. *Plant Cell Environ.* 41: 2654–2667.

Zerai, D.B., Glenn, E.P., Chatervedi, R. et al. (2010). Potential for the improvement of *Salicornia bigelovii* through selective breeding. *Ecol. Eng.* 36: 730–739.

Zhang, W., Wang, C., Xue, R., and Wang, L.J. (2019). Effects of salinity on the soil microbial community and soil fertility. *J. Integr. Agric.* 18: 1360–1368.

Zhao, C., Zhang, H., Song, C. et al. (2020). Mechanisms of plant responses and adaptation to soil salinity. *Innovation* 1: 100017.

Zörb, C., Geilfus, C.M., and Dietz, K.J. (2019). Salinity and crop yield. *Plant Biol.* 21: 31–38.

3

Effects of Salt Stress on Biochemistry of Crop Plants

Poonam Yadav and Durgesh Kumar Jaiswal

Institute of Environment and Sustainable Development, Banaras Hindu University, Varanasi, Uttar Pradesh, India

3.1 Introduction

Soil salinity is one of the challenging threats to agricultural productivity and sustainability. The salinity problem is worldwide and observed under all climatic and geographical conditions. It occurs both naturally and anthropologically. Saline soils are mainly reported from arid and semiarid regions where rainfall is limited to meet the crop water demand, and mineral salts are leached out of the root region. It has been threatening us for centuries. A fine example is observed from Mesopotamia (now Iraq) and the Viru valley of Peru, where well-flourished societies based on irrigated agriculture have failed due to man-made salinization (Shahid et al. 2018). Soluble salt species, such as Ca^{2+}, Na^+, Mg^{2+} and $SO_4{}^{2-}$, $HCO_3{}^-$, Cl^- along with K^+, $CO_3{}^{2-}$, and $NO^-{}_3$, contribute to the soil salinity. If electrical conductivity (EC) of a soil is reported as ≥4 dS/m (equivalent to ≈40 mM NaCl) along with an osmotic pressure of approximately 0.2 MPa, a soil can be termed as saline. Thus, salinity can also be defined as the soil condition when the EC is sufficient to cause growth and yield reduction of most crops. Generally, pH range between 7 and 8.5 is regarded as the pH of saline soils (Mengel et al. 2001). Soil salinity is caused chiefly by over-irrigation, excessive use of fertilizers, flooding, silting, seepage, and a rising water table. Till now, no exact statistics on global salinity exist; however, scientific community reported the various soil differently, for e.g., 10% of the total cultivated land is saline and sodic, sodic soils cover one billion ha of land, and between 30 and 40% of irrigated lands are saline and unproductive. Globally, salt-affected soils are 954 million ha and as per the FAO in 1988, salt-affected soils were 932 million ha, nearly 32 million of ≈1500million ha dryland agriculture is salt affected. Estimations show that as much as 12 billion US$ will be lost globally each year due to the reduction in agricultural production from salt-affected areas (Qadir et al. 2008; Flowers et al. 2010).

In India, many states such as Rajasthan, Punjab, Haryana, Maharashtra, Orissa, West Bengal delta region, Madhya Pradesh, and Karnataka have so far been affected by soil salinity problems. Depending upon the plants and abiotic/biotic stresses, an imbalance in

Physiology of Salt stress in Plants: Perception, Signalling, Omics and Tolerance Mechanism, First Edition.
Edited by Pratibha Singh, Madhulika Singh, Rajiv Kumar Singh, and Sheo Mohan Prasad.

reactive oxygen species (ROS) and reactive nitrogen species (RNS) metabolisms can occur. This leads to nitro-oxidative stress in plants' cell, and hence antioxidant systems (enzymatic and nonenzymatic), plants' redox status (NADPH, NADH, and GSH/GSSG), and other responsive/defensive systems (hormones and calcium ion signaling) being affected differently (Asgher et al. 2017; Yadav et al. 2020).

Environmental stresses such as heavy metal(loids) toxicity, drought, salinity, chilling, and UV-B radiation as well as pathogens attack lead to enhanced production of ROS in plants due to disruption of cellular homeostasis (Han et al. 2009; Srivastava and Dubey 2011; Sharma et al. 2012; Poonam and Srivastava 2019). ROS generates in low concentration in plants under normal functioning of cells, but all the forms of ROS are extremely harmful to plants and organisms when produced at high concentrations during any kind of stress. When the level of ROS exceeds the optimum concentration and beyond the control of defense machinery, a cell is said to be in a state of "oxidative stress." The enhanced production of ROS during stresses can pose a serious threat to cells by causing damage to nucleic acids, oxidation of proteins, peroxidation of lipids, enzyme inhibition, protein metabolism, amino acid (AA) synthesis, photosynthesis, damage to protein hence to AA and DNA, enzymatic imbalance, redox homeostasis, and activation of programmed cell death (PCD) pathway, ultimately leading to the death of the cells. Some other observable effects can also be witnessed clearly in the plant as chlorosis, necrosis, and curling of leaf margin, and these all result in reduced and stunted growth of plants under the salinity or other stress.

There is a wide range of ROS and RNS including $^1O_2$, $O^{2\bullet-}$, H_2O_2, $^{\bullet}OH$, NO^{2-}, NO^-, and others. Two unpaired electrons with a parallel spin in oxygen in its ground state make it harmless as it makes it paramagnetic and, hence, unlikely to participate in reactions with organic molecules unless it is activated (Apel and Hirt 2004). Activation of oxygen follows two possible mechanisms viz., first, absorption of sufficient energy to reverse the spin on one of the unpaired electrons by absorbing the energy by an unpaired electron itself leading to the formation of $^1O_2$ and, second, stepwise monovalent reduction of oxygen molecule where O_2 is sequentially reduced to $O^{2\bullet-}$, H_2O_2, and $^{\bullet}OH$.

In the presence of salt stress, molecular oxygen gets the high-energy electrons from salts and transfers to the ROS which damages the plant cells. During unusual stress conditions, the rate of ROS production get enhanced and the scavenging capability of the cellular defense system also gets activated, resulting in oxidative pressure (Ahanger et al. 2017). Salt stress influences the plant quality, which adversely affects the productiveness of crops and the nature of water stockpiling. Over a span of development and improvement, plants have created numerous physiological and biochemical mechanisms to adjust to the environment stress (Zhu et al. 2014; Zhang et al. 2017), and enormous physiological and metabolic responses have been delivered in plants to oblige this modification (Apel and Hirt 2004; Lata and Prasad 2011). A significant degree of effect under ROS can influence numerous functions of plant cells by damaging nucleic acids only (Imlay and Linn 1988).

ROS-initiated damage to nucleic acid creates genotoxic stress in plants (Tuteja et al. 2009). In parallel, on sensing the stress at the root level, plant sends a signal to its distant organs for preparedness of defense systems. These signaling molecules are H_2O_2 calcium ion (Ca^{2+}) signaling, hormonal signaling, etc.

3.2 Effects of Salt Stress on Lipid Metabolism

Plants have a well-developed barrier and defense machinery to fight the stress. Antioxidants play a wide role in fighting the stress and promoting the growth and better health of the plant. In the process of lipid peroxidation, free radicals attack unsaturated fatty acids in the lipid membrane of the cell. These free radicals remove an H atom from the lipid, and thus, a lipid radical is formed, and an oxygen-mediated chain reaction starts in the cell membrane. This leaves the cell membrane damaged with lots of lipid hydroperoxides (nonradical intermediates). These lipid hydroperoxides severely hamper membrane functionality by way of increased rigidity, reduced photosynthetic process, accumulation of toxic substances, allowing ions like calcium to leak across the membrane, etc.

Increased lipid peroxidation of cell membrane is one of the negative effects of oxidative damage under salinity (Hernández and Almansa 2002) and other stresses in plants. Malondialdehyde (MDA) is one of the end products of lipid peroxidation of the cell membrane and is used to measure the level of damage caused by oxidative stress in plants. MDA is a final decomposition product of lipid peroxidation (Poonam and Srivastava 2019). Under the salinity stress, a gradual increase in MDA content in leaves and roots has been reported (You et al. 2019). Similar trends have been reported in crop plants (rice) exposed to other stresses (Poonam and Srivastava 2019; Yadav et al. 2020).

The effect of salinity stress on lipid metabolism varies with the sensitivity of the crops (HediyeSekmen et al. 2007). In a study by Koca et al. (2006), a wild and tolerant variety of tomato (*Lycopersicon pennellii*) showed higher superoxide dismutase (SOD) and peroxidase (POX) activities and lower level of lipid peroxidation than in *Lycopersicon esculentum* (sensitive variety). Wild tolerant variety showed a better defense system for oxidative damage, thus protecting the membrane under short and long durations of stress. Similar results were reported in other studies, showing less damage to lipid membrane with an increased level of SOD and POX in salt-stressed plants, cotton, and senna (*Cassia angustifolia Vahl.*), respectively (Agarwal and Pandey 2004). Controlled lipid peroxidation level in tolerant varieties under salinity stress can be due to the production of an increased level of SOD and POX that help in reducing the oxidative stress and thus lowering the lipid membrane damage by oxidative molecules. Other plants, *Plantago maritima*, maize, pea, rice, etc., have shown a high level of MDA content as a marker of lipid peroxidation (Hernández and Almansa 2002; Hediye Sekmen et al. 2007; AbdElgawad et al. 2016).

AbdElgawad et al. (2016) showed that root of maize seedlings under varying salt stress accumulated more Na^+ ions in comparison with above-ground parts. High oxidative toxicity as a result of high salinity was reported in roots and mature leaves compared with younger leaves which showed increased electrolyte leakage (EL) and MDA (Hernández and Almansa 2002). Under the increased salt stress and lipid peroxidation, glutathione helps in the reduction of stress by electron donor and chelation of products of peroxidation in GSH-S-transferase (GST) catalyzed reaction in the cell cytoplasm. This results in a reduced level of GSH and an increased level of GST under the stress (Marrs 1996). GSH is the predominant cellular nonprotein thiol (NP-SH) found in the plants and regulates sulfur uptake at the root level. It is used by the GSTs in the detoxification of peroxides generated due to metalloid-induced oxidative damage and xenobiotics and is also a precursor of the phytochelatins, which are crucial in controlling the cellular heavy metal concentration. GSH acts as an antioxidant and redox buffer during any kind of stress (Hasan et al. 2016).

3.3 Effects of Salt Stress on the Amino Acids Synthesis and Nitrogen Metabolism

AAs have a very sensitive role in protecting the plants under the stress. They are nonenzymatic components of the biomolecules pool. Glutamate, methionine, glutamine, γ-aminobutyrate, alanine, proline, and aspartate are some of the important AAs playing a critical role in the stress-fighting machinery of the plants. Nitrogen is one of the main binding components of the AAs with other elements and thus plays highly diverse and essential roles and is very essential for plant developmental processes. In addition, AAs are also secreted from the roots as they are the basic element of all the living systems (Chaparro et al. 2013; Mondal et al. 2013) and under the stress have also been observed to be regulated to reduce the stress effects of plants. Thus, AA in root secretion contributes to growth, progression, and stress resistance in plants (Pavlikova et al. 2014). Plants' root either assimilates the nitrogen into AAs or takes up AAs from the soil directly (Tegeder and Rentsch 2010). AAs and nitrogen are the building blocks of enzymes and proteins, thus contributing to the structure and metabolism of the plant (Tegeder and Ward 2012). Osmotic disturbances are the key effects of salt stress, and plant accumulates a number of compatible substances (proteins, AAs, and organic acids) to maintain the cell's natural condition.

Proline, an α AA, is found in very large quantities in plants. It helps mainly in the regulatory process, osmolytes, and stress signaling, thus increasing especially under the stress conditions (Fougere et al. 1991). It also works as an active osmoprotectant (organic solute) found mainly in the roots of the plants. The amount of proline produced is time- and concentration-dependent of the stress. It also varies with plant ecotypes, varieties, and types of stress. The effect of time and variation in the concentration of salt stress on AAs' level has also been discussed in the literature. Xie et al. (2020) reported an increment in AAs level in *Phragmites australis*, at 1% salinity stress, but it was found to decrease at 3 and 6% salinity stress. In addition, the increased level of AAs was mainly due to the increment in proline content. Even, proline content is used as a marker of salt stress in *P. australis*, grown in saltwater wetlands (Xie et al. 2020). Whereas, Zhang et al. (2011) discussed the levels of pyrimidine and purine in tobacco and showed that these two AAs decreased under 50 mM of NaCl salt stress. Additionally, an increment in the levels of proline and aromatic AAs was reported under 500 mM concentration of NaCl. Thus, the level of AAs has an important role in stress mitigation and is marked as an indicator of salt stress in plants.

Salinity is also reported to affect the level of AAs' composition in xylem by increasing the concentration of *Ser, Ala, Gaba,* and *Pro*, and decreasing the *Asn*. A similar effect was noticed for *Asn* and *Ser* in roots with a high increment in *Gaba*. It is hypothesized that in the root, Asn converts to *Ala* and *Gaba* through glutamate (*Glu*). *Pro* and *Gaba* promote the salinity adaptation in the plant (Bouché et al. 2003; Ashraf and Harris 2004; Queiroz et al. 2012; Xie et al. 2020). GABA is also reported to alleviate the salt stress effect during seed germination of white clover (Cheng et al. 2018).

However, the addition of NH_4^+ (ATP-dependent) to the γ-carboxyl group of glutamate is catalyzed by glutamine synthetase (GS) to produce glutamine, the center for nitrogen metabolic pathway in plants. Then, GS further catalyzes the conversion of glutamine and 2-oxoglutarate to produce two molecules of glutamate. One molecule gets involved in further NH_4^+ assimilation through GS, while the other donates nitrogen (reduced) for nitrogen-containing biomolecules of the whole system (Lea and Ireland 1999). GS was observed to increase in many salt-tolerant plants, for example, *Salicornia europaea* (Wang et al. 2009), *Thellungiella salsuginea* (Pang et al. 2010), and rice roots (Yan et al. 2005). Enhanced glutamate ammonia ligase was found in *Sorghum bicolor* in response to salt stress that is important for nitrogen assimilation and AA biosynthesis (Swami et al. 2011).

GABA has been suggested to accumulate in cells under salt stress due to polyamine degradation observed in roots of the soybean cultivar *Suxie-1* (Xing et al. 2007). GABA accumulation and gene expression need more focus with action mechanism under salt stress, as it is reported to be highly regulating protein under other stress. You et al. 2019 reported that lysine catabolism is done by the saccharopine pathway. They further added that some of the saccharopine pathway genes (LKR/SDH and AASADH) were upregulated in osmotic and/or salt-stressed plants.

Stress generated by excess ROS can also affect the protein biosynthesis in the cell by affecting the enzymes of nitrogen metabolism. Nitrate reductase (NR) enzyme is very important in N-cycle and plays a very sensitive role in nitrogen uptake by plants. NR is also very sensitive to high NaCl concentration. Salt trace works as a barrier in nitrate uptake and transport, therefore required assimilation of N is reduced in the amount needed for protein biosynthesis (Maaroufi-Dguimi et al. 2011; Qu et al. 2012; Carillo et al. 2019). It was reported decades back that N_2 fixation by roots decreased under salinity stress (Fougere et al. 1991). Nitrate reduction was also reported in cowpea (Silveira et al. 2001) and cashew (Silveira et al. 2003). This reduction can be related to reduced uptake of nitrate due to interferences caused by Cl^- to membrane transport machinery (Silveira et al. 2001; Silveira et al. 2003). In this support, it is proposed that Cl^- competes and binds to the nitrate transporters sites, hence nitrate uptake by transporters is blocked by Cl^- ion (Carillo et al. 2005). This leads to decreased AAs concentration in root and supported the finding of Da Costa et al. (2011) reported for cowpea. Thus, reduced AAs concentration in the root can directly be related to the reduced nitrate, consequently resulting in reduced N uptake and assimilation, so the biosynthesis of AAs (Silveira et al. 2001; Flores et al. 2002; Rubinigg et al. 2004; Carillo et al. 2005).

A proline–proline cycle is also proposed, stating that it helps in scavenging $OH^•$ radicals at a significant level under the action of pyrroline-5-carboxylate reductase and NADPH (Signorelli et al. 2015; Abbas et al. 2018). Proline dehydrogenase is the main controlling enzyme behind the process of proline degradation and accumulation in plants. It is also observed that stress-tolerant plants accumulate proline in excess that helps to overcome the plant water stress by osmotic management (Yadu et al. 2016; Chandrakar et al. 2017). So, it shows that proline is an essential bio-component for the plant to fight all types of stress and promote the growth of plants.

3.4 Effects of Salt Stress on Protein Biosynthesis

Salinity and other stresses firstly increase the production of ROS as the first response of stress. Being highly reactive, they interrupt the entire cellular metabolism through oxidative damages. Ionic stress results in ROS and Na^+ accumulation in cells that damage protein by denaturing protein content, thus protein metabolism of plants also gets disturbed (Dat et al. 2000; Imlay 2003). Jiang et al. (2014) showed that tolerance to salinity in the transgenic gene was enhanced by overexpression of gene *GmDREB1*, whereas leaf protein response was also improved under high salt stress. In addition, proteins related to osmotic and oxidative stress were upregulated in leaves of transgenic wheat under salinity stress conditions.

The protein exhibits a number of structures and folding, therefore differences in their activity level exist, which causes differential responses of plant species and varieties (tolerant and sensitive) to variable levels of stresses (Inan et al. 2004). The imbalance caused by salinity also affects the nutrients involved in protein and photosynthetic pigments synthesis, which can lead to the inhibition of related processes (Santos 2004), as well as degradation of photosynthetic pigments (Di Martino et al. 2003).

Furthermore, an active stress acclimation and acquisition of salinity tolerance processes are also done by an enhanced biosynthesis of several novel proteins. A novel role of pea G-protein in high-salinity stress tolerance has been observed by Misra et al. (2007) and authors reported that the pea Gα and Gβ transcripts were also upregulated by hydrogen peroxide (H_2O_2) treatment of pea seedlings. In addition, increased assimilation of N and AAs biosynthesis is responsible for upregulated protein biosynthesis in salinity tolerance in halophytes and glycophytes. In contrast, a decreased relative abundance of some proteins involved in protein and AA biosynthesis, for example,mRNA-binding protein and GS, has been observed in salt-sensitive plants like potato (Aghaei et al. 2008). Salt stress to *Paulowinia imperialis* also resulted in decreased protein concentration at higher salt levels (Ayala-Astorga and Alcaraz-Meléndez 2010).

3.5 Effect of Salt Stress on the Oxidation of Membrane Proteins

Membrane, being the primary structure of the cell, goes through the adjustment under the salinity stress and various changing conditions; so the membrane proteins. There is compartmentalization of the various proteins and lipids in the membrane that help in signaling, developments and survival under the changing conditions. Membrane proteins are responsible for various biological actions and functions. Various functions such as transportation of ions and metabolites, cell differentiation and structural role is carried out by a number of membrane proteins. ROS produced under salt stress damages the cell membrane and its proteins by oxidizing it. Thus various functions related to theses protiens also gets disturbed. ROS are the significant players behind cell toxicity under salinity or any other stress. Hence, growth and yield of various crops decreases under these stress and toxicity conditions due to the disturbed plant energetics and various metabolic pathways (Ahmad et al. 2009; Hasanuzzaman et al. 2013).

3.6 Effect of ROS on the Nucleic Acid Formation

Formation and elimination of ROS need to be carefully controlled in the cell in order to keep away from oxidative damage. At the same time, the degree of ROS is enhancedand the defense mechanisms of cells are supposed to be in a condition of "oxidative stress." Notwithstanding, the harmony among generation and scavenging of ROS is bothered under various stress conditions, for example, salinity, drought, excessive light, the toxicity of metals (Yadav and Srivastava 2020), pathogens, etc. These responses can change intrinsic cell membrane characterization like smoothness, transport of ion, decrease catalyst action, cross-linking of protein, restraint of protein synthesis, DNA damage, etc., at last bringing about cell death (Sharma et al. 2012). Grune et al. (1997) observed that ROS can cause oxidative damage to nuclear, chloroplast, and mitochondrial DNA. Due to the lack of protective nature of histone protein, chloroplast and mitochondrial DNA were more susceptible to oxidative damage than nuclear DNA (Richter 1992). Oxidative stress on DNA brings about oxidation of deoxyribose, breakage of the nucleic strand, evacuation of nucleotides, and DNA-protein cross-links (DPC). In addition, deviations in the nucleotides of nucleic acid in one strand can bring about mismatches with other strands of the nucleotides and finally leading to the subsequent mutations. The component of nucleotide, such as sugar, and nitrogen base moieties are more vulnerable to oxidation by ROS. Dizdaroglu (1993) observed that oxidative stress to a nitrogenous base of DNA, commonly addition of OH radical, occurs between double bonds of nucleotides, whereas damage to DNA sugar mainly results from removal of the hydrogen atom from the C4 position of deoxyribose, which leads to the production of deoxyribose radical that further attacksnucleic acid and finally damages DNA (Evans et al. 2004). In vivo condition, the toxicity of ROS acts more similar to the consequence of Fenton reaction, and when $^{\bullet}OH$ radical attacks on either nucleic acid or its associated proteins, as a result, DPC are formed (Oleinick et al. 1987). Thus, nucleic acid plays role as cell's genetic and hereditary material. If any damage to the nucleic acid happens, it can bring about changes/modifications in the encoded proteins, which can prompt denaturation or complete inactivation of the encoded proteins (Sharma et al. 2012). Many studies observed that degradation of DNA increased when plants were exposed to salinity stress (Liu et al. 2000a; Joseph and Jini 2010; Khan and Hemalatha 2016).

3.7 DNA–Protein Cross-links

DPC have been shown to cause an extreme DNA harm, as they disturb quintessential processes, such as DNA replication and transcription, because of their massiveness. DPC have been described through a covalent linkage between a protein and DNA, and it is exceptionally huge because of their proteinaceous part that disables numerous chromatin-based procedures by steric block. To keep up cell viability and access to significant genetic regions, plants have advanced mechanically, and three independent pathways have been developed to fix this sort of exceptionally toxic lesion (Confalonieri et al. 2014; Hacker et al. 2020). DPC varied in their size, physicochemical properties, and the nature of the cross-link itself. In general, DPC can be partitioned into four distinct classes established on the presence and kind of DNA breaks adjacent to the DPC (Hacker et al. 2020). RNA-binding proteins

(RBPs) are also critical players in post-transcriptional gene regulation. They are also known to protect the plant cell under stress by enhancing plant immunity.

3.7.1 Type 1 DPC

Type 1 DPC is not related to DNA breaks; rather, the protein is connected to a flawless DNA strand. This kind of DPC happens all the time, as theoretically any type of protein near the DNA can shape an adduct affected by different ecological, endogenous, or synthetic agents (Ide et al. 2011). ROS are produced as a by-product of a number of metabolic methods like photosynthesis or the demethylation of histones that are the endogenous sources of the development of Type 1 DPC (Turrens 2003; Tsukada et al. 2006; Enderle et al. 2019). Type 1 has both enzymatic and nonenzymatic formation of DPC, and Type 2 and 4 showed the only enzymatic reaction to the DPC (Ide et al. 2018; Hacker et al. 2020).

3.7.2 Type 2 DPC

Type 2 DPC is characterized to a DNA-adduct close to an AP site (apurinic/apyrimidinic site, a site in DNA where no purine or pyrimidine is found), which emerges during base excision repair (BER) either with the poly (ADP-ribose) polymerase (PARP1) or with the DNA polymerase Polβ stabilized at the DNA (Ide etal. 2018). The cross-link among PARP1 and the DNA may be a Schiff base at the 3'-end, which can emerge suddenly when there is a stabilized AP destination in the DNA. Alkylating agents like MMS (methyl methane sulfonate) initiated the crosslink between PARP1 and DNA, but some other additional cross-linkers agents like *Olaparib*, MK4827, and *veliparib* are reported to contribute the formation of Type 2 DPC (Prasad et al. 2014). Plants have three homologs of PARP1, to be specific PARP1, PARP2, and PARP3. For PARP1 and PARP2, it could as of now be demonstrated that they most likely add to plant BER (Babiychuk et al. 1998; Klemm et al. 2017).

Subsequently, PARP1 DPC may likewise be naturally applicable in plants. The Polβ DPC are framed under the presence of oxidizing agents that are causing the arrangement of 2-desoxyribonolactone (dL) AP site. This type of injury can be proficiently fixed by long-patch BER. But, during the attempt to short patch BER of the 2-desoxyribonolactone (dL) lesion, DNA polymerase beta (Polβ) turns out to be covalently caught at the 5'-end of the AP site. Therefore, these events result in the harmful cross-links (DeMott et al. 2002; Quiñones et al. 2015). In any case, plants donot harbor a Polβ homolog, and Polλ, which is the main individual from the X polymerase family in plants, that is not able to sidestep 2-dL lesions (Amoroso et al. 2011; Mentegari et al. 2016; Roldán-Arjona et al. 2019). Besides, it isnot yet realized whether plants perform short-patch BER (Roldán-Arjona et al. 2019). Thus, the significance of this type of DPC probably remains obsolete in plants (Hacker et al. 2020).

3.7.3 Type 3 & 4 DPC

Type 3 and 4 DPC comprise of caught topoisomerase cleavage complexes, which can happen immediately and informs where DNA must be loosened up or unraveled, for example, DNA replication, recombination, interpretation, or chromosome isolation (Ide et al. 2018). DNA topoisomerases are equipped for relaxing the DNA by making transient DNA breaks.

Topoisomerase 1 (TOP1) produces single-strand breaks (SSBs), while topoisomerase 2 (TOP2) produces double-strand breaks (DSBs). Their capacity depends on a functional tyrosyl residue in the reactive center of the catalyst, with which they can attack the phosphate of the DNA backbone in a reversible way. But, the reaction intermediate can be caught under specific conditions (Wigley 1995). Along these, DNA defects, for example, mismatched bases, SSBs, AP sites, or massive DNA adducts, encourage the formation of both Type 3 and Type 4 DPC (Connelly and Leach 2004). In spite of the fact that the roots of both types of DPC are fundamentally the same but chemically very distinct, as a result, the consequences for genome integrity vary significantly. In Type 3 DPC, TOP1 is associated with the DNA by a tyrosyl phosphodiester bond at the 3'-end of a SSB. Conversely, in Type 4 DPC, the caught TOP2 is related to the DNA by two tyrosyl-phosphodiester bonds at the 5'-end of a DSB (Wigley 1995). Thus, Type 4 DPC is characterized as a potential hazard to DNA damage compared with Type 3 DPC (Hacker et al. 2020). The activity of DNA topoisomerases is highly conserved among all living kingdoms, thus Type 3 and Type 4 DPC are additionally of high biological pertinence in plants (Singh et al. 2004).

DPC is profoundly harmful injuries, and in this way, represents a significant threat to genome respectability in plants. So as to guarantee the viability of cells, this damage must be recognized and fixed in a convenient way (Hacker et al. 2020).

3.8 Effect of Salt Stress on Dephosphorylation of RNA and DNA

Generally, the molecular level responses and related mechanisms are relatively more complex as compared with the physiological and biochemical processes under any stress. Noncoding RNAs (ncRNAs) and protein-coding polyadenylated mRNAs are the main components of transcriptome of eukaryotes. But, they are supposed to have nonsignificant protein-coding potential (Liu et al. 2000a). Various environmental stress stimuli and developmental factors regulate the expression of ncRNA and mRNA. ncRNA has had some unique functions for growth and stress responses (Heo and Sung 2011). The detailed role of salt stress on DNA phosphorylation is discussed in Section 3.7. Additionally, calcium-binding proteins cascade (SOS1, SOS2, SOS3, SOS4, and SOS5), sense the calcium signaling, and decode the signals provided to the distant part. These decoded messages are sent downstream that activate the phosphorylation cascade. Phosphorylation cascade activation further regulates the gene expression and some ion pumps.

3.9 Future Advances and Conclusion

Crop plants show reductions in agricultural production and yield grown in the salt-affected agricultural field. In a crop field multiple stress occur and interact. To improve the yield and production on marginal saline land, more exploration of salt-tolerant variety, as well as improvement in mechanisms behind the tolerance of crops are needed in the present scenario. Overall, the high-salinity stress effect on ROS, signaling, and activation of various metabolic pathways is an important area of research with respect to increased crop productivity under optimal conditions. A detailed study on metabolomics and salinity stress-responsive genes for tolerance and sensitivity adaptation will help to uncover the complexity of salinity tolerance at the molecular level and will be a useful tool for breeders to develop a salt-tolerant variety. This can be achieved by the fast transfer of required traits into the desired crop variety. Development of genetically modified or transgenic plants with transformation using the key genes in salinity-tolerance pathways can also be a promising approach in improving agricultural productions under salt stress. While talking about the molecular aspects, research in DPC repair in plants has little information so it is to explore and identify a pathway that controls the DPC damage. It will be an energizing assignment later on to disentangle further factors involved in plant DPC repair and to comprehend the complicated regulation network of the individual proteins, as our present information knowledge represents only the tip of the iceberg.

References

Abbas, G., Murtaza, B., Bibi, I. et al. (2018). Arsenic uptake, toxicity, detoxification, and speciation in plants: physiological, biochemical, and molecular aspects. *Int. J. Environ. Res. Public Health* 15: 59.

AbdElgawad, H., Zinta, G., Hegab, M.M. et al. (2016). High salinity induces different oxidative stress and antioxidant responses in maize seedlings organs. *Front. Plant Sci.* 8: 276.

Agarwal, S. and Pandey, V. (2004). Antioxidant enzyme responses to NaCl stress in *Cassia angustifolia*. *Biol. Plant.* 48: 555–560.

Aghaei, K., Ehsanpour, A.A., and Komatsu, S. (2008). Proteome analysis of potato under salt stress. *J. Proteome Res.* **7** (11): 4858–4868.

Ahanger, M.A., Tomar, N.S., Tittal, M. et al. (2017). Plant growth under water/salt stress: ROS production; antioxidants and significance of added potassium under such conditions. *Physiol. Mol. Biol. Plants* 23: 731–744.

Ahmad, P., Jaleel, C.A., Azooz, M.M., and Nabi, G. (2009). Generation of ROS and non-enzymatic antioxidants during abiotic stress in plants. *Bot. Res. Int.* 2: 11–20.

Amoroso, A., Concia, L., Maggio, C. et al. (2011). Oxidative DNA damage bypass in *Arabidopsis thaliana* requires DNA polymerase λ and proliferating cell nuclear antigen 2. *Plant Cell* 23: 806–822.

Apel, K. and Hirt, H. (2004). Reactive oxygen species: metabolism, oxidative stress, and signal transduction. *Annu. Rev. Plant Biol.* 55: 373–399.

Asgher, M., Per, T.S., Masood, A. et al. (2017). Nitric oxide signaling and its crosstalk with other plant growth regulators in plant responses to abiotic stress. *Environ. Sci. Pollut. Res.* 24: 2273–2285.

Ashraf, M.P. and Harris, P.J. (2004). Potential biochemical indicators of salinity tolerance in plants. *Plant Sci.* 166: 3–16.

Ayala-Astorga, G.I. and Alcaraz-Meléndez, L. (2010). Salinity effects on protein content, lipid peroxidation, pigments, and proline in *Paulownia imperialis* (Siebold&Zuccarini) and *Paulownia fortunei* (Seemann&Hemsley) grown in vitro. *Electron. J. Biotechnol.* 13: 13–14.

Babiychuk, E., Cottrill, P.B., Storozhenko, S. et al. (1998). Higher plants possess two structurally different poly (ADP-ribose) polymerases. *Plant J.* 15: 635–645.

Bouché, N., Lacombe, B., and Fromm, H. (2003). GABA signaling: a conserved and ubiquitous mechanism. *Trends Cell Biol.* 13: 607–610.

Carillo, P., Mastrolonardo, G., Nacca, F., and Fuggi, A. (2005). Nitrate reductase in durum wheat seedlings as affected by nitrate nutrition and salinity. *Funct. Plant Biol.* 32: 209–219.

Carillo, P., Raimondi, G., Kyriacou, M.C. et al. (2019). Morpho-physiological and homeostatic adaptive responses triggered by omeprazole enhance lettuce tolerance to salt stress. *Sci. Hortic.* 249: 22–30.

Chandrakar, V., Yadu, B., Meena, R.K. et al. (2017). Arsenic-induced genotoxic responses and their amelioration by diphenyleneiodonium, 24-epibrassinolide and proline in *Glycine max L. Plant Physiol. Biochem.* 112: 74–86.

Chaparro, J.M., Badri, D.V., Bakker, M.G. et al. (2013). Root exudation of phytochemicals in *Arabidopsis* follows specific patterns that are developmentally programmed and correlate with soil microbial functions. *PLoS One* 8: e55731.

Cheng, B., Li, Z., Liang, L. et al. (2018). The γ-aminobutyric acid (GABA) alleviates salt stress damage during seeds germination of white clover associated with Na+/K+ transportation, dehydrins accumulation, and stress-related genes expression in white clover. *Int. J. Mol. Sci.* 19: 2520.

Confalonieri, M., Fae, M., Balestrazzi, A. et al. (2014). Enhanced osmotic stress tolerance in *Medicagotruncatula* plants overexpressing the DNA repair gene MtTdp2α (tyrosyl-DNA phosphodiesterase 2). *Plant Cell, Tissue and Organ Cult. (PCTOC)* 116: 187–203.

Connelly, J.C. and Leach, D.R. (2004). Repair of DNA covalently linked to protein. *Mol. Cell* 13: 307–316.

Da Costa, R.C.L., Lobato, A.K., da Silveira, J.A., and Laughinghouse, H.D. IV (2011). ABA-mediated proline synthesis in cowpea leaves exposed to water deficiency and rehydration. *Turk. J. Agric. For.* 35: 309–317.

Dat, J., Vandenabeele, S., Vranova, E.V. et al. (2000). Dual action of the active oxygen species during plant stress responses. *Cell. Mol. Life Sci.* 57: 779–795.

DeMott, M.S., Beyret, E., Wong, D. et al. (2002). Covalent trapping of human DNA polymerase β by the oxidative DNA lesion 2-deoxyribonolactone. *J. Biol. Chem.* 277: 7637–7640.

Di Martino, C., Delfine, S., Pizzuto, R. et al. (2003). Free amino acids and glycine betaine in leaf osmoregulation of spinach responding to increasing salt stress. *New Phytol.* 158: 455–463.

Dizdaroglu, M. (1993). Chemistry of free radical damage to DNA and nucleoproteins. In: *DNA and Free Radicals* (eds. B. Halliwell and O.I. Aruoma), 19–39. London, UK: Ellis Horwood.

Enderle, J., Dorn, A., and Puchta, H. (2019). DNA-and DNA-protein-crosslink repair in plants. *Int. J. Mol. Sci.* 20: 4304.

Evans, M.D., Dizdaroglu, M., and Cooke, M.S. (2004). Oxidative DNA damage and disease: induction, repair and significance. *Mutat. Res. Rev. Mutat. Res.* 567: 1–61.

Flores, P., Angeles, B.M., Martínez, V., and Cerdá, A. (2002). Response to salinity of tomato seedlings with a split-root system: nitrate uptake and reduction. *J. Plant Nutr.* 25: 177–187.

Flowers, T.J., Gaur, P.M., and Gowda, C.L. (2010). Salt sensitivity in chickpea. *Plant Cell Environ. 33*: 490–509.

Fougere, F., Le Rudulier, D., and Streeter, J.G. (1991). Effects of salt stress on amino acid, organic acid, and carbohydrate composition of roots, bacteroids, and cytosol of alfalfa (Medicago sativa L.). *Plant Physiol.* 96 (4): 1228–1236.

Grune, T., Reinheckel, T., and Davies, K.J. (1997). Degradation of oxidized proteins in mammalian cells. *FASEB J.* 11 (7): 526–534.

Hacker, L., Dorn, A., and Puchta, H. (2020). Repair of DNA-protein crosslinks in plants. *DNA Repair* 87: 102787.

Han, D., Lerner, A.G., Walle, L.V. et al. (2009). IRE1α kinase activation modes control alternate endoribonuclease outputs to determine divergent cell fates. *Cell* 138: 562–575.

Hasan, M.K., Liu, C., Wang, F. et al. (2016). Glutathione-mediated regulation of nitric oxide, S-nitrosothiol and redox homeostasis confers cadmium tolerance by inducing transcription factors and stress response genes in tomato. *Chemosphere* 161: 536–545.

Hasanuzzaman, M., Nahar, K., and Fujita, M. (2013). Plant response to salt stress and role of exogenous protectants to mitigate salt-induced damages. In: *Ecophysiology and Responses of Plants Under Salt Stress* (eds. P. Ahmad, M. Azooz and M. Prasad), 25–87. New York, NY: Springer.

Hediye Sekmen, A., Türkan, İ., and Takio, S. (2007). Differential responses of antioxidative enzymes and lipid peroxidation to salt stress in salt-tolerant *Plantago maritima* and salt sensitive *Plantago media*. *Physiol. Plant.* **131** (3): 399–411.

Heo, J.B. and Sung, S. (2011). Vernalization-mediated epigenetic silencing by a long intronic noncoding RNA. *Science* 331 (6013): 76–79.

Hernández, J.A. and Almansa, M.S. (2002). Short-term effects of salt stress on antioxidant systems and leaf water relations of pea leaves. *Physiol. Plant* 115 (2): 251–257.

Ide, H., Shoulkamy, M.I., Nakano, T. et al. (2011). Repair and biochemical effects of DNA–protein crosslinks. *Mutat. Res.* 711 (1–2): 113–122.

Ide, H., Nakano, T., Salem, A.M., and Shoulkamy, M.I. (2018). DNA–protein cross-links: formidable challenges to maintaining genome integrity. *DNA Repair* 71: 190–197.

Imlay, J.A. (2003). Pathway of oxidative damage. *Annu. Rev. Microbiol.* 57: 395–418.

Imlay, J.A. and Linn, S. (1988). DNA damage and oxygen radical toxicity. *Science* 240: 1302–1309.

Inan, G., Zhang, Q., Li, P. et al. (2004). Salt cress. A halophyte and cryophyte Arabidopsis relative model system and its applicability to molecular genetic analyses of growth and development of extremophiles. *Plant Physiol.* 135 (3): 1718–1737.

Jiang, Q., Hu, Z., Zhang, H., and Ma, Y. (2014). Overexpression of GmDREB1 improves salt tolerance in transgenic wheat and leaf protein response to high salinity. *Crop. J.* 2 (2–3): 120–131.

Joseph, B. and Jini, D. (2010). Salinity induced programmed cell death in plants: challenges and opportunities for salt-tolerant plants. *J. Plant Sci.* 5: 376–390.

Khan, M.S. and Hemalatha, S. (2016). Biochemical and molecular changes induced by salinity stress in Oryza sativa L. *Acta. Physiol. Plant.* 38 (7): 167.

Klemm, T., Mannuß, A., Kobbe, D. et al. (2017). The DNA translocase RAD 5A acts independently of the other main DNA repair pathways, and requires both its ATP ase and RING domain for activity in *Arabidopsis thaliana*. *Plant J.* 91 (4): 725–740.

Koca, H., Ozdemir, F., and Turkan, I. (2006). Effect of salt stress on lipid peroxidation and superoxide dismutase and peroxidase activities of *Lycopersicon esculentum* and *L. pennellii*. *Biol. Plant.* 50: 745–748.

Lata, C. and Prasad, M. (2011). Role of drebs in regulation of abiotic stress responses in plants. *J. Exp. Bot.* 62: 4731–4748.

Lea, P.J. and Ireland, R.J. (1999). Nitrogen metabolism in higher plants. In: *Plant Amino Acids: Biochemistry and Biotechnology (Chapter 1)* (ed. B.K. Singh). CRC Press, Taylor & Francis Group.

Liu, T., Van, S.J., and Cress, W.A. (2000). Salinity induced nuclear and DNA degradation in meristematic cells of soybean (*Glycine max* L.) roots. *Plant Growth Regul.* 30 (1): 49–54.

Maaroufi-Dguimi, H., Debouba, M., Gaufichon, L. et al. (2011). An *Arabidopsis* mutant disrupted in ASN2 encoding asparagine synthetase 2 exhibits low salt stress tolerance. *Plant Physiol. Biochem.* 49 (6): 623–628.

Marrs, K.A. (1996). The functions and regulation of Glutathione-S -transferase in plants. *Annu. Rev. Plant Physiol.* 47 (127): 158.

Mengel, K., Kirkby, E.A., Kosegarten, H., and Appel, T. (2001). Nitrogen. In: *Principles of Plant Nutrition* (eds. K. Mengel and E.A. Kirkby), 397–434. Dordrecht: Springer.

Mentegari, E., Kissova, M., Bavagnoli, L. et al. (2016). DNA polymerases λ and β: the double-edged swords of DNA repair. *Genes* 7 (9): 57.

Misra, S., Wu, Y., Venkataraman, G. et al. (2007). HeterotrimericGprotein complex and Gproteincoupled receptor from a legume (*Pisumsativum*): role in salinity and heat stress and crosstalk with phospholipase C. *Plant J.* 51 (4): 656–669.

Mondal, M.F., Asaduzzaman, M., Kobayashi, Y. et al. (2013). Recovery from autotoxicity in strawberry by supplementation of amino acids. *Sci. Hortic.* 164: 137–144.

Oleinick, N.L., Chiu, S.M., Ramakrishnan, N., and Xue, L.Y. (1987). The formation, identification, and significance of DNA-protein cross-links in mammalian cells. *Br. J. Cancer* Supplement.8: 135.

Pang, Q., Chen, S., Dai, S. et al. (2010). Comparative proteomics of salt tolerance in *Arabidopsis thaliana* and *Thellungiella halophila*. *J. Proteome Res.* 9 (5): 2584–2599.

Pavlikova, D., Zemanova, V., Prochazkova, D. et al. (2014). The long-term effect of zinc soil contamination on selected free amino acids playing an important role in plant adaptation to stress and senescence. *Ecotoxicol. Environ. Saf.* 100: 166–170.

Poonam and Srivastava, S. (2019). Assessing the role of glutathione in arsenic toxicity amelioration in rice (*Oryza sativa* L.) at early seedling growth. *Res. J. Chem. Environ.* 23 (7): 1–6.

Prasad, R., Horton, J.K., Chastain, P.D. et al. (2014). Suicidal cross-linking of PARP-1 to AP site intermediates in cells undergoing base excision repair. *Nucleic Acids Res.* 42 (10): 6337–6351.

Qadir, M., Tubeileh, A., Akhtar, J. et al. (2008). Productivity enhancement of salt affected environments through crop diversification. *Land Degrad. Dev.* 19 (4): 429–453.

Qu, C., Liu, C., Gong, X. et al. (2012). Impairment of maize seedling photosynthesis caused by a combination of potassium deficiency and salt stress. *Environ. Exp. Bot.* 75: 134–141.

Queiroz, H.M., Sodek, L., and Haddad, C.R. (2012). Effect of salt on the growth and metabolism of Glycine max. *Braz. Arch. Biol. Technol.* 55 (6): 809–817.

Quiñones, J.L., Thapar, U., and Yu, K. (2015). Enzyme mechanism-based, oxidative DNA–protein cross-links formed with DNA polymerase β in vivo. *Proc. Natl. Acad. Sci.* 112 (28): 8602–8607.

Richter, C. (1992). Reactive oxygen and DNA damage in mitochondria. *Mut. Res./DNAging* 275 (3–6): 249–255.

Roldán-Arjona, T., Ariza, R.R., and Córdoba-Cañero, D. (2019). DNA base excision repair in plants: an unfolding story with familiar and novel characters. *Front. Plant Sci.* 10: 1055.

Rubinigg, M., Wenisch, J., Elzenga, J.T., and Stulen, I. (2004). NaCl salinity affects lateral root development in Plantago maritima. *Funct. Plant Biol.* 31 (8): 775–780.

Santos, C.V. (2004). Regulation of chlorophyll biosynthesis and degradation by salt stress in sunflower leaves. *Sci. Hortic.* 103 (1): 93–99.

Shahid, S.A., Zaman, M., and Heng, L. (2018). Soil salinity: historical perspectives and a world overview of the problem. In: *Guideline for Salinity Assessment, Mitigation and Adaptation Using Nuclear and Related Techniques* (eds. M. Zaman, S.A. Shahid and L. Heng), 43–53. Cham: Springer.

Sharma, P., Jha, A.B., Dubey, R.S., and Pessarakli, M. (2012). Reactive oxygen species, oxidative damage, and antioxidative defense mechanism in plants under stressful conditions. *J. Bot.* 2012: 1–26.

Signorelli, S., Dans, P.D., Coitiño, E.L. et al. (2015). Connecting proline and γ-aminobutyric acid in stressed plants through non-enzymatic reactions. *PLoS One* 10 (3): e0115349.

Silveira, J.A., Melo, A.R., Viégas, R.A., and Oliveira, J.T. (2001). Salinity-induced effects on nitrogen assimilation related to growth in cowpea plants. *Environ. Exp. Bot.* 46 (2): 171–179.

Silveira, J.A.G., De Almeida, V.R., da Rocha, I.M.A. et al. (2003). Proline accumulation and glutamine synthetase activity are increased by salt-induced proteolysis in cashew leaves. *J. Plant Physiol.* 160 (2): 115–123.

Singh, B.N., Sopory, S.K., and Reddy, M.K. (2004). Plant DNA topoisomerases: structure, function, and cellular roles in plant development. *CRC Crit. Rev. Plant Sci.* 23 (3): 251–269.

Srivastava, S. and Dubey, R.S. (2011). Manganese-excess induces oxidative stress, lowers the pool of antioxidants and elevates activities of key antioxidative enzymes in rice seedlings. *Plant Growth Regul.* 64 (1): 1–6.

Swami, A.K., Alam, S.I., Sengupta, N., and Sarin, R. (2011). Differential proteomic analysis of salt stress response in *Sorghum bicolor* leaves. *Environ. Exp. Bot.* 71 (2): 321–328.

Tegeder, M. and Rentsch, D. (2010). Uptake and partitioning of amino acids and peptides. *Mol. Plant* 3 (6): 997–1011.

Tegeder, M. and Ward, J.M. (2012). Molecular evolution of plant AAP and LHT amino acid transporters. *Front. Plant Sci.* 3: 21.

Tsukada, Y.I., Fang, J., Erdjument-Bromage, H. et al. (2006). Histone demethylation by a family of JmjC domain-containing proteins. *Nature* 439 (7078): 811–816.

Turrens, J.F. (2003). Mitochondrial formation of reactive oxygen species. *J. Physiol.* 552 (2): 335–344.

Tuteja, N., Ahmad, P., Panda, B.B., and Tuteja, R. (2009). Genotoxic stress in plants: shedding light on DNA damage, repair and DNA repair helicases. *Mutat. Res. Rev. Mutat. Res.* 681 (2–3): 134–149.

Wang, X., Fan, P., Song, H. et al. (2009). Comparative proteomic analysis of differentially expressed proteins in shoots of *Salicornia europaea* under different salinity. *J. Proteome Res.* 8 (7): 3331–3345.

Wigley, D.B. (1995). Structure and mechanism of DNA topoisomerases. *Annu. Rev. Biophys. Biomol. Struct.* 24 (1): 185–208.

Xie, E., Wei, X., Ding, A. et al. (2020). Short-term effects of salt stress on the amino acids of *Phragmites australis* root exudates in constructed wetlands. *Water* 12 (2): 569.

Xing, S.G., Jun, Y.B., Hau, Z.W., and Liang, L.Y. (2007). Higher accumulation of γ-aminobutyric acid induced by salt stress through stimulating the activity of diamine oxidases in *Glycine max* (L.) Merr. roots. *Plant Physiol. Biochem.* 45 (8): 560–566.

Yadav, P. and Srivastava, S. (2020). Co-culturing *Hydrillaverticillata* with rice (*Oryza sativa*) plants ameliorates arsenic toxicity and reduces arsenic accumulation in rice. *Environ. Technol. Innov.* 18: 100722.

Yadav, P., Srivastva, S., Patil, T. et al. (2020). Tracking the time-dependent and tissue-specific processes of arsenic accumulation and stress responses in rice (*Oryza sativa* L.). *J. Hazard. Mater.* 406: 124307.

Yadu, B., Chandrakar, V., and Keshavkant, S. (2016). Responses of plants to fluoride: an overview of oxidative stress and defense mechanisms. *Fluoride* 49: 293–302.

Yan, S., Tang, Z., Su, W., and Sun, W. (2005). Proteomic analysis of salt stress responsive proteins in rice root. *Proteomics* 5 (1): 235–244.

You, J., Zhang, Y., Liu, A. et al. (2019). Transcriptomic and metabolomic profiling of drought-tolerant and susceptible sesame genotypes in response to drought stress. *BMC Plant Biol.* 19 (1): 1–6.

Zhang, J., Zhang, Y., Du, Y. et al. (2011). Dynamic metabonomic responses of tobacco (*Nicotianatabacum*) plants to salt stress. *J. Proteome Res.* 10 (4): 1904–1914.

Zhang, Z., Mao, C., Shi, Z., and Kou, X. (2017). The amino acid metabolic and carbohydrate metabolic pathway play important roles during salt-stress response in tomato. *Front. Plant Sci.* 8: 1231.

Zhu, M., Chen, G., Zhang, J. et al. (2014). The abiotic stress-responsive NAC-type transcription factor SlNAC4 regulates salt and drought tolerance and stress-related genes in tomato (*Solanumlycopersicum*). *Plant Cell Rep.* 33 (11): 1851–1863.

4

Salt Ion Transporters in Crop Plants at Cellular Level

Ria Khare[1,2], Gurpreet Sandhu[1,2], Aruba Khan[1], and Prateek Jain[3]

[1] *CSIR-National Botanical Research Institute (CSIR-NBRI), Lucknow, India*
[2] *Academy of Scientific and Innovative Research (AcSIR), New Delhi, India*
[3] *Department of Biology, University of North Carolina at Chapel Hill, Chapel Hill, NC, USA*

4.1 Introduction

Plants are exposed to various types of biotic and abiotic stresses in their natural environment. Salt toxicity is one of the chief constraints on the agriculture sector as it hinders an array of metabolic processes and lowers the crop yield and productivity. Areas with low precipitation and high evaporation including arid and semiarid zones are maximally exaggerated due to salt stress. Globally, about 800 million hectares of land including 20% of cultivated area is severely affected by salinity stress (Ghosh et al. 2016). This necessitates the selection and development of salt-tolerant plants for agriculture's growth and welfare. Short-term salinity stress leads to an increase in soil osmotic potential that lowers the root water uptake (thus causing osmotic stress), whereas long-term salt exposure causes nutrient imbalance in cell cytosol and leads to ionic stress, which ultimately hinders the plant growth and development (Acosta-Motos et al. 2017; Gupta and Huang 2014). Preliminary effects of osmotic stress comprise membrane disruption, nutrient imbalance, ruins reactive oxygen species (ROS) detoxification machinery, affect antioxidant enzymes, reduce photosynthetic activity, and decrease stomatal aperture (Wu 2018). Osmotic homeostasis is mediated by the movement of osmolytes, which regulates the osmotic stress, eventually influencing stomatal opening and cellular water transport in plants (Munns and Tester 2008; Rahnama et al. 2011). Under high-salt (NaCl) condition, plants accumulate Na^+ and Cl^- ions, which interferes with cytosolic K^+ and instigate ionic stress in plants (Isayenkov 2012). The hydrated Na^+ ions have similar radii to K^+, which accelerate their uptake through K^+ channel and cause Na^+ toxicity and K^+ deficiency under saline environment (Keisham et al. 2018). Na^+ is nonessential for plants, and its toxicity inhibits enzyme activity by binding to their inhibition sites or by replacing K^+ from activation sites while K^+ mediates essential metabolic processes necessary for plant growth and development (Keisham

Physiology of Salt stress in Plants: Perception, Signalling, Omics and Tolerance Mechanism, First Edition.
Edited by Pratibha Singh, Madhulika Singh, Rajiv Kumar Singh, and Sheo Mohan Prasad.

et al. 2018). Thus, a lower Na^+/K^+ ratio is a requisite for cellular homeostasis in plants to alleviate salt stress and hence enhance plant productivity.

Previous years have seen a huge growth in studies related to salinity stress in crop plants. Several important genes regulating salt assimilation have been identified using genotypes with varied response toward salinity stress in crop plants (Hanin et al. 2016; Ismail and Horie 2017; Morton et al. 2019). Plants are categorized as glycophytes (unable to withstand salinity) or halophytes (able to cope with salt stress) on the basis of their ability to survive in saline environment (Cheeseman 2015). Although owing difference in their growth under saline condition both glycophytes and halophytes cannot withstand higher cytoplasmic salt concentration. Monocot crops such as rice (*Oryza sativa*) and wheat (*Triticum aestivum*) possess lesser salt tolerance than barley (*Hordeum vulgare*). Contrastingly, dicots such as *Arabidopsis thaliana* show severe sensitivity toward salinity compared with halophytes including *Atriplex* sp., *Thellungiella salsuginea* (previously known as *Thellungiella halophila*), etc (Munns and Tester 2008; Pang et al. 2010). The smaller life cycle and immense studies on model plant *A. thaliana* have helped in elucidation of genes associated with salt transportation and detoxification (Jossier et al. 2010). Further transcriptomic, proteomic, metabolomic, and ionomic studies helped to understand the effects of salt on molecular pathways in plants (Hernández 2019; Jha et al. 2019). To counter the excessive salt conditions with a lower Na^+ concentration and higher cytosolic K^+/Na^+ ratio, following strategies are used by plants: (i) lowering Na^+ uptake (reducing influx or increasing efflux via root cells) from roots, (ii) regulating long-distance Na^+ transport, (iii) vacuolar compartmentalization of Na^+, and (iv) secretion of Na^+ from leaves (Gupta and Huang 2014). Understanding the role of salt ions in sensing and signaling under salt stress could provide a broader picture of molecular aspects associated with salt tolerance in plants. Thus, to have a deep understanding of salinity stress in plants, this chapter will discuss the progress in salt ion transport and related transporters at the cellular level.

4.2 Absorption of Na^+ from Soil and Its Compartmentalization in Plant Cell

Under high-salt concentration, plant roots take up Na^+ from soil by passive transport through nonselective cation channels (NSCC), comprising two families: cyclic nucleotide-gated channels (CNGCs) and glutamate receptors (GLRs) (Brini and Masmoudi 2012). Plant CNGCs possess a cytoplasmic C-terminal calmodulin-binding domain (CaMBD), a N-terminal hexa-transmembrane (TM), and a cyclic nucleotide-binding domain (CNB). The cation transportation (K^+, Na^+, Ca^+) is known to be facilitated by the TM domain (Saand et al. 2015). CNGC genes were first time reported in barley (Schuurink et al. 1998) and were further explored in other plant species including *Arabidopsis* (Bridges et al. 2005), rice (Bridges et al. 2005; Nawaz et al. 2014), and maize (Hao and Qiao 2018). These channels are regulated by various salt-induced indicators, including calcium, ROS, and 3′,5′-cyclic guanosine monophosphate (cGMP) (Duszyn et al. 2019). Plant species have variable number of CNGC genes with 20 members in *A. thaliana*, 16 and 18 members in rice and tomato, respectively. In *Arabidopsis*, out of 20 CNGC genes, five *AtCNGCs* (*AtCNGC1, 2, 3, 4,* and *10*) have been characterized. *AtCNGC1* and *AtCNGC4* possess equal permeability to K^+ and Na^+ and are activated by cyclic nucleotides, whereas *AtCNGC2* and *AtCNGC10* are known to be K^+ selective (Brini and Masmoudi 2012). Functional complementation studies in yeast as well as characterization of *Arabidopsis* T-DNA knockout mutants revealed that *AtCNGC3* plays a role in Na^+ influx as seedlings of the mutant had improved growth against high NaCl concentrations (Gobert et al. 2006). Transcriptomic analysis in *Arabidopsis* roots has depicted that the expressions of *AtCNGC1, AtCNGC19,* and *AtCNGC20* are increased under salinity, whereas *AtCNGC19, AtCNGC3,* and *AtCNGC8* were higher in shoots. Another NSCC, glutamate receptors (GLRs) family comprises 20 members in *Arabidopsis*. GLRs in *Arabidopsis* consist of a putative ligand-binding domain, three transmembrane domains, and an ion pore, which together form the functional tetramer (Wang et al. 2019). They form an interaction with glutamate and form cation channel, which is permeable for various ions including Na^+ and Ca^{2+} (Price et al. 2012). Na^+ is transported symplastically across the cellular membranes with exception of few plant species including rice, which has apoplastic Na^+ transport pathway from the soil root interface (Almeida et al. 2017; Kronzucker and Britto 2011). Earlier studies suggest involvement of other channels and transporters for Na^+ influx in plant roots. In monocots, Na^+ uptake is mediated by another important transporter known as high-affinity K^+ transporters (HKT). HKTs are conserved among bacteria, fungi, and plants and classified into two classes: class I HKTs (Na^+-specific) and class II HKTs (Na^+–K^+ cotransporter) (Brini and Masmoudi 2012). *A. thaliana* contains only single-class 1HKT isoform AtHKT1, which regulates long-distance transport of Na^+ by its reabsorption from the xylem sap into the root cells compared with monocots with multiple HKT isoforms (Berthomieu et al. 2003; Rus et al. 2004). Recently, few aquaporins including *AtPIP2; 1* and *AtPIP2; 2*, (most abundant channels on the plasma membrane of root epidermis) are reported to be involved in Na^+ uptake in plants (Byrt et al. 2017).

After following either the symplastic or apoplastic route, Na^+ finally adopts the symplastic pathway across the endodermis. From here, it is released from xylem parenchyma cells into the apoplastic xylem. Acquired Na^+ ions later accumulate in cytosol and instigate cellular ion toxicity (Almeida et al. 2017). The intracellular sequestration of Na^+ ions is necessitated to circumvent the generated osmotic stress due to salinity stress. Plants use tonoplast-localized Na^+/H^+ exchanger (NHX) to remove Na^+ stress in plant cytosol via its vacuolar compartmentalization (Qiu et al. 2016). It belongs to a large super family of monovalent cation/proton antiporters (CPAs), divided into two subgroups, CPA1 and CPA2. The CPA1 family consists of NHX-type transporters, omnipresent in all eukaryotes while CPA2 family includes cation/H^+ exchangers (CHXs) and K^+ efflux antiporters (KEA) (Bassil et al. 2012). *Arabidopsis* encodes eight NHX genes among which the AtNHX1 and AtNHX2 account for the maximum vacuolar transport activity (Barragán et al. 2012). However, rice possesses six NHX-type antiporters with distinct locations in plant cell (Fukuda et al. 2011). Overexpression of NHX1 transporters improves the salt tolerance in *Arabidopsis* (Apse et al. 2003), wheat (Xu et al. 2008), rice (Fukuda et al. 2004), and tomato (Zhang and Blumwald 2001). Transcript levels of *AtNHX 1, 2,* and *5* and *OsNHX1, 2, 3,* and *5* are reported to be enhanced by salt stress (NaCl) suggesting substantial involvement of NHX genes in salt stress response (Fukuda et al. 2011). NHX members characterized till date are known to transport both K^+ and Na^+ and are involved in mediating K^+/H^+ exchange rather than only Na^+/H^+ exchange depending on the significant accumulation of

Na^+ ions in the cytosol (Almeida et al. 2017; Jiang et al. 2010; Maathuis et al. 2014). Captivatingly, individual reports on overexpression of NHX proteins (tomato LeNHX2 in tomato and wheat TaNHX2 in alfalfa plants) depict the essential role of this transporter for maintaining K^+ homeostasis and increasing K^+ absorption during salt stress under ample K^+ supply, thus providing salt tolerance to the transgenic plants (Huertas et al. 2013; Xu et al. 2013; Zhang et al. 2015). These studies prioritize the role of NHX family as the regulator of cellular pH and K^+ homeostasis under salinity stress in plants. In addition to NHX, salt ions are regulated by various signal transducers described in the next section, which are activated in response to salt stress and mediate ionic homeostasis.

4.3 Salt Ions Regulation in Plant Cells and Tissues

Salt stress is known to be perceived as a consequence of osmotic imbalance in plants. Osmotic stress exalts the secondary messengers such as Ca^{2+} in plant cytosol, which is believed to be a receptor for salt signaling (Maathuis 2014). Captivatingly, Jiang et al. 2019 reported the role of glycosyl inositol phosphorylceramide (GIPC), a plasma membrane sphingolipid in Na^+ sensing that leads to calcium influx in plant cytosol. This membrane lipid was identified to play a central role in regulating sodium flux in plants (Steinhorst and Kudla 2019). In response to ionic stress, elevated Ca^{2+} mediates the activation of salt overly sensitive (SOS) pathway to regulate Na^+ movement in plant cytosol (Brini and Masmoudi 2012; Yang and Guo 2018). The three main components of this pathway (SOS1, SOS2, and SOS3) were identified in a genetic screening for mutants identification showing root growth hypersensitivity under salt stress (Ji et al. 2013). SOS1 is controlled through protein phosphorylation via SOS2/SOS3 protein kinase complex. In cellular context, SOS signaling pathway has been known as a key mechanism for Na^+ exclusion and ion homeostasis control in salt-stressed plants (Zhu 2000). Sodium efflux from root cells restricts the accumulation of toxic levels of Na^+ in cytosol and further transport to shoot. Expectedly, overexpression of SOS pathway genes confers tolerance against salinity stress in *Arabidopsis* and in other plant species (Shi et al. 2003; Yang et al. 2009; Zhang and Blumwald 2001).

Plants exposed to salt stress also have increased level of cGMP and ROS as they are also reported to be rapidly enhanced under salt stress (Maathuis 2014). In *Arabidopsis* root protoplast, cyclic nucleotides such as cGMP inhibit Na^+ uptake by targeting NSCC channels and are reported to regulate unidirectional Na^+ influx in *Arabidopsis* and mature pepper plants (Maathuis and Sanders 2001). Salinity stress also fosters the ROS production in plants specifically of hydrogen peroxide (H_2O_2), which is either produced at the apoplast by Ca^{2+} signaling or originates through activation of plant NADPH oxidases (NOXs) also referred to as respiratory burst oxidase homologs (RBOHs) (Suzuki et al. 2011). NOX in turn generates ROS signal, which activates copious signaling transduction pathways that enhance adaptive abiotic stress response against salinity (Ijaz et al. 2017). When exposed to salt stress, ROS hampers proteins, lipids, DNA, and carbohydrates synthesis in plants (Miller et al. 2010). Various enzymatic and nonenzymatic antioxidant enzymes comes in play to protect plant from damaging ROS, which were reported to impart salt tolerance in plants (Yang and Guo 2018). There is not much information on the cross talk between ROS and salt stress; however, MAP kinase has been depicted to be involved in regulating downstream targets of ROS pathway (Livanos et al. 2014; Miller et al. 2010). The specificity of ROS to salt stress is debated as these toxicants are produced in plants in response to various types of biotic and abiotic stresses. Cellular ionic imbalance is also regulated by the osmolytes generated in response to osmotic stress in plants (Ketehouli et al. 2019). Osmolytes include compounds such as polyols, sugars, amino acids, and betaines, which generate a suction force for water absorption by keeping cellular water potential low inside the plant (Park et al. 2016). Moreover, transgenic studies in plants have revealed the role of solutes such as glycine betaine, sorbitol, mannitol, proline, and trehalose in improving salt tolerance (Gupta and Huang 2014).

Salinity stress-mediated hormonal regulation of salt ions has been also elucidated, where abscisic acid (ABA) was known to augment the vacuolar sequestration of Na^+ by enhancing the expression of AtNHX1 (Shi et al. 2002). Salt-mediated stress also evokes synthesis of phytohormones such as ethylene, brassinosteroids (BR), and salicylic acid (SA) in various plants. In *A. thaliana*, ethylene works downstream of cGMP, which induces the PM H^+-ATPase activity, thereby regulating ion homeostasis under salt stress (Li et al. 2014). Similarly, SA enhances K^+ retention by restoring membrane potential and preventing salt-induced K^+ loss via a guard cell outward rectifying K^+(GORK) channel, thus mediating tolerance against salt stress in *Arabidopsis* (Jayakannan et al. 2013). On the other hand, BR application increases the activity of antioxidant enzymes (SOD, POX, APX, and GPX) and accumulation of nonenzymatic antioxidant compounds (tocopherol, ascorbate, and reduced glutathione), which in turn provides tolerance against salt stress (El Mahi et al. 2019).

Ion flux must be strictly controlled to maintain ionic homeostasis in plants under salinity stress. As described above, several channels and transporters including HKT (Na^+ uptake and translocation), NHX (vacuolar sequestration), and SOS (Na^+ efflux) regulate salt transport in plants (Keisham et al. 2018; Wu 2018). To have in-depth understanding of these transporters, next section will be dedicated to the channels and transporters involved in ion fluxing in crop plants.

4.4 Role of Ion Channels and Salt Ion Transporter in Crop Plants at Cellular Level

Salinity mediates ionic disparity (high Na^+/K^+) in plant cytosol, which impairs various metabolic activities eventually reducing plant growth and yield. A high cytosolic K^+/Na^+ ratio is desirable for preserving cellular metabolism, as K^+ deficit disrupts several vital processes in plants including protein synthesis, carbohydrate metabolism, stomatal conductance, ROS detoxification, etc (Hasanuzzaman et al. 2018). Additionally, plants exposed to salinity stress (high NaCl) have increased Cl^- accumulation in cells (Wu and Li 2019). Cl^- is an essential micronutrient and serves a role in regulating turgor and osmotic potential in plants (Franco-Navarro et al. 2016). Although it is predominantly present in saline soils, mechanism for Cl^- ion transport is not much understood (Teakle and Tyerman 2010). Previous studies have exquisitely revealed the role of ion channels and transporters, which facilitates ionic transport (Na^+, K^+ and Cl^-) to maintain ion homeostasis in crop plants (Figure 4.1). As previously described, NSCCs are the ion channels, which govern the

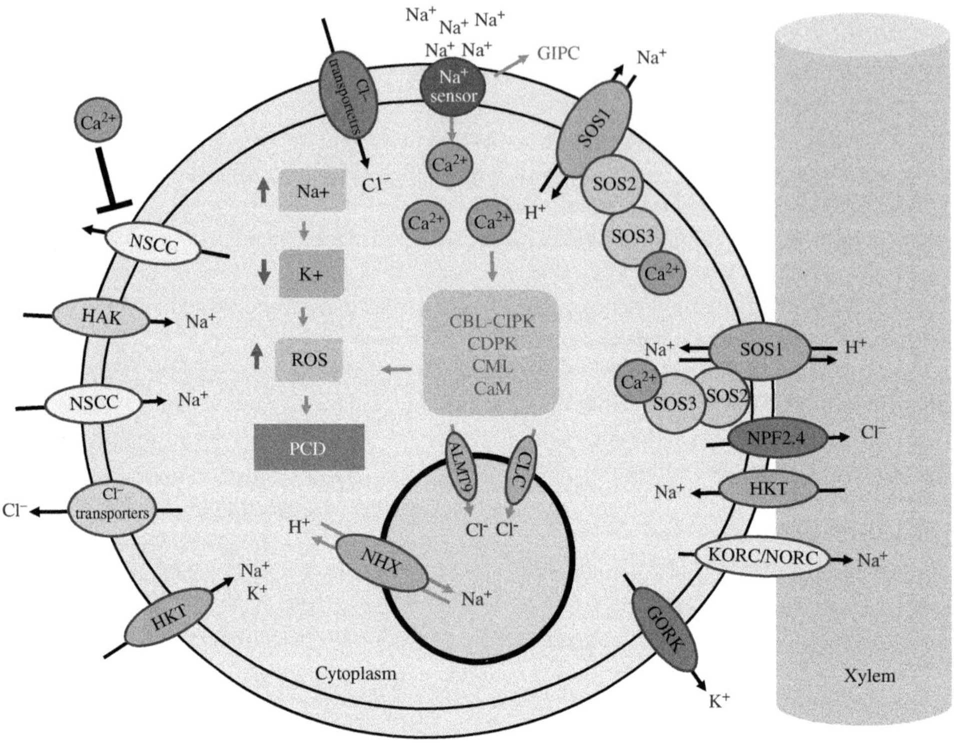

Figure 4.1 Schematic diagram of salt ions regulation and their transport under salt stress.

nonselective cation uptake in plants. Based on their voltage dependency, they are categorized as: voltage-insensitive NSCCs (VI-NSCCs), depolarization-activated NSCCs (DA-NSCCs), and hyperpolarization-activated NSCCs (HA-NSCCs) of which most of the Na^+ influx into the roots is mediated by VI-NSCCs in plants (Keisham et al. 2018). Additionally, K^+ deficiency is encouraged by DA-NSCC in plants along with one other K^+ efflux channel, thereby facilitating K^+ efflux under salt stress in *A. thaliana* and wheat (Wu et al. 2013). Similarly, depolarization-activated outward rectifying K^+ channels (e.g. GORK) lead to K^+ efflux in response to Na^+ accumulation in cytosol (Adams and Shin 2014). As described earlier, NSCCs are comprised of CNGC and GLR in plants. Cyclic nucleotides such as cAMP and cGMP function as signaling molecule modulating the expression of the CNGCs in *A. thaliana*. Moreover, sensitivity of net Na^+ influx to cyclic nucleotides validates CNGCs association in Na^+ transportation in plants (Jha et al. 2016). Crop plants including rice, wheat, and barley possess 15, 47, and 9 CNGCs genes respectively, participating in salt transport along with their role in various additional biotic and abiotic stresses (Guo et al. 2018; Ma and Berkowitz 2011). Expression of these transporters varies in salt-sensitive and -tolerant genotypes in plants, e.g. under salt stress conditions reduced expression of *OsCNGC1* was observed in salt-tolerant cultivar as compared with salt-sensitive cultivar (Senadheera et al. 2009). Alike NSCCs, a low-affinity cation transporter 1 (LCT1) was identified in wheat, which enables Na^+ transport. Later, studies on TaLCT1 disapproved the direct role of LCT1 in Na^+ assimilation, which was implicated to induce transportation mediated via native transporters in yeast membrane (Plett et al. 2010). Na^+ and Ca^{2+} influx in *Arabidopsis* roots is also mediated by another NSCC subfamily, known as GLRs (Plett et al. 2010). Previous studies have indicated that expression of *AtGLR3.7* in the *Xenopus laevis* oocytes led to Na^+ permeable conductance (Roy et al. 2008) together with its participation in salinity stress response in *Arabidopsis* (Wang et al. 2019). Overexpression of *AtGluR2* exhibited high sensitivity to increased level of external Na^+ and K^+ (Kim et al. 2001). Similar results were also found in case of *AtGLR1.1* and *AtGLR1.4* whose ion pore domains facilitate Na^+ transportation (Tapken and Hollmann 2008).

Role of carriers known as KT/HAK/KUP transporters, which are one of the largest K^+ transporter families, has been reported in salinity stress in plants (Assaha et al. 2017). Plants thriving on land are classified into six clusters or clades (I–VI) with clade I including transporters involved in K^+ uptake from the soil. Clade I is clustered into a distinct subgroup clade Ia, known as HAK1-like transporters due to resemblance with the fungal counterparts (Santa-María et al. 2018). Clade Ia includes HAKs from various crop plants including barley HvHAK1 (Gierth and Mäser 2007), *Arabidopsis* AtHAK5 (Gierth et al. 2005), rice OsHAK1 and OsHAK5 (Banuelos et al. 2002), tomato LeHAK5 (Nieves-Cordones et al. 2007), etc. *AtHAK5* expression is inhibited by Na^+ (Nieves-Cordones et al. 2007) whereas HvHAK1 is known to facilitate Na^+ influx (Santa-María et al. 1997). Similarly, *OsHAK5* expression in tobacco (*Nicotiana tabacum*) BY2 cells confers salt tolerance through K^+ influx with no effect on the concentration of Na^+ (Horie et al. 2011). Na^+ homeostasis is crucial for plant health, and regulating its distribution in various tissues is one of the essential approaches to prevent salt toxicity. Understanding the high-affinity K^+ transporters (HKT transporters) is extremely essential in this context as this family of transporters control transportation of Na^+ flux by functioning either as Na^+ selective channels (Class I HKTs) or as Na^+/K^+ symporters (Class II HKTs), thereby having key roles in salinity tolerance in plants. Class I HKTs are found in both monocots and dicots with SGGG (S – Serine, G – Glycine) pore-forming motif while Class II HKTs are found only in monocots with GGGG pore-forming motif. *A. thaliana* has only one HKT homolog, i.e. *AtHKT1; 1,* which obstructs Na^+ overaccumulation in shoot tissues by retrieving it from xylem (Davenport et al. 2007). *AtHKT1; 1* expression in rice plants led to low Na^+ accumulation and improved growth under the salinity stress in shoots (Plett et al. 2010). In durum wheat, Nax1(TmHKT1; 4-A2) and Nax2 (TmHKT1; 5-A) decrease the leaf blade Na^+ concentration by 50 % and 30 % respectively, while together both reduced the accumulation by 60% (James et al. 2011). *TmHKT1; 5-A* is a Na^+-selective transporter, which facilitates Na^+ withdrawal from xylem, thereby reducing Na^+ transportation to leaves (Munns et al. 2012). Overexpression of *HvHKT2; 1* transporter in barley resulted in increased uptake of Na^+, higher concentration of Na^+ in xylem sap, and increased Na^+ translocation in leaves (Mian et al. 2011). In rice, HKT represents a multigene family with nine different members out of which five belong to Class I and four belong to Class II. *OsHKT2; 1* and *OsHKT2; 2* led to salt tolerance in rice, whereas *OsHKT1; 5* has role in Na^+ reabsorption from xylem (Oomen et al. 2012). Moreover, previous studies in plant species including wheat, barley, and *Brassica* emphasized the association of salt tolerance with K^+ retention ability of roots (Chakraborty et al. 2016; Wu et al. 2015). This association in roots of *Brassica* reflects the role of membrane depolarization-activated, GORK (outward-rectifying potassium selective) channels in facilitating K^+ efflux under salinity stress (Véry et al. 2014). Plants exposed to salinity stress, compartmentalize excess Na^+ into vacuoles as one of the detoxification mechanisms to reduce toxic effects of Na^+ in cytosol. As described earlier, this compartmentalization of Na^+ into cytosol occurs with the aid of vacuolar Na^+/H^+ exchangers (NHXs). Based on their differential subcellular location, NHXs have been classified into vacuolar-localized type I NHXs and type II NHXs located at endosome, trans-Golgi Network (TGN)/Golgi, and prevacuolar compartments (Chanroj et al. 2012). Similar to rice OsNHX1–OsNHX4, Arabidopsis AtNHX1–AtNHX4 belong to type I category (AtNHX5 and AtNHX6) while type II includes OsNHX5 (Andrés et al. 2014; Barragán et al. 2012; Bassil et al. 2012). NHX proteins play an important role in K^+ ion homeostasis, Na^+ compartmentalization, pH regulation, salinity tolerance, osmotic adjustment, vesicular trafficking, and flowering (Huertas et al. 2013; Zhang et al. 2015). Previous studies indicated that overexpression of *AtNHX1* and *DmNHX1* in *A. thaliana* led to enhanced salt tolerance (Apse et al. 1999). Moreover, expression of *AtNHX1* in kiwifruit (*Actinidia deliciosa*) led to improved salt tolerance compared with wild-type plants (Tian et al. 2011). The transgenic alfalfa (*Medicago sativa* L.) expressing *TaNHX2* had lower Na^+ concentration and more K^+ accumulation in leaves (Zhang et al. 2015). These transgenic plants also showed higher K^+/Na^+ ratio in leaves and cell sap (Zhang et al. 2015). Overexpression of K^+/H^+ antiporter *LeNHX2* in tomato resulted in improved K^+ homeostasis (high K^+ uptake velocity and lower K^+ activity in cytosol) and increased salt tolerance (Huertas et al. 2013). Lower concentration of NaCl (25 mM) in halophytes such as *Puccinellia tenuiflora* induces the expression of *PtNHX1,* which is further enhanced in shoots on exposure to higher concentration of NaCl (150 mM) (Zhang et al. 2017). Recently, another major Na^+/H^+ antiporter regulating Na^+ homeostasis in plants known as SOS transporter, which provides salt tolerance in various plant species by mediating Na^+ exclusion from plant cytosol (Wu 2018) (This transporter will be discussed in detail in the next section).

Plant growing on highly saline (NaCl) condition also experiences Cl^- toxicity, which necessitates understanding Cl^- assimilation under salt stress (Gong et al. 2011). Cl^- is a micronutrient, which stimulates plant growth by regulating activities including enzymatic reactions, turgor pressure, osmotic potential, and also functions as a cofactor in photosynthesis (Franco-Navarro et al. 2016). Under salinity stress, Cl^- accumulation negatively affects photosynthesis and plant growth (Tavakkoli et al. 2011). Plant species such as soybean and woody plants, e.g. citrus are unable to exclude accumulation of Cl^- from leaves and thus possess greater Cl^- toxicity (Munns and Tester 2008). Previous studies reported contribution of several transporters including slow anion channel and associated homologs (SLAC/SLAH), chloride channels (CLC), nitrate transporter 1/peptide transporter family (NPF), aluminum-activated malate transporter (ALMT), cation chloride cotransporters (CCC), etc. for Cl^- transportation in plants (Li et al. 2016; Shelke et al. 2019). Many of them have been implicated in mediating salt tolerance in plants (Teakle and Tyerman 2010; Wu and Li 2019). Cl^- influx into plants is facilitated by *AtNPF2.4* member of NPF, popularly known to mediate low-affinity active NO_3^- transport (Corratgé-Faillie et al. 2017). *AtNPF2.4* loads Cl^- into the root xylem, and therefore, its overexpression increases Cl^- accumulation in shoots (Li et al. 2016). Another member *AtNPF2.5* encodes Cl^--permeable transporter that is expressed in root cortical cells and is induced in response to salinity stress (Li et al. 2017). Mutation of the *AtNPF2.5* gene in *Arabidopsis* disrupts Cl^- exclusion from roots, hence raising Cl^- accumulation in shoots (Li et al. 2017). Thus, *AtNPF2.5* could be regarded as salt-responsive genes and its overexpression might enhance salt tolerance in plants. Moreover, advancement in studies to understand Cl^- signaling under salinity stress helped in the identification of Cl^- transporters, such as the chloride channel proteins (CLCs), which plays an essential role in Cl^- homeostasis in plants exposed to salt stress. CLC protein family are ubiquitous channels, which were first identified as CLC-0 in torpedo fish (Nguyen et al. 2016). *A. thaliana* encodes seven CLC genes, out of which AtCLCa, AtCLCb, AtCLCc, AtCLCg, Cl^- channels are present in tonoplast and are possibly involved in vacuolar sequestration of Cl^- under salt stress (Wei et al. 2017). Chloride channel-c (CLCc) in plants is known to hinder transportation of Cl^- in shoot in saline condition (Jossier et al. 2010). Previously, first identified tonoplast-localized ALMT, AtALMT9, a malate-activated Cl^- channel, was showed to be involved in vacuolar Cl^- sequestration and needed for stomatal opening in *Arabidopsis* (De Angeli et al. 2013). It is transcriptionally induced under salt stress, and its knockout mutant (*almt9*) possess reduced shoot Cl^- accumulation (Baetz et al. 2016). Altogether, the above studies suggest that vacuolar Cl^- sequestration might be another component for overall plant salt tolerance. Beside these, Cation Chloride Cotransporters (CCCs) act as potential chloride transporters as loss of *AtCCC* results in increased Cl^- uptake in plants (Henderson et al. 2015).

4.5 Transport of Na^+ Through SOS Signal Transduction Pathway: At Cellular Level

Exposure of salinity stress exerts both ionic and osmotic stress in plants. Therefore, plants adopt various strategies to cope with high Na^+ concentration inside the cytosol. One of these strategies involves increase in Na^+ efflux out of the cell, which in turn maintains ion homeostasis. After perception of salt stress, increase in Ca^{2+} level is sensed by the SOS pathway regulating Na^+ flux in plant cytosol (Hilleary and Gilroy 2018). SOS signaling is mediated by three key components – SOS1, SOS2, and SOS3. The increase in cytosolic Ca^{2+} levels occurs either through intracellular stores or apoplastic region by activation of phospholipase C (PLC) (White and Broadley 2003). There are mechanosensitive (MS) channels, which sense osmotic shock and are Ca^{2+}-permeable (Keisham et al. 2018). The targets of these MS channels include kinases such as Calcineurin B-like proteins (CBLs), calcium-dependent protein kinases (CDPKs), and CBL-interacting protein kinases (CIPKs). Excess Na^+ inside the cell leads to increase in cytosolic Ca^{2+}, which in turn is sensed by SOS3/Calcineurin B-like 4 (CBL4) (Figure 4.1). SOS3 is a myristoylated calcium-binding protein with three elongation factor hands in Ca^{2+} binding, expressed prominently in roots, and acts as a primary Ca^{2+} sensor (Ji et al. 2013; Keisham et al. 2018). Plants having mutant SOS3 are hypersensitive to salinity stress, and this response can be partially reversed by addition of Ca^{2+} (Ye et al. 2013). Ca^{2+} binds SOS3 and activates SOS2/CIPK24 (Ji et al. 2013). Interaction of SOS3 with SOS2 occurs in the roots (Quan et al. 2007). SOS2 is a Ser/Thr protein kinase, which belongs to SnRK3 (sucrose non-fermenting protein kinase-3) family, has a role in maintaining Na^+ and K^+ homeostasis (Ji et al. 2013). SOS2 protein structure consists of N-terminal kinase domain and a C-terminal regulatory domain. SOS3 binds to the FISL motif present in the regulatory domain, thereby liberating its kinase domain from inactivation (Gong et al. 2004). Expression of SOS2 occurs in both roots and shoots but is induced by salt stress specifically in roots (Liu et al. 2000). An alternative regulator of SOS2 that functions in shoots in place of SOS3 is SCaBP8/CBL10 (Quan et al. 2007). In vivo studies indicated that SOS2 phosphorylates SCaBP8 at its C-terminal, and this event is induced under salt stress (Lin et al. 2009). CBL10 plays important role in salinity tolerance by activating Na^+ compartmentalization into the vacuoles. Upon activation by SOS3, SOS2 is recruited to the plasma membrane where it phosphorylates and therefore activates SOS1, a Na^+/H^+ antiporter, in order to prevent toxic accumulation of Na^+ (Quintero et al. 2011). Upon salt stress, the SOS2–SOS3 complex phosphorylates the SOS1 C-terminal auto inhibitory domain, hereby making SOS1 free of its auto inhibition (Quintero et al. 2011). Several SOS genes were identified in a genetic screen to elucidate components regulating salt tolerance in *A. thaliana* (Wu et al. 1996). *AtSOS1* gets upregulated in both root and shoot upon salt stress (Shi et al. 2002). Overexpression of *AtSOS1* leads to reduced Na^+ concentration in both root and shoot (Shi et al. 2003). Studies on *sos1* mutant suggest that SOS1 is responsible for controlled uptake of Na^+ in roots and its long-distance transport to shoots (El Mahi et al. 2019). *Atsos1* mutants and *sos1* mutants in rice accumulate higher Na^+ concentration with impaired loading of Na^+ into the xylem (El Mahi et al. 2019). Previously, the photoperiodical and circadian clock switch Gigantea (GI) was identified to negatively regulate the SOS pathway. GI inhibits SOS1 phosphorylation in the absence of salt stress by preventing interaction of SOS2–SOS3. Upon salt stress, GI is degraded, thereby allowing interaction between SOS2 and SOS3. Plants overexpressing GI showed enhanced salt sensitivity while *gi* mutants showed increased salt tolerance (Kim et al. 2013).

4.6 Role of Salt Tolerance Responsive Genes in Transport of Na^+ and Cl^- Ions

Cellular Na^+ and Cl^- homeostasis is indispensable in terms of salinity stress as substantial accumulation of these ions raises toxicity in plant cells. Previous studies on halophytes including *Aeluropus, Mesembryanthemum, Suaeda, Atriplex, Thellungiella*, etc. have greatly assisted in the identification of genes imparting salt tolerance in plants (Mishra and Tanna 2017). As discussed earlier, salt detoxification is mediated by well known salt transporters such as NHX, SOS, HKTs; however, there are various other genes that regulate salt tolerance in plants by modulating distribution of Na^+, K^+, and Cl^- ions (Hernández 2019). Here, we will revisit the role of the chief salt-responsive genes, which improves plant growth under salt stress. Cytosolic Na^+ concentration is majorly controlled by Na^+/H^+ antiporters such as NHX1 and SOS1 (also known as NHX7) in plants. Vacuolar sequestration of Na^+ by NHX antiporters relaxes Na^+ toxicity in plant cytosol, which is evident from tolerance developed by overexpression *AtNHX1* and its orthologs in *Arabidopsis*, maize, rice, and tomato plants (Ismail and Horie 2017). In addition to providing tolerance against Na^+ accumulation under salt stress, overexpression of *AtNHX1 in* tomato also encourages K^+ transport from root to shoot along with its vacuolar sequestration, thereby regulating pH homeostasis in plant cell (Leidi et al. 2010). Interestingly, NHX1 efficiency is controlled by vacuolar H^+ ATPase (VATPase) and vacuolar PPase, which establishes H^+ gradient across tonoplast. Overexpression of vacuolar PPase, AVP1 improved salt tolerance in transgenic *Arabidopsis* (Gaxiola et al. 2001). Similarly, under greenhouse condition, *AtAVP1* expression in barley causes the improvement in shoot biomass and grain yield and also provided higher salt tolerance (Schilling et al. 2014). Activity of NHX1 is also controlled by SOS1, the only known Na^+/H^+ antiporter of the SOS pathway, which facilitates Na^+ efflux under salinity stress (Park et al. 2016). Similarly, SOS1 regulates Na^+ loading in xylem under mild salt stress. Overexpression of *SOS1* in tobacco increases xylem Na^+ loading and hence increases salt tolerance (Oh et al. 2010). Moreover, SOS pathway governs salt tolerance in plants via SOS2, which binds to regulatory subunits of VATPase and enhances its activity (Batelli et al. 2007). Although SOS1 does not transport K^+, *sos1*, *sos2,* and *sos3* mutants are hypersensitive to K^+ deficiency (Quintero et al. 2002). Captivatingly, few mutations and deletions in Na^+-transporter HKT1 restrain the Na^+ accumulation and thus suppress the salt-hypersensitive phenotype of *sos* mutants (Rus et al. 2004). As described previously, HKT restricts Na^+ toxicity in shoots by mediating it efflux from root xylem tissue, thus maintaining Na^+ flux in plants. *AtHKT1; 1*, and its rice ortholog *OsHKT1; 5,*relieves Na^+ accumulation in leaves by transporting it from xylem sap into the surrounding parenchyma cells and thus protecting leaves from Na^+ toxicity (Kobayashi et al. 2017). In contrast to other HKT1; 5, a recent study revealed negative role of barley *HvHKT1; 5* in salt stress as *HvHKT1; 5* knockdown barley lines displayed enhanced tolerance toward salt stress by increasing Na^+ translocation from root to shoots and thus maintaining high K^+/Na^+ ratio in the cytosol (Huang et al. 2020).

In addition to Na^+, presence of high salt (NaCl) concentration enhances Cl^- concentration in plants, which was an underlooked micronutrient in context of salinity stress; however, recent studies have revealed its involvement in salt toxicity (Wu and Li 2019). Mechanosensitive channels including MSL8 and MSL10 were previously identified to mediate Cl^- efflux and showed greater preference for Cl^- over Na^+ (approx. 6: 1) under salt stress (Maksaev and Haswell 2012). Another detoxification mechanism operates in soyabean, *Lotus,* and grapevine, which eliminates Cl^- from shoots and imparts salt tolerance to these plants (Teakle and Tyerman 2010). This shoot-mediated Cl^- detoxification leads to salt tolerance in *Arabidopsis*, identified to be aided by the anion channel SLAC1 and its homolog SLAH1 (Qiu 2016). Similar to the function of NHX in vacuolar compartmentation of Na^+, tonoplast-localized CLC plays an important role in Cl^- homeostasis, which is distinct from earlier studies in which overexpression of *AtCLCc and GmCLC1* provided salt tolerance in *Arabidopsis* and soybean (Hu et al. 2017). Recently, a salinity induced GsCLC-c2 channel (identified from *Glycine soja* accession of soyabean) was recognized to sequester Cl^- in root vacuoles, thereby preventing Cl^- toxicity in shoots under salt stress (Wei et al. 2019). These studies accentuate need of further exploring role of Cl^- in salinity stress, which would help in identification of novel genes associated with salt tolerance in plants.

4.7 Role of Ions in Salt Stress Tolerance

Na^+ accumulation and K^+ deficiency under salinity stress in plants foster Na^+ toxicity as it is nonessential for plant growth and metabolism. Plants get tolerant against salinity by maintaining higher K^+/Na^+ ratio in plant cytosol. Being the most abundant cation, K^+ plays an important role in various enzymatic reactions, regulates pH and membrane potential, and provides ionic homeostasis (Maathuis 2014). Maintaining a more negative potential and K^+ retention inside the cell prevents membrane depolarization facilitated by salt stress in plants (Falhof et al. 2016). Previous studies in plants including *Arabidopsis*, barley, cotton, and brassica have exquisitely linked K^+ retention in plants with salt tolerance (Chakraborty et al. 2016; Wu 2018; Wu et al. 2015). Na^+ influx into plant cell depolarizes the membrane potential below resting potential, which activates K^+ extrusion from the cell mediated by outward rectifier channels, such as GORK (guard cell outward rectifying K^+ channel) (Isayenkov and Maathuis 2019). K^+ efflux from leaf mesophyll cells in wheat genotypes was significantly higher in salt-sensitive varieties than salt-tolerant under salinity stress (Wu et al. 2013). Salt stress also induces the ROS production in plant cell, which in turn facilitates K^+ efflux by activating outward rectifying channels such as GORK, SKOR, etc, leading to programmed cell death (PCD) (Figure 4.1). Decrease in K^+ efflux and hence, salt tolerance was reported by overexpressing animal antiapoptotic CED-9 gene in tobacco mesophyll cells (Shabala et al. 2007). Additionally, slower development in salt stress-mediated PCD was observed in *A. thaliana gork1-1* mutant. Thus, K^+ plays an important role in alleviating salt toxicity by an interplay of K^+ channels, which regulates cellular homeostasis (Demidchik et al. 2010). Alternatively, K^+ uptake by high-affinity KT/HAK/KUP family under salt stress could also relieve the K^+ deficit induced by high Na^+ concentration under salt stress. For example, on exposure to salt stress OsHAK5, overexpression lines show significantly improved shoot K^+ accumulation and biomass as compared with WT rice while OsHAK1-D mutation resulted in significant decrease in root and shoot K^+ content and biomass (Chen et al. 2015). Interestingly, higher Na^+ concentration under salt stress elevates Ca^{2+} flux in plant cytosol, which stimulates calcineurin B-like (CBL) calcium sensors, e.g. CBL4 and CBL10 and thus CIPKs (Manishankar et al. 2018). K^+ uptake mediated by HAK5 is activated by CBL10–CIPK23 complex, suggesting that K^+ signaling is also influenced by the presence of Ca^{2+} in response to salinity (Ragel et al. 2015). In addition to affecting K^+ flux, increased extracellular Ca^{2+} also obstructs Na^+ influx by inhibiting Na^+-permeable NSCCs (Tester and Davenport 2003). Former study on *Arabidopsis* root epidermis showed that extracellular Ca^{2+} has been reported to block one half of the Na^+ influx current via NSCCs, which is further enhanced at higher concentration of Ca^{2+} (2–20 mm Ca^{2+} led to 75–80% inhibition) (Demidchik and Tester 2002).

As previously described, salt-induced Ca^{2+} flux in cytosol also activates SOS pathway, which is crucial for effluxing accumulated Na^+, thus alleviating salt toxicity (Gupta and Huang 2014; Keisham et al. 2018). However, Na^+ extrusion mediated by SOS1 (also known as NHX7, a Na^+-H^+ antiporter) depends on plasma membrane H^+-ATPase activity indispensable for repolarization of membrane voltage (after depolarization by Na^+ influx). Ample Ca^{2+} stimulates H^+-ATPase activity through Ca^{2+}-calmodulin-dependent protein kinases and regulates membrane integrity and ionic homeostasis under salinity stress in plants (Hasegawa 2013). Additionally, Ca^{2+} sensor, CBL10 interacts and suppresses the activity of K^+ uptake channel AKT1, thus assuring it functions as an interconnecting regulator of SOS1 and AKT1 and regulating Na^+/K^+ ratio in the cell (Almeida et al. 2017; Ketehouli et al. 2019). Thus Ca^{2+} functions as a major salt stress regulator in plants, and there must be greater focus on its effect on other unidentified components of salt signaling, which could help in elucidating new genes providing salt tolerance under salinity stress.

4.8 Reaching the Current Status and Conclusion

Salinity is one of the major abiotic stresses that hinders agricultural productivity by inducing ionic and osmotic imbalance in plant cell. Ion homeostasis is crucial for the functioning of various metabolic processes required for plant survival. High salt (NaCl) leads to accumulation of Na^+ and Cl^- ions in plant, which negatively regulates the uptake of K^+ ions. Adequate K^+ is required to prevent salt toxicity in plants caused by Na^+ accumulation under salinity stress (Figure 4.1). Salinity induces formation of various secondary messengers including Ca^{2+}, cyclic nucleotides, and ROS, which regulate plant growth and metabolism. The genotypic variation, transcriptomic, proteomic, and metabolomic approaches have helped to gather ample information about genes related to salt tolerance in plants (Gupta and Huang 2014; Hernández 2019). Plants adopted several strategies including impeding Na^+ uptake, enhancing Na^+ efflux, and vacuolar sequestration of Na^+ to limit cytosolic Na^+ (Yang and Guo 2018). Various Na^+ and K^+ transporters including NSCCs, HKTs, NHXs, SOS, etc. work under salinity stress for cellular homeostasis (Assaha et al. 2017). Ca^{2+} plays an essential role in salt sensing and detoxification in plants as it is a major component of SOS (Na^+ efflux) pathway. Although Na^+ uptake and signaling have been well explored in plants, the sensors, which sense Na^+ toxicity, are not much known. Recently, GIPC has been identified to sense Na^+, which then binds to Ca^{2+} ions and activates SOS pathway in plants exposed to salt stress (Jiang et al. 2019). However, additional information is required for Na^+ sensors, which could be helpful in regulating salt uptake and thus will provide tolerance against salinity. As research on salinity is progressing, studies on Cl^- toxicity in salt stress helped to identify the role of Cl^- transporters involved in salt tolerance in plants. Thus, in-depth understanding of the cross talk of ions (Na^+, Cl^-, K^+ and Ca^{2+}) is necessitated to identify unexplored genes regulating cellular salt ion accumulation to develop salt tolerant crops.

References

Acosta-Motos, J.R., Ortuño, M.F., Bernal-Vicente, A. et al. (2017). Plant responses to salt stress: adaptive mechanisms. *Agronomy* 7 (1): 18.

Adams, E. and Shin, R. (2014). Transport, signaling, and homeostasis of potassium and sodium in plants. *J. Integr. Plant Biol.* 56 (3): 231–249.

Almeida, D.M., Oliveira, M.M., and Saibo, N.J. (2017). Regulation of Na^+ and K^+ homeostasis in plants: towards improved salt stress tolerance in crop plants. *Genet. Mol. Biol.* 40 (1): 326–345.

Andrés, Z., Pérez-Hormaeche, J., Leidi, E.O. et al. (2014). Control of vacuolar dynamics and regulation of stomatal aperture by tonoplast potassium uptake. *Proc. Natl. Acad. Sci.* 111 (17): E1806–E1814.

Apse, M.P., Aharon, G.S., Snedden, W.A., and Blumwald, E. (1999). Salt tolerance conferred by overexpression of a vacuolar Na+/H+ antiport in Arabidopsis. *Science* 285 (5431): 1256–1258.

Apse, M.P., Sottosanto, J.B., and Blumwald, E. (2003). Vacuolar cation/H^+ exchange, ion homeostasis, and leaf development are altered in a T DNA insertional mutant of AtNHX1, the Arabidopsis vacuolar Na^+/H^+ antiporter. *Plant J.* 36 (2): 229–239.

Assaha, D.V., Ueda, A., Saneoka, H. et al. (2017). The role of Na+ and K+ transporters in salt stress adaptation in glycophytes. *Front. Physiol.* 8: 509.

Baetz, U., Eisenach, C., Tohge, T. et al. (2016). Vacuolar chloride fluxes impact ion content and distribution during early salinity stress. *Plant Physiol.* 172 (2): 1167–1181.

Banuelos, M.A., Garciadeblas, B., Cubero, B., and Rodrıguez-Navarro, A. (2002). Inventory and functional characterization of the HAK potassium transporters of rice. *Plant Physiol.* 130 (2): 784–795.

Barragán, V., Leidi, E.O., Andrés, Z. et al. (2012). Ion exchangers NHX1 and NHX2 mediate active potassium uptake into vacuoles to regulate cell turgor and stomatal function in Arabidopsis. *Plant Cell* 24 (3): 1127–1142.

Bassil, E., Coku, A., and Blumwald, E. (2012). Cellular ion homeostasis: emerging roles of intracellular NHX Na^+/H^+ antiporters in plant growth and development. *J. Exp. Bot.* 63 (16): 5727–5740.

Batelli, G., Verslues, P.E., Agius, F. et al. (2007). SOS2 promotes salt tolerance in part by interacting with the vacuolar H^+-ATPase and upregulating its transport activity. *Mol. Cell. Biol.* 27 (22): 7781–7790.

Berthomieu, P., Conéjéro, G., Nublat, A. et al. (2003). Functional analysis of AtHKT1 in Arabidopsis shows that Na^+ recirculation by the phloem is crucial for salt tolerance. *EMBO J.* 22 (9): 2004–2014.

Bridges, D., Fraser, M.E., and Moorhead, G.B. (2005). Cyclic nucleotide binding proteins in the *Arabidopsis thaliana* and *Oryza sativa* genomes. *BMC Bioinf.* 6 (1): 6.

Brini, F. and Masmoudi, K. (2012). Ion transporters and abiotic stress tolerance in plants. *ISRN Mol. Biol.* 2012: 1–13.

Byrt, C.S., Zhao, M., Kourghi, M. et al. (2017). Non selective cation channel activity of aquaporin AtPIP2; 1 regulated by Ca^{2+} and pH. *Plant Cell Environ.* 40 (6): 802–815.

Chakraborty, K., Bose, J., Shabala, L., and Shabala, S. (2016). Difference in root K+ retention ability and reduced sensitivity of K^+-permeable channels to reactive oxygen species confer differential salt tolerance in three Brassica species. *J. Exp. Bot.* 67 (15): 4611–4625.

Chanroj, S., Wang, G., Venema, K. et al. (2012). Conserved and diversified gene families of monovalent cation/H+ antiporters from algae to flowering plants. *Front. Plant Sci.* 3: 25.

Cheeseman, J.M. (2015). The evolution of halophytes, glycophytes and crops, and its implications for food security under saline conditions. *New Phytol.* 206 (2): 557–570.

Chen, G., Hu, Q., Luo, L. et al. (2015). Rice potassium transporter O s HAK 1 is essential for maintaining potassium-mediated growth and functions in salt tolerance over low and high potassium concentration ranges. *Plant Cell Environ.* 38 (12): 2747–2765.

Corratgé-Faillie, C., Ronzier, E., Sanchez, F. et al. (2017). The Arabidopsis guard cell outward potassium channel GORK is regulated by CK 33. *FEBS Lett.* 591 (13): 1982–1992.

Davenport, R.J., Munoz Mayor, A., Jha, D. et al. (2007). The Na^+ transporter AtHKT1; 1 controls retrieval of Na^+ from the xylem in *Arabidopsis*. *Plant Cell Environ.* 30 (4): 497–507.

De Angeli, A., Zhang, J., Meyer, S., and Martinoia, E. (2013). AtALMT9 is a malate-activated vacuolar chloride channel required for stomatal opening in Arabidopsis. *Nat. Commun.* 4 (1): 1–10.

Demidchik, V. and Tester, M. (2002). Sodium fluxes through nonselective cation channels in the plasma membrane of protoplasts from Arabidopsis roots. *Plant Physiol.* 128 (2): 379–387.

Demidchik, V., Cuin, T.A., Svistunenko, D. et al. (2010). Arabidopsis root K+-efflux conductance activated by hydroxyl radicals: single-channel properties, genetic basis and involvement in stress-induced cell death. *J. Cell Sci.* 123 (9): 1468–1479.

Duszyn, M., Świeżawska, B., Szmidt-Jaworska, A., and Jaworski, K. (2019). Cyclic nucleotide gated channels (CNGCs) in plant signalling–current knowledge and perspectives. *J. Plant Physiol.* 241: 153035.

El Mahi, H., Pérez-Hormaeche, J., De Luca, A. et al. (2019). A critical role of sodium flux via the plasma membrane Na^+/H^+ exchanger SOS1 in the salt tolerance of rice. *Plant Physiol.* 180 (2): 1046–1065.

Falhof, J., Pedersen, J.T., Fuglsang, A.T., and Palmgren, M. (2016). Plasma membrane H+-ATPase regulation in the center of plant physiology. *Mol. Plant* 9 (3): 323–337.

Franco-Navarro, J.D., Brumós, J., Rosales, M.A. et al. (2016). Chloride regulates leaf cell size and water relations in tobacco plants. *J. Exp. Bot.* 67 (3): 873–891.

Fukuda, A., Nakamura, A., Tagiri, A. et al. (2004). Function, intracellular localization and the importance in salt tolerance of a vacuolar Na^+/H^+ antiporter from rice. *Plant Cell Physiol.* 45 (2): 146–159.

Fukuda, A., Nakamura, A., Hara, N. et al. (2011). Molecular and functional analyses of rice NHX-type Na^+/H^+ antiporter genes. *Planta* 233 (1): 175–188.

Gaxiola, R.A., Li, J., Undurraga, S. et al. (2001). Drought-and salt-tolerant plants result from overexpression of the AVP1 H^+-pump. *Proc. Natl. Acad. Sci.* 98 (20): 11444–11449.

Ghosh, B., Md, N.A., and Gantait, S. (2016). Response of rice under salinity stress: a review update. *Rice Res. Open Access* 4 (2): 1–8.

Gierth, M. and Mäser, P. (2007). Potassium transporters in plants–involvement in K^+ acquisition, redistribution and homeostasis. *FEBS Lett.* 581 (12): 2348–2356.

Gierth, M., Mäser, P., and Schroeder, J.I. (2005). The potassium transporter AtHAK5 functions in K^+ deprivation-induced high-affinity K^+ uptake and AKT1 K^+ channel contribution to K^+ uptake kinetics in Arabidopsis roots. *Plant Physiol.* 137 (3): 1105–1114.

Gobert, A., Park, G., Amtmann, A. et al. (2006). *Arabidopsis thaliana* cyclic nucleotide gated channel 3 forms a non-selective ion transporter involved in germination and cation transport. *J. Exp. Bot.* 57 (4): 791–800.

Gong, D., Guo, Y., Schumaker, K.S., and Zhu, J.K. (2004). The SOS3 family of calcium sensors and SOS2 family of protein kinases in Arabidopsis. *Plant Physiol.* 134 (3): 919–926.

Gong, X., Chao, L., Zhou, M. et al. (2011). Oxidative damages of maize seedlings caused by exposure to a combination of potassium deficiency and salt stress. *Plant Soil* 340 (1–2): 443–452.

Guo, J., Islam, M.A., Lin, H. et al. (2018). Genome-wide identification of cyclic nucleotide-gated ion channel gene family in wheat and functional analyses of TaCNGC14 and TaCNGC16. *Front. Plant Sci.* 9: 18.

Gupta, B. and Huang, B. (2014). Mechanism of salinity tolerance in plants: physiological, biochemical, and molecular characterization. *Int. J. Genomics* 2014: 1–18.

Hanin, M., Ebel, C., Ngom, M. et al. (2016). New insights on plant salt tolerance mechanisms and their potential use for breeding. *Front. Plant Sci.* 7: 1787.

Hao, L. and Qiao, X. (2018). Genome-wide identification and analysis of the CNGC gene family in maize. *Peer J.* 6 (e): 5816.

Hasanuzzaman, M., Bhuyan, M., Nahar, K. et al. (2018). Potassium: a vital regulator of plant responses and tolerance to abiotic stresses. *Agronomy* 8 (3): 31.

Hasegawa, P.M. (2013). Sodium (Na^+) homeostasis and salt tolerance of plants. *Environ. Exp. Bot.* 92: 19–31.

Henderson, S.W., Wege, S., Qiu, J. et al. (2015). Grapevine and Arabidopsis cation-chloride cotransporters localize to the Golgi and trans-Golgi network and indirectly influence long-distance ion transport and plant salt tolerance. *Plant Physiol.* 169 (3): 2215–2229.

Hernández, J.A. (2019). *Salinity Tolerance in Plants: Trends and Perspectives*. Multidisciplinary Digital Publishing Institute.

Hilleary, R. and Gilroy, S. (2018). Systemic signaling in response to wounding and pathogens. *Curr. Opin. Plant Biol.* 43: 57–62.

Horie, T., Kaneko, T., Sugimoto, G. et al. (2011). Mechanisms of water transport mediated by PIP aquaporins and their regulation via phosphorylation events under salinity stress in barley roots. *Plant Cell Physiol.* 52 (4): 663–675.

Hu, R., Zhu, Y., Wei, J. et al. (2017). Overexpression of PP2AC5 that encodes the catalytic subunit 5 of protein phosphatase 2A in Arabidopsis confers better root and shoot development under salt conditions. *Plant Cell Environ.* 40 (1): 150–164.

Huang, L., Kuang, L., Wu, L. et al. (2020). The HKT transporter HvHKT1; 5 negatively regulates salt tolerance. *Plant Physiol.* 182 (1): 584–596.

Huertas, R., Rubio, L., Cagnac, O. et al. (2013). The K+/H+ antiporter LeNHX2 increases salt tolerance by improving K^+ homeostasis in transgenic tomato. *Plant Cell Environ.* 36 (12): 2135–2149.

Ijaz, R., Ejaz, J., Gao, S. et al. (2017). Overexpression of annexin gene Ann Sp2, enhances drought and salt tolerance through modulation of ABA synthesis and scavenging ROS in tomato. *Sci. Rep.* 7 (1): 1–14.

Isayenkov, S. (2012). Physiological and molecular aspects of salt stress in plants. *Cytol. Genet.* 46 (5): 302–318.

Isayenkov, S.V. and Maathuis, F.J. (2019). Plant salinity stress: many unanswered questions remain. *Front. Plant Sci.* 10: 80.

Ismail, A.M. and Horie, T. (2017). Genomics, physiology, and molecular breeding approaches for improving salt tolerance. *Annu. Rev. Plant Biol.* 68: 405–434.

James, R.A., Blake, C., Byrt, C.S., and Munns, R. (2011). Major genes for Na^+ exclusion, Nax 1 and Nax 2 (wheat HKT1; 4 and HKT1; 5), decrease Na^+ accumulation in bread wheat leaves under saline and waterlogged conditions. *J. Exp. Bot.* 62 (8): 2939–2947.

Jayakannan, M., Bose, J., Babourina, O. et al. (2013). Salicylic acid improves salinity tolerance in *Arabidopsis* by restoring membrane potential and preventing salt-induced K^+ loss via a GORK channel. *J. Exp. Bot.* 64 (8): 2255–2268.

Jha, S.K., Sharma, M.K., and Pandey, G. (2016). Role of cyclic nucleotide gated channels in stress management in plants. *Curr. Genomics* 17 (4): 315–329.

Jha, U.C., Bohra, A., Jha, R., and Parida, S.K. (2019). Salinity stress response and 'omics' approaches for improving salinity stress tolerance in major grain legumes. *Plant Cell Rep.* 38 (3): 255–277.

Ji, H., Pardo, J.M., Batelli, G. et al. (2013). The Salt Overly Sensitive (SOS) pathway: established and emerging roles. *Mol. Plant* 6 (2): 275–286.

Jiang, X., Leidi, E.O., and Pardo, J.M. (2010). How do vacuolar NHX exchangers function in plant salt tolerance? *Plant Signal. Behav.* 5 (7): 792–795.

Jiang, Z., Zhou, X., Tao, M. et al. (2019). Plant cell-surface GIPC sphingolipids sense salt to trigger Ca 2^+ influx. *Nature* 572 (7769): 341–346.

Jossier, M., Kroniewicz, L., Dalmas, F. et al. (2010). The Arabidopsis vacuolar anion transporter, AtCLCc, is involved in the regulation of stomatal movements and contributes to salt tolerance. *Plant J.* 64 (4): 563–576.

Keisham, M., Mukherjee, S., and Bhatla, S.C. (2018). Mechanisms of sodium transport in plants—progresses and challenges. *Int. J. Mol. Sci.* 19 (3): 647.

Ketehouli, T., Carther, I., Foka, K. et al. (2019). Adaptation of plants to salt stress: characterization of Na^+ and K^+ transporters and role of CBL gene family in regulating salt stress response. *Agronomy* 9 (11): 687.

Kim, S.A., Kwak, J., Jae, S.K. et al. (2001). Overexpression of the AtGluR2 gene encoding an *Arabidopsis* homolog of mammalian glutamate receptors impairs calcium utilization and sensitivity to ionic stress in transgenic plants. *Plant Cell Physiol.* 42 (1): 74–84.

Kim, W.Y., Ali, Z., Park, H.J. et al. (2013). Release of SOS2 kinase from sequestration with GIGANTEA determines salt tolerance in Arabidopsis. *Nat. Commun.* 4 (1): 1–13.

Kobayashi, N.I., Yamaji, N., Yamamoto, H. et al. (2017). OsHKT1; 5 mediates Na^+ exclusion in the vasculature to protect leaf blades and reproductive tissues from salt toxicity in rice. *Plant J.* 91 (4): 657–670.

Kronzucker, H.J. and Britto, D.T. (2011). Sodium transport in plants: a critical review. *New Phytol.* 189 (1): 54–81.

Leidi, E.O., Barragán, V., Rubio, L. et al. (2010). The AtNHX1 exchanger mediates potassium compartmentation in vacuoles of transgenic tomato. *Plant J.* 61 (3): 495–506.

Li, J., Jia, H., and Wang, J. (2014). cGMP and ethylene are involved in maintaining ion homeostasis under salt stress in *Arabidopsis* roots. *Plant Cell Rep.* 33 (3): 447–459.

Li, B., Byrt, C., Qiu, J. et al. (2016). Identification of a stelar-localized transport protein that facilitates root-to-shoot transfer of chloride in Arabidopsis. *Plant Physiol.* 170 (2): 1014–1029.

Li, B., Qiu, J., Jayakannan, M. et al. (2017). *AtNPF2.5* modulates chloride (Cl^-) efflux from roots of *Arabidopsis thaliana*. *Front. Plant Sci.* 7: 2013.

Lin, H., Yang, Y., Quan, R. et al. (2009). Phosphorylation of SOS3-LIKE CALCIUM BINDING PROTEIN8 by SOS2 protein kinase stabilizes their protein complex and regulates salt tolerance in Arabidopsis. *Plant Cell* 21 (5): 1607–1619.

Liu, J., Ishitani, M., Halfter, U. et al. (2000). The Arabidopsis thaliana SOS2 gene encodes a protein kinase that is required for salt tolerance. *Proc. Natl. Acad. Sci.* 97 (7): 3730–3734.

Livanos, P., Galatis, B., Gaitanak, i.C., and Apostolakos, P. (2014). Phosphorylation of a p38-like MAPK is involved in sensing cellular redox state and drives atypical tubulin polymer assembly in angiosperms. *Plant Cell Environ.* 37 (5): 1130–1143.

Ma, W. and Berkowitz, G.A. (2011). Ca^{2+} conduction by plant cyclic nucleotide gated channels and associated signaling components in pathogen defense signal transduction cascades. *New Phytol.* 190 (3): 566–572.

Maathuis, F.J. (2014). Sodium in plants: perception, signalling, and regulation of sodium fluxes. *J. Exp. Bot.* 65 (3): 849–858.

Maathuis, F.J. and Sanders, D. (2001). Sodium uptake in Arabidopsis roots is regulated by cyclic nucleotides. *Plant Physiol.* 127 (4): 1617–1625.

Maathuis, F.J., Ahmad, I., and Patishtan, J. (2014). Regulation of Na^+ fluxes in plants. *Front. Plant Sci.* 5: 467.

Maksaev, G. and Haswell, E.S. (2012). MscS-Like10 is a stretch-activated ion channel from Arabidopsis thaliana with a preference for anions. *Proc. Natl. Acad. Sci.* 109 (46): 19015–19020.

Manishankar, P., Wang, N., Köster, P. et al. (2018). Calcium signaling during salt stress and in the regulation of ion homeostasis. *J. Exp. Bot.* 69 (17): 4215–4226.

Mian, A., Oomen, R.J., Isayenkov, S. et al. (2011). Over expression of an Na^+ and K^+ permeable HKT transporter in barley improves salt tolerance. *Plant J.* 68 (3): 468–479.

Miller, G., Suzuki, N., Ciftci-Yilmaz, S., and Mittler, R. (2010). Reactive oxygen species homeostasis and signalling during drought and salinity stresses. *Plant Cell Environ.* 33 (4): 453–467.

Mishra, A. and Tanna, B. (2017). Halophytes: potential resources for salt stress tolerance genes and promoters. *Front. Plant Sci.* 8: 829.

Morton, M.J., Awlia, M., Al-Tamimi, N. et al. (2019). Salt stress under the scalpel–dissecting the genetics of salt tolerance. *Plant J.* 97 (1): 148–163.

Munns, R. and Tester, M. (2008). Mechanisms of salinity tolerance. *Annu. Rev. Plant Biol.* 59: 651–681.

Munns, R., James, R.A., Xu, B. et al. (2012). Wheat grain yield on saline soils is improved by an ancestral Na^+ transporter gene. *Nat. Biotechnol.* 30 (4): 360.

Nawaz, Z., Kakar, K.U., Saand, M.A., and Shu, Q.Y. (2014). Cyclic nucleotide-gated ion channel gene family in rice, identification, characterization and experimental analysis of expression response to plant hormones, biotic and abiotic stresses. *BMC Genomics* 15 (1): 853.

Nguyen, C.T., Agorio, A., Jossier, M. et al. (2016). Characterization of the chloride channel-like, AtCLCg, involved in chloride tolerance in Arabidopsis thaliana. *Plant Cell Physiol.* 57 (4): 764–775.

Nieves-Cordones, M., Martínez-Cordero, M.A., Martínez, V., and Rubio, F. (2007). An $NH4^+$-sensitive component dominates high-affinity K+ uptake in tomato plants. *Plant Sci.* 172 (2): 273–280.

Oh, D.H., Lee, S.Y., Bressan, R.A. et al. (2010). Intracellular consequences of SOS1 deficiency during salt stress. *J. Exp. Bot.* 61 (4): 1205–1213.

Oomen, R.J., Benito, B., Sentenac, H. et al. (2012). HKT2; 2/1, a K^+ permeable transporter identified in a salt tolerant rice cultivar through surveys of natural genetic polymorphism. *Plant J.* 71 (5): 750–762.

Pang, H.C., Li, Y.Y., Yang, J.S., and Liang, Y.S. (2010). Effect of brackish water irrigation and straw mulching on soil salinity and crop yields under monsoonal climatic conditions. *Agric. Water Manag.* 97 (12): 1971–1977.

Park, H.J., Kim, W.Y., and Yun, D.J. (2016). A new insight of salt stress signaling in plant. *Mol. Cell* 39 (6): 447.

Plett, D., Safwat, G., Gilliham, M. et al. (2010). Improved salinity tolerance of rice through cell type-specific expression of AtHKT1; 1. *PLoS One* 5 (9): e12571.

Price, M.B., Jelesko, J., and Okumoto, S. (2012). Glutamate receptor homologs in plants: functions and evolutionary origins. *Front. Plant Sci.* 3: 235.

Qiu, Q.S. (2016). Plant endosomal NHX antiporters: activity and function. *J. Exp. Bot.* 11 (5): e0144716.

Qiu, J., Henderson, S.W., Tester, M. et al. (2016). SLAH1, a homologue of the slow type anion channel SLAC1, modulates shoot Cl− accumulation and salt tolerance in *Arabidopsis thaliana*. *J. Exp. Bot.* 67 (15): 4495–4505.

Quan, R., Lin, H., Mendoza, I. et al. (2007). SCABP8/CBL10, a putative calcium sensor, interacts with the protein kinase SOS2 to protect *Arabidopsis* shoots from salt stress. *Plant Cell* 19 (4): 1415–1431.

Quintero, F.J., Ohta, M., Shi, H. et al. (2002). Reconstitution in yeast of the Arabidopsis SOS signaling pathway for Na^+ homeostasis. *Proc. Natl. Acad. Sci.* 99 (13): 9061–9066.

Quintero, F.J., Martinez-Atienza, J., Villalta, I. et al. (2011). Activation of the plasma membrane Na/H antiporter Salt-Overly-Sensitive 1 (SOS1) by phosphorylation of an auto-inhibitory C-terminal domain. *Proc. Natl. Acad. Sci.* 108 (6): 2611–2616.

Ragel, P., Ródenas, R., García-Martín, E. et al. (2015). The CBL-interacting protein kinase CIPK23 regulates HAK5-mediated high-affinity K^+ uptake in *Arabidopsis* roots. *Plant Physiol.* 169 (4): 2863–2873.

Rahnama, A., Munns, R., Poustini, K., and Watt, M. (2011). A screening method to identify genetic variation in root growth response to a salinity gradient. *J. Exp. Bot.* 62 (1): 69–77.

Roy, S., Gilliham, M., Berger, B. et al. (2008). Investigating glutamate receptor-like gene co expression in *Arabidopsis thaliana*. *Plant Cell Environ.* 31 (6): 861–871.

Rus, A., Lee, B.-h., Muñoz-Mayor, A. et al. (2004). AtHKT1 facilitates Na^+ homeostasis and K^+ nutrition in plants. *Plant Physiol.* 136 (1): 2500–2511.

Saand, M.A., Xu, Y.P., Munyampundu, J.P. et al. (2015). Phylogeny and evolution of plant cyclic nucleotide-gated ion channel (CNGC) gene family and functional analyses of tomato CNGCs. *DNA Res.* 22 (6): 471–483.

Santa-María, G.E., Rubio, F., Dubcovsky, J., and Rodríguez-Navarro, A. (1997). The HAK1 gene of barley is a member of a large gene family and encodes a high-affinity potassium transporter. *Plant Cell* 9 (12): 2281–2289.

Santa-María, G.E., Oliferuk, S., and Moriconi, J.I. (2018). KT-HAK-KUP transporters in major terrestrial photosynthetic organisms: a twenty years tale. *J. Plant Physiol.* 226: 77–90.

Schilling, R.K., Marschner, P., Shavrukov, Y. et al. (2014). Expression of the Arabidopsis vacuolar H^+ pyrophosphatase gene (AVP 1) improves the shoot biomass of transgenic barley and increases grain yield in a saline field. *Plant Biotechnol.* 12 (3): 378–386.

Schuurink, R.C., Shartzer, S.F., Fath, A., and Jones, R.L. (1998). Characterization of a calmodulin-binding transporter from the plasma membrane of barley aleurone. *Proc. Natl. Acad. Sci.* 95 (4): 1944–1949.

Senadheera, P., Singh, R., and Maathuis, F.J. (2009). Differentially expressed membrane transporters in rice roots may contribute to cultivar dependent salt tolerance. *J. Exp. Bot.* 60 (9): 2553–2563.

Shabala, S., Cuin, T.A., Prismall, L., and Nemchinov, L.G. (2007). Expression of animal CED-9 anti-apoptotic gene in tobacco modifies plasma membrane ion fluxes in response to salinity and oxidative stress. *Planta* 227 (1): 189–197.

Shelke, D., Nikalje, G., Nikam, T. et al. (2019). Chloride (Cl^-) uptake, transport, and regulation in plant salt tolerance. In: *Molecular Plant Abiotic Stress: Biology and Biotechnology* (eds. A. Roychoudhury and D. Tripathi), 241–268. John Wiley & Sons, Ltd.

Shi, H., Quintero, F.J., Pardo, J.M., and Zhu, J.K. (2002). The putative plasma membrane Na^+/H^+ antiporter SOS1 controls long-distance Na^+ transport in plants. *Plant Cell* 14 (2): 465–477.

Shi, H., Kim, Y., Guo, Y. et al. (2003). The Arabidopsis SOS5 locus encodes a putative cell surface adhesion protein and is required for normal cell expansion. *Plant Cell* 15 (1): 19–32.

Steinhorst, L. and Kudla, J. (2019). *How Plants Perceive Salt*. Nature Publishing Group.

Suzuki, N., Miller, G., Morales, J. et al. (2011). Respiratory burst oxidases: the engines of ROS signaling. *Curr. Opin. Plant Biol.* 14 (6): 691–699.

Tapken, D. and Hollmann, M. (2008). *Arabidopsis thaliana* glutamate receptor ion channel function demonstrated by ion pore transplantation. *J. Mol. Biol.* 383 (1): 36–48.

Tavakkoli, E., Fatehi, F., Coventry, S. et al. (2011). Additive effects of Na^+ and Cl^- ions on barley growth under salinity stress. *J. Exp. Bot.* 62 (6): 2189–2203.

Teakle, N.L. and Tyerman, S.D. (2010). Mechanisms of Cl^- transport contributing to salt tolerance. *Plant Cell Environ.* 33 (4): 566–589.

Tester, M. and Davenport, R. (2003). Na^+ tolerance and Na^+ transport in higher plants. *Ann. Bot.* 91 (5): 503–527.

Tian, N., Wang, J., and Xu, Z. (2011). Overexpression of Na^+/H^+ antiporter gene AtNHX1 from *Arabidopsis thaliana* improves the salt tolerance of kiwifruit (Actinidia deliciosa). *S. Afr. J. Bot.* 77 (1): 160–169.

Véry, A.A., Nieves-Cordones, M., Daly, M. et al. (2014). Molecular biology of K^+ transport across the plant cell membrane: what do we learn from comparison between plant species? *J. Plant Physiol.* 171 (9): 748–769.

Wang, P.H., Lee, C.E., Lin, Y.S. et al. (2019). The glutamate receptor-like protein GLR3. 7 interacts with 14-3-3ω and participates in salt stress response in *Arabidopsis thaliana*. *Front. Plant Sci.* 10: 1169.

Wei, Q., Luo, Q., Wang, R. et al. (2017). A wheat R2R3-type MYB transcription factor TaODORANT1 positively regulates drought and salt stress responses in transgenic tobacco plants. *Front. Plant Sci.* 8: 1374.

Wei, P., Che, B., Shen, L. et al. (2019). Identification and functional characterization of the chloride channel gene, GsCLC-c2 from wild soybean. *BMC Plant Biol.* 19 (1): 121.

White, P.J. and Broadley, M.R. (2003). Calcium in plants. *Ann. Bot.* 92 (4): 487–511.

Wu, H. (2018). Plant salt tolerance and Na^+ sensing and transport. *Crop J.* 6 (3): 215–225.

Wu, H. and Li, Z. (2019). The importance of Cl^- exclusion and vacuolar Cl^- sequestration: revisiting the role of Cl^- transport in plant salt tolerance. *Front. Plant Sci.* 10: 1418.

Wu, S.J., Ding, L., and Zhu, J.K. (1996). SOS1, a genetic locus essential for salt tolerance and potassium acquisition. *Plant Cell* 8 (4): 617–627.

Wu, H., Shabala, L., Barry, K. et al. (2013). Ability of leaf mesophyll to retain potassium correlates with salinity tolerance in wheat and barley. *Physiol. Plant.* 149 (4): 515–527.

Wu, H., Zhu, M., Shabala, L. et al. (2015). K^+ retention in leaf mesophyll, an overlooked component of salinity tolerance mechanism: a case study for barley. *J. Integr. Plant Biol.* 57 (2): 171–185.

Xu, H., Jiang, X., Zhan, K. et al. (2008). Functional characterization of a wheat plasma membrane Na+/H+ antiporter in yeast. *Arch. Biochem. Biophys.* 473 (1): 8–15.

Xu, Y., Zhou, Y., Hong, S. et al. (2013). Functional characterization of a wheat NHX antiporter gene TaNHX2 that encodes a K^+/H^+ exchanger. *PLoS One* 8 (11): e78098.

Yang, Y. and Guo, Y. (2018). Elucidating the molecular mechanisms mediating plant salt stress responses. *New Phytol.* 217 (2): 523–539.

Yang, Q., Chen, Z.Z., Zhou, X.F. et al. (2009). Overexpression of SOS (Salt Overly Sensitive) genes increases salt tolerance in transgenic *Arabidopsis*. *Mol. Plant* 2 (1): 22–31.

Ye, J., Zhang, W., and Guo, Y. (2013). *Arabidopsis* SOS3 plays an important role in salt tolerance by mediating calcium-dependent microfilament reorganization. *Plant Cell Rep.* 32 (1): 139–148.

Zhang, H.X. and Blumwald, E. (2001). Transgenic salt-tolerant tomato plants accumulate salt in foliage but not in fruit. *Nat. Biotechnol.* 19 (8): 765–768.

Zhang, Y.M., Zhang, H.M., Liu, Z.H. et al. (2015). The wheat NHX antiporter gene TaNHX2 confers salt tolerance in transgenic alfalfa by increasing the retention capacity of intracellular potassium. *Plant Mol. Biol.* 87 (3): 317–327.

Zhang, W.D., Wang, P., Bao, Z. et al. (2017). SOS1, HKT1; 5, and NHX1 synergistically modulate Na^+ homeostasis in the halophytic grass Puccinellia tenuiflora. *Front. Plant Sci.* 8: 576.

Zhu, J.K. (2000). Genetic analysis of plant salt tolerance using *Arabidopsis*. *Plant Physiol.* 124 (3): 941–948.

5

Salt Ion and Nutrient Interactions in Crop Plants: Prospective Signaling

Ria Khare[1,2] *and Prateek Jain*[3]

[1] *CSIR-National Botanical Research Institute (CSIR-NBRI), Lucknow, India*
[2] *Academy of Scientific and Innovative Research (AcSIR), New Delhi, India*
[3] *Department of Biology, The University of North Carolina at Chapel Hill, Chapel Hill, NC, USA*

5.1 Introduction

Salinity stress is one of major environmental constraint that deleteriously affect plant metabolism and productivity. Soil salinity is a serious threat to crop yield, responsible for the conversion of larger agricultural land into uncultivatable wasteland and obstructing worldwide food security (Ahanger et al. 2019). Irrigation under higher-salt conditions (NaCl) and poor drainage is a major factor causing salinity (Deinlein et al. 2014). Despite NaCl, an array of dissolved salts are also present in soil including Na_2SO_4, $MgSO_4$, $CaSO_4$, $MgCl_2$, KCl, and Na_2CO_3 with a little contribution in salinity stress (Tavakkoli et al. 2011). Saline soil exposes plant roots to higher NaCl concentration, which is assimilated in the form of Na^+ and Cl^- ion via various ion channels and transporters (Almeida et al. 2017). Na^+ compared with Cl^-, leads to greater cytotoxicity and lowers the plant yield (Isayenkov and Maathuis 2019). Na^+ accumulation leads to osmotic and ionic (higher Na^+/K^+ ratio) inequity in cytosol that facilitates membrane depolarization and eventually leads to K^+ extrusion (Keisham et al. 2018). Since Na^+ and K^+ have similar physicochemical properties, they can enter through common channels and transporters, which comprise the low-affinity (e.g. nonselective cation channels (NSCC) and high-affinity (e.g. KUP and high-affinity K^+ transporter (HKT)) transporters (Assaha et al. 2017). Maintaining cytoplasmic K^+ homeostasis is crucial as it is one of the important macronutrients, which regulate various biochemical (protein synthesis, enzyme activation, and carbohydrate metabolism) and physiological processes (stomatal regulation and photosynthesis) essential for proper functioning of plant cell (Hasanuzzaman et al. 2018). K^+ deficiency under salinity stimulates reactive oxygen species (ROS), which harms the cellular machinery by damaging biomolecules such as pigments, proteins, lipids, carbohydrates, and DNA, eventually leading to programmed cell death (PCD) (Wang et al. 2013). Captivatingly, Ca^{2+}

Physiology of Salt stress in Plants: Perception, Signalling, Omics and Tolerance Mechanism, First Edition.
Edited by Pratibha Singh, Madhulika Singh, Rajiv Kumar Singh, and Sheo Mohan Prasad.

influx inside the cell under salt stress serves as an integral part of salt overly sensitive (SOS) pathway detoxification machinery, which mediates Na^+ efflux and mitigates cytosolic salt toxicity (Ketehouli et al. 2019). In addition to K^+ and Ca^{2+}, salt stress also affects uptake and assimilation of various nutrients, which regulate plant growth and development (Guo et al. 2018; Tang et al. 2019). Nutrient homeostasis is obligatory since its deficiency limits plant growth and yield. Previous studies using several plant species including *Arabidopsis*, rice, wheat, and barley revealed the nutrient (N, P, S, and K) importance in salt stress regulation (Ahanger et al. 2019; Zayed et al. 2017). Here we will update the information regarding interplay of nutrients under salt stress in plants, which could help in unmining genes providing tolerance against salinity stress and eventually improve crop yield.

5.2 Effects of Salt Stress on Nutrient Absorption

Salinity affects the uptake of various essential nutrients with most severe impact on K^+ since higher Na^+ lowers the cytosolic K^+/Na^+ ratio and causes K^+ deficit in plants. K^+ uptake is a cumulative effort of high and low-affinity transporters present on cell membrane. High-affinity uptake (HAT) of K^+ is encouraged by members of KT/HAK/KUP family including HAK5 and KUP7, whereas the low-affinity uptake (LAT) of K^+ is facilitated by AKT1 K^+ channels (Assaha et al. 2017; Nieves-Cordones et al. 2016). In few cases, AKT1 has been also reported in high-affinity K^+ uptake in plants (Gierth and Mäser 2007). HAK operates at very low extracellular salt concentration, and its (K^+/H^+ symporters) activity is suppressed by combinatorial effects of higher membrane depolarization and reduced activity of plasma membrane H^+ ATPases (Bacha et al. 2015). Consequently, K^+ uptake via HAK could be improved by regulating membrane depolarization (by limiting Na^+ entry/promoting Na^+ efflux) and amplifying H^+-ATPase activity under salt stress. Interestingly, single amino acid substitutions in AtHAK5 and HvHAK enhance K^+ assimilation conferring salt tolerance in *Arabidopsis* and barley respectively (Alemán et al. 2014). However, alternative mechanism for augmenting K^+ uptake relies on plasma membrane-localized Na^+/H^+ antiporter SOS1, a component of SOS pathway (Qi and Spalding 2004). SOS1 facilitates Na^+ efflux and plays a critical role in salt detoxification in plants. Evidently, high cytoplasmic Na^+ concentration in *sos1* mutants of *Arabidopsis* hinders the K^+ uptake by the AKT1 channel (Qi and Spalding 2004). Salt stress induces K^+ deficiency and further stimulates Ca^{2+} signaling in plant roots, which positively modulates cytoplasmic K^+ flux, thus alleviating salt toxicity (Li et al. 2014). However, Ca^{2+} induction is not a thumb rule for other nutrients, for example, nitrate (NO_3^-) assimilation, which is a major source for N available in the soil (Riveras et al. 2015). Ca^{2+}-dependent phosphorylation negatively regulates NO_3^- uptake activity as it affects the functioning of NRT1.1/NPF6.3, an important transporter that facilitates NO_3^- uptake in plants (Bouguyon et al. 2016). Additionally, salt toxicity mediated by Cl^- also obstructs nitrate absorption in plants. Cl^- toxicity hampers NO_3^- absorption from the soil and strengthens NO_3^- translocation from shoots to roots, thereby generating NO_3^- deficiency in aerial parts of the plants such as leaves (Chen et al. 2012; Gallegos-Cedillo et al. 2016). Root growth is drastically impacted under salt stress in plants. Similarly, among all essential nutrients required for proper functioning of plant cellular machinery, P stress severely affects root development (Amtmann and

Shahzad 2017). Inorganic phosphate (Pi) forms insoluble precipitates in saline soil, which limits P assimilation by plant roots and thus retards plant growth (Naidu and Rengasamy 1993). Salinity decreases Pi uptake in plant species such as cotton (*Gossypium hirsutum*), melon (*Cucumis melo*), and lupin (*Lupinus albus*), whereas contrasting results were witnessed in case of maize (*Zea mays*) (Navarro et al. 2001). In addition to N, P, and K; S associated metabolite, glutathione (GSH) is involved in mitigating salinity stress in plants (Nazar et al. 2011). However, significant information in the context of S absorption in saline soil is still limited, which necessitates the understanding of S uptake regulation in plants under salinity.

5.3 Effects of Salt Stress on Nutrient Cycling in Crop Plants

Nutrients are crucial component for biological processes, which regulate cellular homeostasis in plants. Salinity being one of the major abiotic stress, limits yield and productivity thereby heavily influencing plant nutrient cycling. Increased Na^+ toxicity under salt stress restricts K^+ mobility from xylem to phloem, eventually reducing K^+ flux in above-ground tissues. Xylem to phloem mobility of Na^+ is restricted by a popular class of Na^+ and K^+ symporters, HKT1-type (high-affinity potassium transporter1), which prevents Na^+ toxicity in shoots by salvaging Na^+ from xylem. Previous reports suggest that HKT1 knockout confers salt sensitivity in *Arabidopsis thaliana* (Ali et al. 2019). Moreover, overexpression of AtHKT1 optimizes K^+/Na^+ ratio and leads to better plant biomass under salt stress in tobacco (Wang et al. 2018). Interestingly, QTL analysis to identify genetic component regulating salt tolerance in plant species such as rice and wheat emphasized significance of HKT class 1 genes in monitoring Na^+ buildup in leaves under salinity stress (Kronzucker and Britto 2011). In addition to stimulating K^+ deficit, enhanced Na^+ content under salt stress also leads to Cl^- accumulation in plants (Wu and Li 2019). Bulk of Cl^- in roots competes with NO_3^- channels to access the plant machinery, which inhibits NO_3^- uptake and transport in salt-stressed plants (Li et al. 2017). Evidently, overexpression of *SLAH1* (homolog of the slow-type anion channel SLAC1) known to be involved in Cl^- uptake reduces shoot NO_3^-/Cl^- ratio and also decreases shoot biomass in *Arabidopsis* (Cubero-Font et al. 2016; Qiu et al. 2016). Similarly, higher Cl^- levels in *Brassica rapa* were identified to reduce NO_3^- content in shoots (Reich et al. 2017). Thus, plant requires homeostasis in Na^+/K^+ and NO_3^-/Cl^- ratio to stabilize plant growth in response to salt stress. Apart from Cl^-, a crucial secondary messenger, Ca^{2+}, is induced in response to salinity stress and plays an important role in regulating nitrogen assimilation in plants. Ca^{2+} influx in response to high-salt concentration is sensed by calcium sensor, calcineurin B-like proteins (CBLs) in cytosol, which further activates CBL-interacting protein kinase (CIPK) in plants (Almeida et al. 2017). This Ca^{2+} sensor and kinase complex not only regulates Na^+ level in cell cytosol (mediated by SOS pathway) but also controls NO_3^- uptake in plants. This complex mediates switching of the nitrate transporter NRT1.1/NPF6.3/CHL1 from low affinity to high affinity via phosphorylation, thus assisting N assimilation even at lower concentration in soil (Manishankar et al. 2018). Therefore, a deep understanding of components involved in salt tolerance and nutrient homeostasis is thus necessitated to enlighten knowledge regarding the nutrient cycling under salinity stress in plants.

5.4 Salt Ion and Nutrient Interactions in Crop Plants

Activity of several cytosolic enzymes, membrane potential, and overall cellular osmosis are dependent on the intracellular K^+ and Na^+ homeostasis. Plant cell requires high cytosolic K^+/Na^+ ratio to maintain cellular metabolism and hence plant growth. Na^+ is nonessential for plants, whereas K^+ serves a crucial role in enzyme activation, osmoregulation, protein synthesis, and carbohydrate metabolism. Under salt stress, Na^+ competes with K^+ for uptake via roots and restricts the activity of many enzymes, which require K^+ to perform their functions (Duggleby and Dennis 1973). Effluxing Na^+ out of root cells hinders root to shoot transportation followed by vacuolar compartmentalization that contributes to better growth and higher yield of plant under salinity (Ismail and Horie 2017). Inhibition of Na^+ flux in shoots under salt stress has been correlated with salt stress tolerance in rice and maize (Tester and Davenport 2003). Salt stress due to Na^+ accumulation enhances cytosolic Ca^{2+} flux, which participates in Na^+ detoxification using SOS pathway via SOS1 (Na+/H+ antiporter) involving SOS3/CBL4 (act as a Ca^{2+} sensor) and SOS2/CIPK24. SOS3 uses the kinase CIPK24 (SOS2) to plasma membrane where CIPK24 stimulates SOS1 via phosphorylation for cellular effluxing of excess Na^+ (Liu et al. 2015). Captivatingly, previous studies suggest the role of this complex (CBL-CIPK23) in enhancing K^+ uptake in plants by regulating activity of K^+ transporters. Additionally, Ca^{2+} inhibits Na^+ uptake in plant by checking its entry points for example, NSCCs, which are reported to be involved in Na^+ uptake and K^+ efflux in plants. Ca^{2+} supplementation (5 mM) under K^+ starvation has been studied to reduce Na^+-induced plasma membrane depolarization by obstructing NSCCs in tomato (Bacha et al. 2015). Similar results have been observed in *A. thaliana* suggesting that Ca^{2+} plays an important role in maintaining Na^+/K^+ homeostasis in plants (Shabala et al. 2006). Additionally, Ca^{2+} has been also known to alleviate sulfate (SO_4^-) salinity in *Brassica* (Reich et al. 2017). However, S has been known to reduce ROS production in response to salinity stress, thereby providing salt tolerance in plants (Khan et al. 2013).

In addition to Na^+, Cl^- is also known to impart salt stress in plant species including soybean, grapevine, citrus, and lotus (Wu and Li 2019). However, it is known to be a vital micronutrient for sustaining plant health as it stabilizes the membrane potential and regulates pH and turgor pressure in plant cell (Wu and Li 2019). Although Cl^- transport and detoxification under salinity stress are not so well underpinned as that of Na^+, Cl^- elimination from shoots correlates with the salt tolerance of crop and woody plants (Teakle and Tyerman 2010). Interestingly when talking about the salt ion and nutrient interaction, N is also one among the major players known to alleviate salinity in plants (Mansour 2000). N application increases N-containing compounds, mainly amino acids, upon salt exposure, which are considered as osmoprotectants and are known to reduce oxidative stress by scavenging ROS (Mansour 2000). Presence of augmented NO_3^- in poplar trees and herbaceous species, such as *Pisum sativum*, *Helianthus annuus*, and *Glycine max*, helps in acclimatizing saline conditions (Ashraf and Sultana 2000; Bourgeais-Chaillou et al. 1992; Dluzniewska et al. 2007; Ehlting et al. 2007; Frechilla et al. 2001). Cl^- interacts with NO_3^- through N transporters, which leads to modulation in NO_3^-/Cl^- ratio in plants. Previous studies implicate that increased NO_3^-/Cl^- in shoots governs Cl^-/salt tolerance (Qiu et al. 2016). Hence, improving NO_3^- supply for crop plants can considerably reduce Cl^- accumulation and raises $NO3^-$ level in leaves, thereby enhancing plant salt tolerance (Abdolzadeh et al. 2008).

As mentioned previously, salinity reduces the solubility of macronutrient such as P significantly in the Na^+ rich soil (Martinez et al. 1996). Contrasting result has been observed in different crops for P concentration under salt stress for example, studies on lettuce and spinach showed that soil salinity reduced P concentrations in all plant tissues, whereas in barley shoot and root P concentrations were increased under salt stress. Moreover, increased phosphate content was found to negatively influence salt tolerance in soyabean (Phang et al. 2009). Interestingly, earlier researches suggests availability of P in plants affects Na^+ transportation and hence regulates salinity tolerance in crop plants (Abbas et al. 2018).

5.5 Effect of Salt Stress on Nutrient Transporters

Salt-mediated stress activates cellular signaling modulating expression of various nutrient transporters in plants. Na^+ in the hydrated form possesses similar ionic radii as that of K^+ and therefore, is assimilated via roots through common transporters (Keisham et al. 2018). As discussed previously, salt stress regulates the assimilation of macronutrients including K, N, P, and S. Thus, this section will discuss the effect of salt stress on the K, N, P, and S transporters.

5.5.1 K Transporters

Maintaining cytosolic K^+ flux is of utmost importance since plants exposed to salinity possess a lower K^+/Na^+ ratio, which hampers plant growth and development. This is in turn due to membrane depolarization below resting potential; activates K^+ outward rectifier channels, such as guard cell outward rectifying K^+ (GORK) channel, which eventually mediates K^+ efflux from cell. Therefore, preventing membrane depolarization could be a better strategy to regulate and maintain intracellular negative potential, thereby obstructing activity of the GORK channels and balancing K^+ flux (Assaha et al. 2017; Falhof et al. 2016). K^+ uptake is mediated by the K^+ transporter/high-affinity K^+ transporter/K^+ uptake protein (KT/HAK/KUP), which is one of the chief K^+ assimilation systems in plants. *Arabidopsis* genome encodes 13 genes, whereas rice possesses 27 members belonging to this family (Mäser et al. 2002). Transcriptional activity of the *Mesembryanthemum crystallinum* transporters such as McHAK1, McHAK2, and McHAK3 is significantly upregulated under salt stress (Su et al. 2002). Similarly, enhanced expression of AtKUP6 and AtKUP11 alongside a decrease in transcripts of AtKUP2 under salt stress has been observed in *A. thaliana* (Maathuis 2006). However, AtHAK5 transcriptional activity is reduced under salt stress, but then AtHAK5 also accounts important role in taking up K^+ on encountering salinity stress (Nieves-Cordones et al. 2010). Evidently, OsHAK5 expression in tobacco BY2 cells has been reported to elevate K^+ accumulation, thereby providing tolerance against salinity to cells (Horie et al. 2011). In addition to their prominent role in salinity stress, molecular and physiological aspects of KT/HAK/KUP transporter activity are still debated. Another class of K^+ transporters are the high-affinity K^+ transporters (HKTs), divided into two classes in which class I mediates Na^+ transport, whereas Na^+ and K^+ both are assimilated via Class II HKT transporters (Garciadeblás et al. 2003). Nucleotide substitution of

(G to S) GGGG for class II members (monocots) and SGGG for class I members (both monocots and dicots), in the pore-forming motif of the polypeptide, is the basis of differentiation of two classes (Deinlein et al. 2014). Reportedly, substitution causes steric hindrance in the first pore domain of class I HKTs, thereby restricting K^+ permeability (Cotsaftis et al. 2012). *Arabidopsis* encodes single high-affinity K^+ transporter, AtHKT1;1, which imparts salt tolerance either by transporting Na^+ from root xylem to xylem parenchyma cells or by loading Na^+ from the shoot/leaf xylem to xylem parenchyma cells (Tada 2019). Exceptionally, rice OsHKT2;1, a Class II HKT transporter shows Na^+-selectivity, having SGGG in place of GGGG motif. Hence, overexpression of OsHKT2;1 confers salt sensitivity in rice, whereas its downregulation provided tolerance against salinity stress (Horie et al. 2007). Captivatingly, HKTs of the monocot plants such as wheat (*Triticum aestivum*, TaHKT2;1), barley (*Hordeum vulgare*, HvHKT2;1), and reed (*Phragmites australis*, PhaHKT2;1) are K^+ pervious and thus play an important role in providing salt tolerance by improving cytosolic K^+ concentration (Assaha et al. 2017).

5.5.2 N Transporters

As discussed above, N functions in NO_3^-, K^+ or Cl^- shuttling, thus it owes an essential role in ameliorating salinity tolerance in plants. NO_3^- is the most prevalent form of N in soil, which is taken up and assimilated by various transporters including nitrate transporter 1 (NRT1)/peptide transporter (PTR) family (NPF), NRT2, chloride channel (CLC) family, and slowly activating anion channel and their homologs (SLAHs) (Wang et al. 2018). These transporters are differentially regulated under salinity stress in plants. Previous gene expression analysis postulated salt-mediated stimulation of NO_3^- transporters including *McNRT1* in *M. crystallinum* (Popova et al. 2003). Contrastingly, expressions of N transporters including *NRT1.1*, *NRT2.1*, and *NRT1.5* in 24 hours treatments in *Brassica juncea* and *A. thaliana* were shown to be reduced under salt and osmotic stresses (Taochy et al. 2015). Interestingly, NPF2.3, a NO_3^- efflux transporter, encourages NO_3^- translocation from roots to shoots in plants exposed to salinity stress (Taochy et al. 2015). In addition to NPF2.3, *Arabidopsis* AtNPF2.4 functions in loading Cl^- into the root xylem during salinity stress. *Atnpf2.4* knockdown mutants accumulate lesser Cl^- in shoot, whereas overexpression of AtNPF2.4 led to increased shoot Cl^- accumulation in plants (Zhang et al. 2018). Cl^- efflux from the root is also mediated by AtNPF2.5, the closest homolog to AtNPF2.4, which helps in exclusion of Cl^- from shoot under salt stress, eventually reducing Cl^- toxicity/salinity in *Arabidopsis* (Li et al. 2017). Formerly, CLC has been identified to transport both NO_3^- and Cl^-, hence regulating NO_3^-/Cl^- ratio in cells (O'Brien et al. 2016). The CLC families show a substrate shift between Cl^- and NO_3^- as a selectivity motif that has been identified in protein sequence of various CLC a/e members in plants (Zifarelli and Pusch 2010). Recently two root anion channels in SLAH1 and SLAH3 have been also implicated to regulate shoot NO_3^-/Cl^- ratio in *A. thaliana* (Cubero-Font et al. 2016). However, SLAH3 alone was unable to transport Cl^- but was able to transport NO_3^- with the help of anion-channel-activating protein kinase CIPK21. Interestingly, co-expression of SLAH1 and SLAH3 in *Xenopus* oocytes facilitated both NO_3^- and Cl^- transport (Cubero-Font et al. 2016). Hence, a deeper understanding of these transporters is required to elucidate their role in imparting salt tolerance in plants.

5.5.3 P Transporters

Salinity stress leads to nutrient imbalance in plants including P, which is a major plant macronutrient (Młodzińska and Zboińska 2016). It serves as a structural component for nucleic acids and phospholipids and participates in energy transfer, signal transduction, and enzyme activation reaction (Młodzińska and Zboińska 2016). P is assimilated from soil in the form of Pi, which is acquired and transported in the plant via five families of phosphate transporters including PHT1, PHT2, PHT3, PHT4, and PHT5 (Poirier and Jung 2018). Previously, *Arabidopsis* PHT3 transporters have been reported to enhance salt sensitivity in plants. Analysis implicated overexpression of *AtPHT3;1*, *AtPHT3;2*, and *AtPHT3;3* led to changes in energy status along with gibberellin synthesis, which enhances plant susceptibility toward salt stress (Zhu et al. 2012). Captivatingly, a member of PHT4 family, PHT4;6, has been also known to interact with salinity stress. An insertional mutation was reported to increase the sensitivity of the A*t*PHT4;6 mutant exposed to salt stress that causes decrease in primary root growth, enriched lateral branching, and root tip swelling (Cubero et al. 2009). Further study of PHT4;6, Golgi membrane-localized P transporter revealed its role in protein N-glycosylation and cell wall biosynthesis, which are crucial for salinity tolerance. Certainly, additional exploration of the Pi transporters could help in empathizing the role of unknown Pi transporters in salinity stress in plants.

5.5.4 S Transporters

Sulfur is an important macronutrient as it is involved in several metabolic processes crucial for plant growth and metabolism. It is a constituent of amino acid cysteine and methionine, which are part of coenzymes and prosthetic groups such as iron–sulfur centers, lipoic acid, thiamine, Coenzyme A, etc. thereby regulating structural and biological activity of proteins (Kopriva et al. 2015). S compounds including GSH, phytochelatins (PC), and glucosinolates (GSL) play an essential role in abiotic and biotic stress tolerance respectively (Hasanuzzaman et al. 2018). The major form of S present in soil is sulfate, which is taken up and reduced in plant cells through four groups of sulfate transporters (SULTR) including SULTR1, SULTR2, SULTR3, and SULTR4 (Hawkesford 2017). Earlier studies illustrated the role of SULTRs in regulating drought and salinity stress in plant species such as *A. thaliana* and *Medicago truncatula* (Gallardo et al. 2014). SULTR3 and SULTR4 family of sulfate transporters were identified to be differentially modulated in response to abiotic stress, eventually opening doors for further investigation of this family toward other stresses in plants. Abscisic acid (ABA) biosynthesis is a common component in both drought and salt stress. Interestingly, ABA synthesis is dependent on cysteine (Cys) availability since Abscisic Aldehyde Oxidase 3 (AAO3), a key enzyme in ABA biosynthesis, depends on Cys as the S donor for its molybdenum cofactor sulfuration (Chen et al. 2019). As a consequence, *sultr3;1* possesses reduced AAO3 activity, which is restored by exogenous application of Cys (Cao et al. 2014). Moreover, ABA levels were reported to be lesser in *SULTR3* single mutants under normal and salinity stress and therefore showed hypersensitivity toward exogenous ABA and salt at the germination stage (Cao et al. 2014).

5.6 Role of Nutrient Interactions: Prospective Signaling

Salt stress induces ionic and osmotic imbalance in the cell, which eventually reduces plant growth and yield. It leads to increase in Na^+/K^+ ratio in plant cytosol, stimulating ROS production, which results in PCD in case of severity (Almeida et al. 2017; Gupta and Huang 2014). Maintaining sufficient K^+ is necessary for the proper cellular functioning as it serves as an essential macronutrient in enzyme activation, osmotic regulation, generating turgor pressure, expansion of cells, maintaining membrane electric potential, and pH inside plant cell (Hawkesford 2017; Ragel et al. 2015). Hence, salinity involves interplay between the two monovalent ions in addition to Ca^{2+}, which is induced as a secondary messenger in this signaling. K^+ and Ca^{2+} positively regulate salinity stress, which was evident from earlier study in salt-tolerant alfalfa cultivars that were reported to be capable of absorbing greater K^+ and Ca^{2+} ions and reduce Na^+ absorption under salt stress (Khorshidi et al. 2009). Moreover, Na^+ itself does not govern salinity stress in plants as Cl^- too imposes salt stress in few plant species (Wu and Li 2019). As mentioned previously, Cl^- toxicity disturbs NO_3^- assimilation in plants, thereby reducing N flux required for sustaining plant growth (Shelke et al. 2019). Earlier studies highlighted substantial role of N in imparting salinity tolerance in plants (Ahanger et al. 2019). Thus, plants need to maintain lower Na^+/K^+ and higher NO_3^-/Cl^- inside the cell in order to achieve tolerance against salinity stress. P application has also been identified to modulate plant growth under salt stress in previous studies (Tang et al. 2019). Similarly, S is also an important macronutrient, which helps in coping against salinity by synthesizing reduced S compounds such as GSH. GSH prevents ROS buildup in the cell eventually protecting plant from osmotic stress (Nazar et al. 2011). Presence of S has been also known to control N assimilation in plants, hence understanding the modulation in combined nutritional deficiencies and salinity stress is necessitated.

5.7 Future Prospective and Conclusion

Salinity is one of the major abiotic stresses, which drastically affects agricultural productivity. Numerous genes have been identified in the past, which works to improve plant growth against salt stress. Plant nutrients bear a necessary function in controlling accumulation of toxic Na^+ and Cl^- ions in cell. Usage of conventional and modern molecular techniques revealed several essential components linked with cross talk between nutrient assimilation and salt stress. Although some of nutrients including K, Ca^{2+}, and N are well studied in response to salinity, detailed information related to S and P is still lacking in this respect. Therefore, more focus should be driven toward understanding salinity under different concentrations of S and P also, there should be studies associated with combined nutrient deficiency under salt stress, which could reveal unidentified molecular components providing tolerance against such abiotic stresses.

References

Abbas, G., Chen, Y., Khan, F.Y. et al. (2018). Salinity and low phosphorus differentially affect shoot and root traits in two wheat cultivars with contrasting tolerance to salt. *Agronomy* 8 (8): 155.

Abdolzadeh, A., Shima, K., Lambers, H., and Chiba, K. (2008). Change in uptake, transport and accumulation of ions in Nerium oleander (Rosebay) as affected by different nitrogen sources and salinity. *Ann. Bot.* 102 (5): 735–746.

Ahanger, M.A., Qin, C., Maodong, Q. et al. (2019). Spermine application alleviates salinity induced growth and photosynthetic inhibition in Solanum lycopersicum by modulating osmolyte and secondary metabolite accumulation and differentially regulating antioxidant metabolism. *Plant Physiol. Biochem.* 144: 1–13.

Alemán, F., Caballero, F., Ródenas, R. et al. (2014). The F130S point mutation in the Arabidopsis high-affinity K^+ transporter AtHAK5 increases K^+ over Na^+ and Cs^+ selectivity and confers Na^+ and Cs^+ tolerance to yeast under heterologous expression. *Front. Plant Sci.* 5: 430.

Ali, A., Maggio, A., Bressan, R.A., and Yun, D.-J. (2019). Role and functional differences of HKT1-type transporters in plants under salt stress. *Int. J. Mol. Sci.* 20 (5): 1059.

Almeida, D.M., Oliveira, M.M., and Saibo, N.J. (2017). Regulation of Na^+ and K^+ homeostasis in plants: towards improved salt stress tolerance in crop plants. *Genet. Mol. Biol.* 40 (1): 326–345.

Amtmann, A. and Shahzad, Z. (2017). To respond or not to respond? Natural variation of root architectural responses to nutrient signals. *J. Exp. Bot.* 68 (11): 2636–2640.

Ashraf, M. and Sultana, R. (2000). Combination effect of NaCl salinity and nitrogen form on mineral composition of sunflower plants. *Biol. Plant.* 43 (4): 615–619.

Assaha, D.V., Ueda, A., Saneoka, H. et al. (2017). The role of Na^+ and K^+ transporters in salt stress adaptation in glycophytes. *Front. Physiol.* 8: 509.

Bacha, H., Ródenas, R., López-Gómez, E. et al. (2015). High Ca^{2+} reverts the repression of high-affinity K^+ uptake produced by Na^+ in *Solanum lycopersicum* L.(var. microtom) plants. *J. Plant Physiol.* 180: 72–79.

Bouguyon, E., Perrine-Walker, F., Pervent, M. et al. (2016). Nitrate controls root development through posttranscriptional regulation of the NRT1. 1/NPF6. 3 transporter/sensor. *Plant Physiol.* 172 (2): 1237–1248.

Bourgeais-Chaillou, P., Perez-Alfocea, F., and Guerrier, G. (1992). Comparative effects of N–sources on growth and physiological responses of soyabean exposed to NaCI–stress. *J. Exp. Bot.* 43 (9): 1225–1233.

Cao, M.J., Wang, Z., Zhao, Q. et al. (2014). Sulfate availability affects ABA levels and germination response to ABA and salt stress in A rabidopsis thaliana. *Plant J.* 77 (4): 604–615.

Chen, L., Chen, Y., Jiang, J. et al. (2012). The constitutive expression of Chrysanthemum dichrum ICE1 in Chrysanthemum grandiflorum improves the level of low temperature, salinity and drought tolerance. *Plant Cell Rep.* 31 (9): 1747–1758.

Chen, Z., Zhao, P.X., Miao, Z.Q. et al. (2019). SULTR3s function in chloroplast sulfate uptake and affect ABA biosynthesis and the stress response. *Plant Physiol.* 180 (1): 593–604.

Cotsaftis, O., Plett, D., Shirley, N. et al. (2012). A two-staged model of Na^+ exclusion in rice explained by 3D modeling of HKT transporters and alternative splicing. *PLoS One* 7 (7): e39865.

Cubero, B., Nakagawa, Y., Jiang, X.-Y. et al. (2009). The phosphate transporter PHT4; 6 is a determinant of salt tolerance that is localized to the Golgi apparatus of Arabidopsis. *Mol. Plant* 2 (3): 535–552.

Cubero-Font, P., Maierhofer, T., Jaslan, J. et al. (2016). Silent S-type anion channel subunit SLAH1 gates SLAH3 open for chloride root-to-shoot translocation. *Curr. Biol.* 26 (16): 2213–2220.

Deinlein, U., Stephan, A.B., Horie, T. et al. (2014). Plant salt-tolerance mechanisms. *Trends Plant Sci.* 19 (6): 371–379.

Dluzniewska, P., Gessler, A., Dietrich, H. et al. (2007). Nitrogen uptake and metabolism in Populus× canescens as affected by salinity. *New Phytol.* 173 (2): 279–293.

Duggleby, R.G. and Dennis, D.T. (1973). Pyruvate kinase, a possible regulatory enzyme in higher plants. *Plant Physiol.* 52 (4): 312–317.

Ehlting, B., Dluzniewska, P., Dietrich, H. et al. (2007). Interaction of nitrogen nutrition and salinity in grey poplar (Populus tremula× alba). *Plant Cell Environ.* 30 (7): 796–811.

Falhof, J., Pedersen, J.T., Fuglsang, A.T., and Palmgren, M. (2016). Plasma membrane H^+-ATPase regulation in the center of plant physiology. *Mol. Plant* 9 (3): 323–337.

Frechilla, S., Lasa, B., Ibarretxe, L. et al. (2001). Pea responses to saline stress is affected by the source of nitrogen nutrition (ammonium or nitrate). *Plant Growth Regul.* 35 (2): 171–179.

Gallardo, K., Courty, P.-E., Le Signor, C. et al. (2014). Sulfate transporters in the plant's response to drought and salinity: regulation and possible functions. *Front. Plant Sci.* 5: 580.

Gallegos-Cedillo, V.M., Urrestarazu, M., and Álvaro, J.E. (2016). Influence of salinity on transport of nitrates and potassium by means of the xylem sap content between roots and shoots in young tomato plants. *J. Soil Sci. Plant Nutr.* 16 (4): 991–998.

Garciadeblás, B., Senn, M.E., Bañuelos, M.A., and Rodríguez, N.A. (2003). Sodium transport and HKT transporters: the rice model. *Plant J.* 34 (6): 788–801.

Gierth, M. and Mäser, P. (2007). Potassium transporters in plants–involvement in K^+ acquisition, redistribution and homeostasis. *FEBS Lett.* 581 (12): 2348–2356.

Guo, J., Islam, M.A., Lin, H. et al. (2018). Genome-wide identification of cyclic nucleotide-gated ion channel gene family in wheat and functional analyses of TaCNGC14 and TaCNGC16. *Front. Plant Sci.* 9: 18.

Gupta, B. and Huang, B. (2014). Mechanism of salinity tolerance in plants: physiological, biochemical, and molecular characterization. *Int. J. Genom.* 2014: 1, 701596–18. https://doi.org/10.1155/2014/701596. Academic Editor: Lugi Catuvelli.

Hasanuzzaman, M., Bhuyan, M., Nahar, K. et al. (2018). Potassium: a vital regulator of plant responses and tolerance to abiotic stresses. *Agronomy* 8 (3): 31.

Hawkesford, M.J. (2017). Sulfate transport in plants: a personal perspective. In: *Sulfur Metabolism in Higher Plants-Fundamental, Environmental and Agricultural Aspects* (eds. L.J. De Kok, M.J. Hawkesford, S.H. Haneklaus and E. Schnug), 3–12. Springer.

Horie, T., Costa, A., Kim, T.H. et al. (2007). Rice OsHKT2; 1 transporter mediates large Na^+ influx component into K^+ starved roots for growth. *EMBO J.* 26 (12): 3003–3014.

Horie, T., Brodsky, D.E., Costa, A. et al. (2011). K^+ transport by the OsHKT2; 4 transporter from rice with atypical Na^+ transport properties and competition in permeation of K^+ over Mg^{2+} and Ca^{2+} ions. *Plant Physiol.* 156 (3): 1493–1507.

Isayenkov, S.V. and Maathuis, F.J. (2019). Plant salinity stress: many unanswered questions remain. *Front. Plant Sci.* 10 (80) https://doi.org/10.3389/fpls.2019.00080.

Ismail, A.M. and Horie, T. (2017). Genomics, physiology, and molecular breeding approaches for improving salt tolerance. *Annu. Rev. Plant Biol.* 68: 405–434.

Keisham, M., Mukherjee, S., and Bhatla, S.C. (2018). Mechanisms of sodium transport in plants—progresses and challenges. *Int. J. Mol. Sci.* 19 (3): 647.

Ketehouli, T., Carther, I., Foka, K. et al. (2019). Adaptation of plants to salt stress: characterization of Na^+ and K^+ transporters and role of CBL gene family in regulating salt stress response. *Agronomy* 9 (11): 687.

Khan, M.I.R., Asgher, M., Iqbal, N. et al. (2013). Potentiality of sulphur-containing compounds in salt stress tolerance. In: *Ecophysiology and Responses of Plants Under Salt Stress* (eds. P. Ahmad, M. Azooz and M. Prasad), 443–472. New York, NY: Springer https://doi.org/10.1007/978-1-4614-4747-4_17.

Khorshidi, M., Yarnia, M., and Hassanpanah, D. (2009). Salinity effect on nutrients accumulation in alfalfa shoots in hydroponic condition. *J. Food Agric. Environ.* 7: 787–790.

Kopriva, S., Calderwood, A., Weckopp, S.C., and Koprivova, A. (2015). Plant sulfur and big data. *Plant Sci.* 241: 1–10.

Kronzucker, H.J. and Britto, D.T. (2011). Sodium transport in plants: a critical review. *New Phytol.* 189 (1): 54–81.

Li, J., Jia, H., and Wang, J. (2014). cGMP and ethylene are involved in maintaining ion homeostasis under salt stress in Arabidopsis roots. *Plant Cell Rep.* 33 (3): 447–459.

Li, B., Tester, M., and Gilliham, M. (2017). Chloride on the move. *Trends Plant Sci.* 22 (3): 236–248.

Liu, W., Li, R.-J., Han, T.-T. et al. (2015). Salt stress reduces root meristem size by nitric oxide-mediated modulation of auxin accumulation and signaling in Arabidopsis. *Plant Physiol.* 168 (1): 343–356.

Maathuis, F.J. (2006). The role of monovalent cation transporters in plant responses to salinity. *J. Exp. Bot.* 57 (5): 1137–1147.

Manishankar, P., Wang, N., Köster, P. et al. (2018). Calcium signaling during salt stress and in the regulation of ion homeostasis. *J. Exp. Bot.* 69 (17): 4215–4226.

Mansour, M. (2000). Nitrogen containing compounds and adaptation of plants to salinity stress. *Biol. Plant.* 43 (4): 491–500.

Martinez, C.A., Maestri, M., and Lani, E.G. (1996). in vitro salt tolerance and proline accumulation in Andean potato (Solanum spp.) differing in frost resistance. *Plant Sci.* 116 (2): 177–184.

Mäser, P., Gierth, M., and Schroeder, J.I. (2002). Molecular mechanisms of potassium and sodium uptake in plants. In: *Progress in Plant Nutrition: Plenary Lectures of the XIV International Plant Nutrition Colloquium* (eds. W.J. Horst et al.), 43–54. Dordrecht: Springer https://doi.org/10.1007/978-94-017-2789-1_3.

Młodzińska, E. and Zboińska, M. (2016). Phosphate uptake and allocation–a closer look at Arabidopsis thaliana L. and Oryza sativa L. *Front. Plant Sci.* 7: 1198.

Naidu, R. and Rengasamy, P. (1993). Ion interactions and constraints to plant nutrition in Australian sodic soils. *Soil Res.* 31 (6): 801–819.

Navarro, J.M., Botella, M.A., Cerdá, A., and Martinez, V. (2001). Phosphorus uptake and translocation in salt-stressed melon plants. *J. Plant Physiol.* 158 (3): 375–381.

Nazar, R., Iqbal, N., Masood, A. et al. (2011). Understanding the significance of sulfur in improving salinity tolerance in plants. *Environ. Exp. Bot.* 70 (2–3): 80–87.

Nieves-Cordones, M., Alemán, F., Martínez, V., and Rubio, F. (2010). The Arabidopsis thaliana HAK5 K^+ transporter is required for plant growth and K^+ acquisition from low K^+ solutions under saline conditions. *Mol. Plant* 3 (2): 326–333.

Nieves-Cordones, M., Martínez, V., Benito, B., and Rubio, F. (2016). Comparison between Arabidopsis and rice for main pathways of K^+ and Na^+ uptake by roots. *Front. Plant Sci.* 7: 992.

O'Brien, J.A., Vega, A., Bouguyon, E. et al. (2016). Nitrate transport, sensing, and responses in plants. *Mol. Plant* 9 (6): 837–856.

Phang, T.H., Shao, G., Liao, H. et al. (2009). High external phosphate (Pi) increases sodium ion uptake and reduces salt tolerance of 'Pi-tolerant'soybean. *Physiol. Plant.* 135 (4): 412–425.

Poirier, Y. and Jung, J.Y. (2018). Phosphate transporters. *Annu. Plant Rev.* online: 125–158, 71–83. https://doi.org/10.1023/A:1022376332180.

Popova, O.V., Dietz, K.-J., and Golldack, D. (2003). Salt-dependent expression of a nitrate transporter and two amino acid transporter genes in *Mesembryanthemum crystallinum. Plant Mol. Biol.* 52 (3): 569–578.

Qi, Z. and Spalding, E.P. (2004). Protection of plasma membrane K^+ transport by the salt overly sensitive1 Na^+-H^+ antiporter during salinity stress. *Plant Physiol.* 136 (1): 2548–2555.

Qiu, J., Henderson, S.W., Tester, M. et al. (2016). SLAH1, a homologue of the slow type anion channel SLAC1, modulates shoot Cl^- accumulation and salt tolerance in Arabidopsis thaliana. *J. Exp. Bot.* 67 (15): 4495–4505.

Ragel, P., Ródenas, R., García-Martín, E. et al. (2015). The CBL-interacting protein kinase CIPK23 regulates HAK5-mediated high-affinity K^+ uptake in Arabidopsis roots. *Plant Physiol.* 169 (4): 2863–2873.

Reich, M., Aghajanzadeh, T., Helm, J. et al. (2017). Chloride and sulfate salinity differently affect biomass, mineral nutrient composition and expression of sulfate transport and assimilation genes in Brassica rapa. *Plant Soil* 411 (1–2): 319–332.

Riveras, E., Alvarez, J.M., Vidal, E.A. et al. (2015). The calcium ion is a second messenger in the nitrate signaling pathway of Arabidopsis. *Plant Physiol.* 169 (2): 1397–1404.

Shabala, S., Demidchik, V., Shabala, L. et al. (2006). Extracellular Ca^{2+} ameliorates NaCl-induced K^+ loss from Arabidopsis root and leaf cells by controlling plasma membrane K^+-permeable channels. *Plant Physiol.* 141 (4): 1653–1665.

Shelke, D., Nikalje, G., Nikam, T. et al. (2019). Chloride (Cl^-) uptake, transport, and regulation in plant salt tolerance. Molecular plant abiotic stress. *Biol. Biotechnol.* 12: 241–268.

Su, H., Golldack, D., Zhao, C., and Bohnert, H.J. (2002). The expression of HAK-type K^+ transporters is regulated in response to salinity stress in common ice plant. *Plant Physiol.* 129 (4): 1482–1493.

Tada, Y. (2019). The HKT transporter gene from Arabidopsis, AtHKT1; 1, is dominantly expressed in shoot vascular tissue and root tips and is mild salt stress-responsive. *Plants* 8 (7): 204.

Tang, Y., Bao, X., Zhi, Y. et al. (2019). Overexpression of a MYB family gene, OsMYB6, increases drought and salinity stress tolerance in transgenic rice. *Front. Plant Sci.* 10: 168.

Taochy, C., Gaillard, I., Ipotesi, E. et al. (2015). The Arabidopsis root stele transporter NPF2. 3 contributes to nitrate translocation to shoots under salt stress. *Plant J.* 83 (3): 466–479.

Tavakkoli, E., Fatehi, F., Coventry, S. et al. (2011). Additive effects of Na^+ and Cl^- ions on barley growth under salinity stress. *J. Exp.Bot.* 62 (6): 2189–2203.

Teakle, N.L. and Tyerman, S.D. (2010). Mechanisms of Cl^- transport contributing to salt tolerance. *Plant Cell Environ.* 33 (4): 566–589.

Tester, M. and Davenport, R. (2003). Na^+ tolerance and Na^+ transport in higher plants. *Ann. Bot.* 91 (5): 503–527.

Wang, F., Deng, S., Ding, M. et al. (2013). Overexpression of a poplar two-pore K^+ channel enhances salinity tolerance in tobacco cells. *Plant Cell, Tissue Organ Cult. (PCTOC)* 112 (1): 19–31.

Wang, L., Liu, Y., Feng, S. et al. (2018). AtHKT1 gene regulating K^+ state in whole plant improves salt tolerance in transgenic tobacco plants. *Sci. Rep.* 8 (1): 1–12.

Wu, H. and Li, Z. (2019). The importance of Cl^- exclusion and vacuolar Cl^- sequestration: revisiting the role of Cl^- transport in plant salt tolerance. *Front. Plant Sci.* 10: 1418.

Zayed, B., Abdelaal, M., and Deweedar, G. (2017). Response of rice yield and soil to sulfur application under water and salinity stresses. *Egypt. J. Agron.* 39 (3): 239–249.

Zhang, G.-B., Meng, S., and Gong, J.-M. (2018). The expected and unexpected roles of nitrate transporters in plant abiotic stress resistance and their regulation. *Int. J. Mol. Sci.* 19 (11): 3535.

Zhu, W., Miao, Q., Sun, D. et al. (2012). The mitochondrial phosphate transporters modulate plant responses to salt stress via affecting ATP and gibberellin metabolism in *Arabidopsis thaliana*. *PLoS One* 7 (8): e43530.

Zifarelli, G. and Pusch, M. (2010). CLC transport proteins in plants. *FEBS Lett.* 584 (10): 2122–2127.

6

Effects of Salt Stress on the Morphology, Anatomy, and Gene Expression of Crop Plants

Pragati Kumari[1], Arvind Gupta[2], Harish Chandra[3], Pratibha Singh[4], and Saurabh Yadav[2]

[1] *Scientist Hostel-S-02, Chauras Campus, Srinagar Garhwal, Uttarakhand, India*
[2] *Department of Biotechnology, Hemvati Nandan Bahuguna Garhwal (Central) University, Srinagar, Uttarakhand, India*
[3] *Department of Botany and Microbiology, Gurukula Kangri Vishwavidyalaya, Haridwar, Uttarakhand, India*
[4] *Ranjan Plant Physiology and Biochemistry Laboratory, Department of Botany, University of Allahabad, Prayagraj, Uttar Pradesh, India*

6.1 Introduction

Our society and people living around the world need food to survive. Population is projected to increase around 10 billion in the next 50 years and to meet the food demand, around 50% increase in crop yield is needed (Godfray et al. 2010). On one side, population is increasing but the land needed for agriculture is decreasing about 1–2% per year in arid and semiarid zones around the globe due to salinity increase (Khan 2008). Salinity stress will increase globally due to climate change. Due to illogical crop irrigation practices, excessive fertilization, and excessive plowing, etc., salt stress increases every year (Lin et al. 2018). The sessile nature of the plants makes them very susceptible to get exposed and affected by different type of stresses. Tolerance is multidimensional and there is a complex signaling and concerted efforts to integrate all activities in plants (Cheeseman 2013). To feed the population by next century is a major challenge, so the mechanism underlying salt tolerance in halophytes is important since it withstands harsh salinity (Yang et al. 2017). The plants are classified into sensitive ones as "glycophytes" and tolerant ones as "halophytes" as per the salt stress response. Plants termed as glycophytes have retarded growth and even die under 100–200 mM concentration of salt (Flowers and Colmer 2008; Shabala 2013). Plants termed as halophytes can survive high concentrations of salts in their environments with ≥200 mM NaCl (Flowers 1975). They have important role in restoration of saline soils and ecological protection (Flowers and Colmer 2008).

Physiology of Salt stress in Plants: Perception, Signalling, Omics and Tolerance Mechanism, First Edition.
Edited by Pratibha Singh, Madhulika Singh, Rajiv Kumar Singh, and Sheo Mohan Prasad.

The German botanist Breckle (Breckle 1995) classified halophytes into three major categories on the physiological basis of salt tolerance and transport of ions:

A) *Recretohalophytes* that can secrete salts outside from plant body using salt glands or into salt bladders for temporary storage or scatter from salt bladders on encounter of an external stimuli. Salt thus secreted, reduce the salt concentration of the plant tissues and restore normal growth in salty soils (Kiani-Pouya et al. 2017).
B) *Euhalophytes* and stem succulent euhalophytes absorb salt in the vacuoles of leaves, green tissues, or stems (Chen et al. 2012; Song and Wang 2015). Euhalophytes can absorb salts from soils and compartmentalize excessive salt ions that enter the vacuoles. In another way, this reduces the damage to enzymes and biological macromolecules in the cytoplasm (Chen et al. 2012; Song and Wang 2015).
C) Salt-exclusion halophytes known as the salt excluders which accumulate excess salt in the vacuoles of roots than in the shoots (Matsushita and Matoh 1991).

This salt stress limits the development of salt-sensitive plants and also halophytes (Sobhanian et al. 2011). Salt stress imposes two types of stresses, one when the sodium and chloride ions cause ionic imbalance leading to ionic stress and other osmotic stress caused by reduced water availability due to enhanced osmotic pressure (Flowers and Colmer 2008). The major economical crops like rice, wheat, barley, and maize have suffered ~70% decrease in yield (Acquaah 2009).

The field crops are severely affected by the salt stress in multiple ways. Salt stress inhibits many physiological parameters involving seed germination, seed setting, etc., and also changes the physiological characteristics (Kamal et al. 2012a,b). The increased salt stress affects plants in a multitude of several factors like ion toxicity, metabolic disorders, oxidative stress, cellular toxicity, disorganization of cell membrane, cell division alterations, etc. The biological macromolecules like DNA, RNA, protein, etc., also undergo alteration in structure and function.

To overcome the salt stress, plants must adapt to the increasing salinity, and this adaptation is a long-term and changing process affected by the interplay of many genes which are involved in many processes of plants (Munns and Tester 2008). The model plant *Arabidopsis thaliana* (Ding et al. 2014), rice (Shankar et al. 2016), and soybean (Belamkar et al. 2014) have revealed many salt stress-responsive genes which has shown their effects in many pathways. There are many signaling pathways working under salt stress, like Ca^{2+}-dependent signaling, salt overly sensitive (SOS) pathway, and ABA signaling pathways (Ji et al. 2013). There are many efforts being employed in model plants using molecular and OMICS techniques to understand the salt tolerance mechanisms. The isolation and characterization of some important salt-tolerant genes which help in perception and signaling related to ROS removal, accumulation of osmolytes, and transportation of water across plasma membrane and tonoplast prove significant (Hasegawa et al. 2000; Zhu 2002). The transgenic plants having overexpression of important salt-induced genes show salt stress tolerance (Shi et al. 2003; Hu et al. 2006). Few genes like CDPK, CBL-interacting protein kinase (CIPK)/SOS, and MAPK signaling pathways have been shown to be salt-responsive (Zhu 2002).

The study of salt stress and its deleterious effects on plants lead to growing interest on studying its relation with plant salt metabolism. This will help to make several crops more

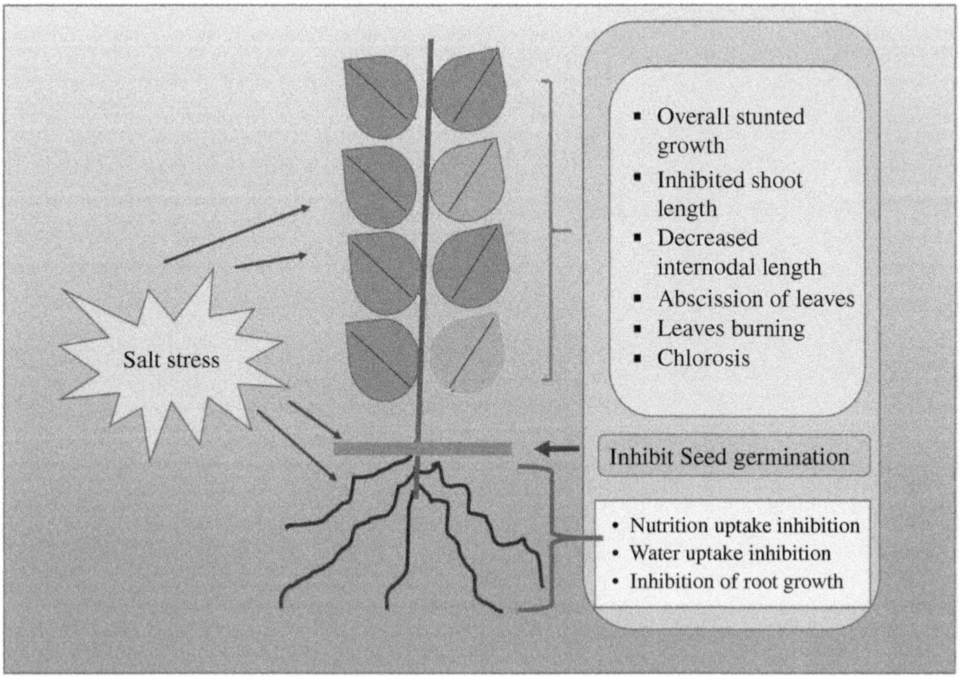

Figure 6.1 Negative effects of salt stress on plants (Above and under the soil).

salt-tolerant by crossing and breeding with wild species and by ectopic expression of genes. These genes may be from halophytes and also from other plants. Plant breeding and plant biotechnology approaches are challenging and some achievements have been made, although slow. In this chapter, we have discussed the correlation of salt stress and its effects on physiological, morphological, yield, yield-related traits, floral attributes, etc. The molecular mechanisms involved in providing stress tolerance exploiting overexpression transgenic studies and other OMICS approaches should surely contribute to alleviate salt stress (Figure 6.1).

6.2 Salt Stress and Effects on Morphology of Plants

The plant growth and development is reduced by salt stress, and plant may suffer from four types of stresses (Greenway and Munns 1980), viz. osmosis-induced water stress, toxicity due to ions, and ROS toxicity. The morphological, physiological, and biochemical responses are altered by salinity stress and impose negative effects on the plant (Figure 6.2). Deleterious effects are symptoms like leaf area reduction, increase in thickness and succulence of leaf, leaf abscission, root and shoot necrosis, and decrease of internodal lengths (Gucci and Tattini 1997; Parida and Das 2005). Seed germination is the first and early response which affects the initial growth stages of plants.

The shoot formation is decreased and thereby reduced biomass production and decline of crop yield and finally plants' death (Pardo 2010). Seedling growth if seriously affected

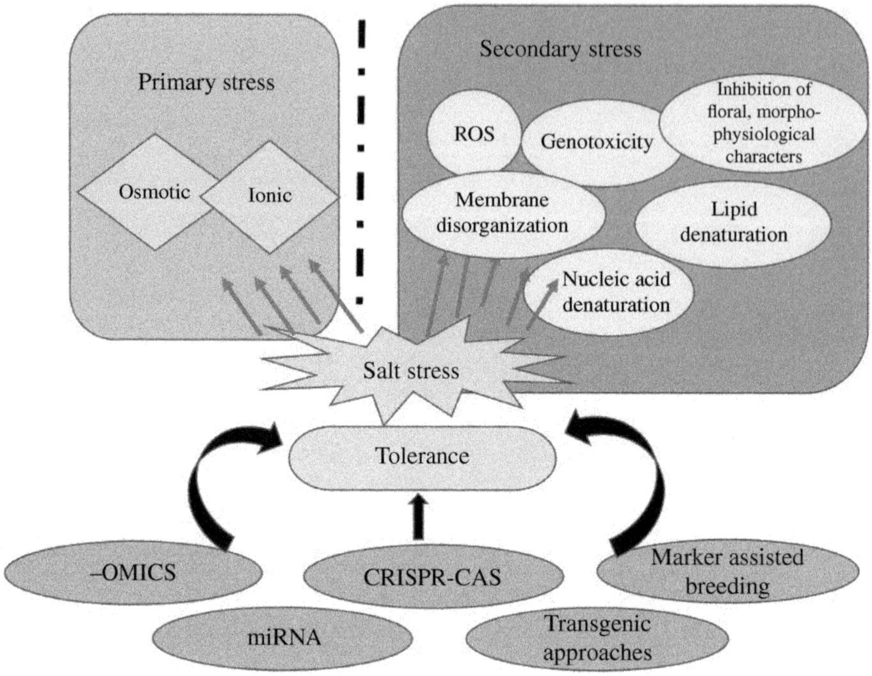

Figure 6.2 Primary and secondary stress under salinity and tolerance strategies.

during the early foundation under stress conditions affects the overall growth of plant, shoot and root dry weight, root and shoot length, and the number of leaves. High concentrations of sodium decreased plant growth in rice seedlings (Mishra et al. 1997), *Vigna unguiculata* L., (Dantas et al. 2005), and *Brassica campestris* L. (Memon et al. 2010). Andriolo et al. (2005) reported that the treatment with salt increased the fresh weight by about 28% in *Lactuca sativa* L. which was in agreement with the results of Dantas et al. (2005) on cowpea (*V. unguiculata* L.) where they reported that 10 mM of sodium chloride increased fresh and dry weights of the shoot of seedlings. Accumulation of toxic salts in the leaf apoplasm leads to turgor loss and dehydration which culminates in death (Marschner 1995). The toxic level of chloride causes leaf burn or drying of leaf tissues, where tips of older leaves are affected then it progresses back along the edges as salt severity increases and the leaves die (Yilmaz and Kina 2008). The leaves number also gets affected as per the study by Ha et al. (2008), etc. The roots are affected by salt as they are responsible to transport nutrients and salts throughout the plant, but the damage is less than shoots (Munns et al. 2002). Some woody plants have their roots capable to accumulate sodium levels as an adaptive mechanism to avoid its toxicity in the shoots (Gucci and Tattini 1997). Thus, this serves to be a tolerance mechanism for controlling root-to-shoot transport of salts. The mild salt stress leads to rapid growth inhibition of leaves and stems, but roots may grow further (Hsiao and Xu 2000). Fresh and dry weights of the shoots are positively or negatively affected by concentration of salts, type of salts, and plant type (Taffouo et al. 2009; Memon et al. 2010).

6.3 Photosynthetic Pigments and Osmolytes Accumulation

Photosynthetic machinery is highly dependent on chlorophyll and is directly related with the growth and development of plants. There is reduction in pigments like chlorophyll and carotenoids under stress and also when there is increase in salt concentrations. It is due to the membrane swelling in chloroplasts and excess sodium and chloride ions in the leaves (Murillo-Amador et al. 2007). We can say that salinity reduces the photosynthetic pigments content in the salt-treated plants.

Osmotic potential outside the cell helps in the accumulation of metabolites which act as compatible solutes without disturbing normal metabolic reactions. The tolerance is provided by the osmolytes which helps in osmotic adjustments by decreasing the internal osmotic potential (McCue and Hanson 1990). There is accumulation of compatible solutes like proline and sugars which help the plants to cope with salt stress (Pattanagul and Thitisaksakul 2008) as these solutes fine-tune and adjust osmotic balance. The sugar accumulation contributes to osmotic balance by providing tolerance to plants as the storage reserves are maintained and modulated (Smeekens 2000), and it has been reported that the total soluble sugar was increased with the increasing salt stress (Muscolo et al. 2003).

6.4 Effect of Saline Stress on Floral Organs

Floral organs are very important as they are involved in seed development, fruit development, and transferring traits to next generation. The fertilization and early seed development from spore mother cells are extremely sensitive to various biotic and abiotic stresses including salts which result in various structural and functional aberrations in reproductive organs, thus leading to distorted fertilization and premature seed abortion. The reproductive development seems to be the most salinity-stress-prone period after seed germination (Salter 1967). Plants have evolved mechanisms to decrease plant fecundity by aborting ovules and/or pollen and then shifting resources from reproductive activities to the cellular metabolic reactions which later aid in stress tolerance. Plants prevent fertilization under salt stress by freezing the male gametophyte and female gametophyte development. Finally, there is premature abortion of the fertilized ovule (Sheoran and Saini 1996).

Under salt stress, the florets are normal, but the growth of the anthers has been reported to be inhibited. The pollen grains shapes are not natural as anthers get shriveled (Sadeghirad et al. 2017). Under salinity stress, the reproductive development of the plant is affected. The microsporogenes is phenomenon, and the elongation of stamen filament elongation get negatively affected (Shrivastava and Kumar 2015). The microspores are sensitive to salt stress during plant growth (Namuco and O'Toole 1986). In *A. thaliana*, under salt stress, microsporocytes are unable to mature to pollen grains and become vacuoles (Sun et al. 2004). The mature pollen grains are unaffected by salinity stress thereby showing that only developmental growth stages of anthers are affected (Sun et al. 2004). The defect in salt stress-affected anthers is related to the metabolic events which are related to the degradation of the pollen development (Saini 1997). Stress results in a reduction of the flower number, faulty pollen grains and ovules development, disturbed fertilization, and a deformed embryo, etc. Salt stress inhibits the cell elongation, and the flower buds are significantly reduced in size, and there are reports about malformed ovule and embryo sac development under stress (Sulpice et al. 2003). Salinity stress negatively affects pollen fertility, and sterile pollen is undersized and due to abortion it becomes cytoplasm-deficient (Rehman et al. 2004). Mahmoodzadeh and Kabiri (2009) observed that under salt stress, the tetrad and young microspores shapes are abnormal and suggested that the tetrad stage and any stage prior to tetrad formation are salt-sensitive. Under salt stress, sporopollenin gets secreted by tapetum cells which results in the formation of improper microspores as this increased sporopollenin content gets accumulated in anther locules. The sporopollenin precursors disturb the osmotic potential of fluids in locules and water gets excreted out of the microspores, leading to collapse of microspores. In *Brassica juncea* (Chauhan and Singh 2005), increased sporopollenin content was reported under salt stress as earlier observed in barley and later reported by Tripathi (2008) in *Helianthus annuus* L. The viability, germination, pollen tube growth, and pod formation in *Cicer arietinum* are affected in plants under salt stress (Samineni et al. 2011).

6.5 Anatomical Features and Salt Stress

Morphophysiological and biochemical perturbations in addition to anatomical changes are reported in salt stress. These are leaf thickening, thickness of epidermis, change in distribution of xylem elements of stem and root, etc. Finally, these above modifications end up adjusting osmotic and ionic status of cells whenever the salt level increases. The transpirational water loss is by stomata and cuticle and both are related to the anatomical architecture of the leaves (Muir 2015). When leaf thickness is more, water loss is reduced under stress conditions. When there is salt stress, there is reduction of stomatal aperture and density in abaxial and adaxial surfaces of leaves (Albaladejo et al. 2017). Since, halophytes are more tolerant, they have developed adaptation strategy to reduce stomatal density with increasing salt concentration (Shabala et al. 2012). In wild-type tomatoes and cultivated species, there were important anatomical differences in leaf mesophyll cells as related to salt stress. In citrus varieties, there were anatomical disturbances due to the chloride salts in sensitive (*Carrizo citrange*) and tolerant (*Cleopatra mandarin*) variety (Romero-Aranda et al. 1998). The photosynthetic efficiency had been linked to the change of leaf anatomical features and in particular increase in the size and succulence of leaves. Anatomical changes have been seen in the leaves of *Arbutus unedo* under salt stress, and a comparative account of control and salt-treated plants showed that cell size of the first layer of palisade tissue is not much altered as compared to the second layer of palisade tissues (Navarro et al. 2007). The henna plants responded by modulating the ratio of leaf mass to leaf area (LSM), increased thickness of leaves, and suggested an increase in leaf thickness and LSM (Fernández-García et al. 2014) under high salt level. This indicates a strategy to adapt to the increasing salt stress. The thickness of leaf is increased as there is an increase in the palisade parenchyma, thereby increasing photosynthetic efficiency in *Viburnum tinus* L. (Gómez-Bellot et al. 2015), zucchini squash (Rouphael et al. 2017), *Myrtus communis* (Acosta-Motos et al. 2015a), and *Eugenia myrtifolia* (Acosta-Motos et al. 2015b). Under high salt stress conditions, the leaves try to maintain very high photosynthetic ability despite of closure of stomata. Trichomes are glandular structures and also provide defense to plants. So, their numbers are increased under abiotic and biotic stresses. Trichome density is found to be increased (Aguirre-Medina et al. 2002). This increase in trichome density is an adaptive mechanism for salt tolerance, and the trichomes help in the excretion of ions under high salt stress (Gucci et al. 1997). Anatomical changes in xylem of stems and thickness of cell in vascular region are more than the control treatment. So, the effect under salt stress is related to concentration-dependent. There is more deposition of lignin in the vascular tissues and xylem. The cell walls become lignified when cell expansion gets decreased under stress (Christensen et al. 1998). In the roots of bean, when NaCl is more there is enhanced lignifications which later inhibit the growth of roots and helps in adaptation to combat salinity stress (Cachorro et al. 1993). Parida et al. (2016) observed that *Salvadora persica* when treated with salt concentrations showed anatomical variations in the architecture of stem, root, and leaf. As compared to the control, seedlings treated with salt show thickening of lower and upper epidermis. Such thickening controls the circulations of sodium, potassium, and other ions which favor the tolerance toward the salt level thereby giving adaptation to salt stress. This thickening is also related to the transpiration rate (Parida et al. 2016). Such kind of thickening of various layers are reduced in the glycophytes under increasing salt concentration (Atabayeva et al. 2013) as observed in *Hordeum vulgare*. So, there exists an evolutionary adaptation mechanism in glycophytes and halophytes, in response to salinity. In stem, the pith area and thickness of hypodermis get increased under the salinity increase and also provide tolerance. The cortical thickness of stem in *S. persica* decreased with increase in salinity (Parida et al. 2016). This anatomical adaptation is helpful to reduce the growth of plants under salt stress conditions and helpful in conservation of energy needed for survival.

6.5.1 Relationship of Structure–Function Operation in Roots

The sodium ions make a direct apoplastic pathway from outside into xylem from the gaps being created by the disruption of suberin deposition in endodermis of roots. This is because suberization is not properly developed in the young roots area, thereby creating the broken patches which get remained in the secondary roots appearance. This makes a discontinuous patch in the endoderm (Flowers et al. 2015). The symplastic pathway and the apoplastic pathway related to the sodium ion transport relate to the anatomy of roots. This anatomy varies from plant to plant and is altered during the salt stress. In rice, the sodium ions which get leaked through apoplastic spaces determine the tolerance toward the salinity response. Also the importance of apoplastic pathway in rice is much more than wheat (Fernández-García et al. 2014). The root anatomy has difference between glycophytes and halophytes. But, the root anatomy of halophytes helps in maintaining the flow of sodium ions through the apoplastic pathway. The casparian strips are thicker than the glycophytes and there is the presence of the second layer of endoderm in halophytes (Tester and Davenport 2003). Salinity also induces and accelerates the formation of the casparian strips and more layers in cotton which is considered to be relatively salt-tolerant (Cameron et al. 2006). So, we can say that there is a correlation between the sodium ions influx into the roots and the salt tolerance levels. Thus, if the apoplastic leakage can be reduced in any sensitive species (e.g. rice), the tolerance may be provided for salt and this may be employed as a strategy for developing salt stress-resistant plants.

6.5.2 Transport Through Xylem

During salt stress, the sodium content in the roots remains undisturbed under the salt stress. This is achieved by the action of root cells to discharge the sodium ions toward external area. This is the effect of the translocation of sodium ions from the xylem vessels to the aerial parts. During the flow of plant sap, the sodium levels of the xylem and phloem get changed. The sodium ions that reach the leaves via the xylem sap can be effectively controlled by the transport in xylem vessels (Franco et al. 2001). In *A. thaliana,* under medium salt stress level, (40 mM) *Sos1* mutants with deleted function of an H^+/Na^+ antiport system accumulate lesser sodium ions in the aerial parts as compared to the wild-type plants (Chrominski et al. 1988; Franco et al. 2011). This observation led to the fact that SOS1 has an important role in the transport of sodium to the upper aerial region from the xylem part. The use of plant biotechnology with the help of a reported gene confirms that the SOS1 protein is located and expressed in the parenchymal cells around the xylem portions (El Mahi et al. 2019). It was observed that SOS1 helps in Na^+ secretion in xylem from the parenchymal cells of stele when there is moderate level of salt stress. The transporters are the proteins which help in the transportation of the salt-related ions from one part of the plant to other part. It has been observed that there is a reduction of sodium accumulation in the aerial parts. This reduction is explained by the process of desalting of xylem sap and it gets excreted from the xylem parts of the roots before the sodium ions reach the foliar parts (Assaha et al. 2017). The correlation of the desalting and the molecular mechanisms involved is related to the two transporters of the HKT gene family. *AtHKT1* of *A. thaliana* and *OsHKT8* of rice are involved in this phenomenon. Also the SKOR potassium channel of the Shaker family in *Arabidopsis* plays an important role in potassium ions secretion in xylem sap (Kim et al. 2015).

6.5.3 Transport into the Phloem

Phloem helps in the distribution of photosynthesis of products synthesized in the aerial portion of the plant and is involved in the long distancing transport. Sodium levels in xylem and phloem get changed throughout the transport in the vessels. There is the circulation of ions like sodium from xylem to phloem which makes it possible to reduce the content of salt in the leaves in few plants, viz. corn (Lohaus et al. 2000), barley (Shabala et al. 2010), and lupin (Munns et al. 1988). There are many plants that have some kind of salt adaptation strategy manifested by their anatomical architecture. A wild-type tomato, *Lycopersicon pennellii*, which is tolerant to salt stress has a remarkable sodium ion translocation in the phloem portion of the plants as compared to their domesticated counterparts (Alfocea et al. 2000).

The flow of sodium ions should be directed to the roots via phloem. Under a gradient setup between phloem and roots, translocation of sodium ions occurs from aerial parts to the roots. The *AtHKT1* of *A. thaliana*, which is expressed in phloem, is correlated with the circulation of sodium ions from leaves to roots via phloem and controls sodium accumulation and is correlated with the salt stress tolerance (An et al. 2017). Out of 18 pomegranate (*Punica granatum* L.) cultivars subjected to 200 mM NaCl for five weeks, some cultivars were found to be tolerant to salt stress to some extent. This was because of their ability to store sodium in roots and transport potassium and calcium in the leaves. So, pomegranate uses such an adaptive strategy to endure salt stress, but further validation is required (Liu et al. 2018).

6.6 Yield and Related Traits

Under salinity, apart from the growth and development, the crop yield also gets severely affected and also the execution of tolerant metabolic reactions occurs. This helps the plant for adaptation, growth, and sustaining stress responses (Munns and Gilliham 2015). The experiments which led to the identification of salt-tolerant lines are invariably related to the yield trait, and mostly the salt-tolerant lines produce good yield of agricultural importance. The stress tolerance index (STI) has been earlier proposed which is related to stress tolerance and yield performance. Munns et al. (2002) gave S/C index which indicates stress tolerance and the plant's performance under non-stress conditions. STI proposed by Fernandez (Fernandez 1992) is a STI, which helps in the identification of those lines which are stress-tolerant and also gives high yield. Saade et al. (2016) gave a new index, i.e. stress-weighted performance (SWP) in which there is a possibility to differentiate the 100 top plants that are high yielding and display tolerance under salt stress. Biomass is an organic material produced by photosynthesis in plants. It is a source of energy for many metabolic reactions that help in the overall development of plants and enduring stress thus making it an important component to be studied. Under high salinity, the biomass yield decreases instantaneously after sensing osmotic stress. The rate of salt absorption and the ability of the plant to compartmentalize the salt in various tissues are some of the factors leading to the membrane injury (Volkmar et al. 1998). So, we can say that the ability of the plant to sustain many abiotic stresses including salinity is the determining factor for plants yield. Like other stresses, salinity involves complex mechanisms and, also various factors determine yield and related traits (Flowers and Flowers 2005; Ashraf and Akram 2009). Regarding salinity stress tolerance, Maas and Hoffman (1977) have told about threshold value or level of salinity after which the crops yield gets reduced in a biphasic manner. This threshold value responds to the environmentally induced factors and is different for different plants (Volkmar et al. 1998). When salt stress was applied to some *P. granatum* cultivars to observe the effects, it was seen that excess salt reduced the yield and quality of the crop (Liu et al. 2018). In rice cultivars, the salt stress acted as a limiting factor and the yield per plant reduced in most of the rice genotypes studied, and only the genotypes, i.e. DM-38-88 and NR1, showed the maximum yield under salt-stressed conditions (Ali et al. 2004). The related yield traits like reduction in the area of leaf and yield components under salt stress conditions have been correlated well with the reduced growth of the plant. This is because of the reduced water uptake, sodium and chloride toxicity, reduced photosynthesis, etc. (Ali et al. 2004).

In *H. vulgare* L. (barley), the evaluation of salt stress done to study the effect on yield parameters showed that the CM67 barley cultivar produced highest and another cultivar UH3 produced lowest biological and grain yield. This study showed that the grain yield, harvest index, grain per ear, weight of grain, height of plant, and overall biological yield were reduced under the saline conditions. Comparing the cultivars used in the study, the CM67 cultivar showed salt stress tolerance to a satisfactory degree and was suitable for cultivation in salinity conditions (Bagheri and Sadeghipour 2009). In wild barley, saline conditions helped to quantify the salinity effects on several yield-related traits. These traits which were affected were modulation in flowering time and harvest index (Saade et al. 2016). Flowering time fluctuations were related to the genes related to the quantitative trait loci (QTLs). In this barley cultivar, yield-related salinity tolerance loci were also found with favorable effects under salinity tolerance. This study helped to improve barley cultivar better suited to high yield under excess salt-stressed conditions (Saade et al. 2016). A class I trehalose phosphate synthetase (TPS) gene related to role in salinity tolerance and subsequent growth of plants from rice (*OsTPS* 8) is related to control yield-related traits. The salinity stress is observed due to the enhanced deposition of suberin in rice (Vishal et al. 2019). When the OsTPS8 is overexpressed, there is no any yield loss and salt tolerance is imparted thus showing promising effects in genetic improvement of rice (Vishal et al. 2019).

6.7 Salt Stress and its Effect on Gene Expression

Indeed being sessile organisms, there are sophisticated mechanisms adopted by plants to combat abiotic and biotic stress. Plants sense the stress and relay properly to the cellular level and complex signaling, thereby imparting stress tolerance. This signaling involves gene expression and molecular players to control and modulate the stress response and roles of early responsive genes (Wei et al. 2013), differentially expressed genes (Liu et al. 2000), and different proteomic responses (Fatehi et al. 2012; Kamal et al. 2012a,b; Ji et al. 2019). This involves many kinds of transcription factors (Hoang et al. 2017). This change of gene expression helps to maintain the cellular homeostasis which has an important role in the survival of plants under unfavorable conditions. Salinity is another very important abiotic stress which leads to modulation at transcription levels (Ibrahimova et al. 2021; Shinozaki and Yamaguchi-Shinozaki 2007). The genes are upregulated or downregulated to maintain the tolerance level, under salinity as seen in soybean. Thus, acting as molecular switches which when needed performs the required modulation. Such factors are transcription factors like bZIP, WRKY, AP2, NAC, C2H2 zinc finger, MYB, and DREB. Thereafter, the transcription factors alter the expression of genes by specific binding in the promoters. In *Limonium bicolor*, overexpression of *DREB* resulted in enhanced tolerance to salt stress (Zhao et al. 2019). The different types of genes having roles in salt stress are *SOS* genes, *NHX1* genes, and *H^+ -ATPase* genes which have the role in ion transport and *SAGs* genes (senescence-associated genes). The *CAMTA* gene family of *Phaseolus vulgaris* L. has a role in salt stress as revealed by their expression profiling (Buyuk et al. 2019). When vacuolar H^+-pyrophosphatase and a B subunit of H^+-ATPase are ectopically expressed after cloning from *Halostachys caspica* (halophyte), the *A. thaliana* plants are salt stress-tolerant (Hu et al. 2013). Heat-shock proteins (HSPs) act as molecular chaperones, and dehydration-related transcription factors (e.g. DREB) participate in osmotic stress. The SOS transcription family of genes has a role in ion homeostasis which imparts salt tolerance (Hasegawa et al. 2000). Schmidt et al. (2013) have shown that ROS scavenging and osmotic-regulating genes are upregulated by salinity-tolerant cultivars. Chakraborty et al. (2012) have observed that *SOS1* and *SOS2* genes of *B. juncea* functions as plasma membrane Na^+/K^+ antiporters upregulators after 24 hours in roots and/or in leaves in 25 mM NaCl. A root-specific salt-responsive ERF1 transcription factor of *Oryza sativa* has shown induction upon salt and prooxidant (H_2O_2) treatment (Schmidt et al. 2013). It has been also observed that ERF1 mutant lines are susceptible to salt tolerance as compared to wild type, but ectopic expression of ERF1 provides salinity tolerance. Many kinases which regulate the transcription factors prove their role in plant adaptation to salt stress. In *O. sativa*, *OsRMC* gene encodes a receptor-like kinase and negatively regulates salt stress responses in rice (Serra et al. 2013). *bZIP* genes are upregulated in sensitive cultivar of wheat when grown under saline stress and downregulated in tolerant variety (Johnson et al. 2002). OsNAC5 and ZFP179 are the transcription factors which are upregulated under salinity stress, and the synthesis, regulation and accumulation of proline, sugar, and late embryogenesis abundant (LEA) proteins are modulated which later exert their role in stress tolerance (Song et al. 2011). WRKY8 gene in *Arabidopsis* gets upregulated and it directly binds to the promoter of RD29A and has a role in salt stress (Hu et al. 2013). A MYB gene *MYB73* from cotton when overexpressed in *A. thaliana* induced salinity tolerance and many genes related to osmotic stress like *NHX1, P5CS1,* and *SOS3* were induced (Zhao et al. 2019). Overexpression of caffeic acid *O*-methyltransferase (COMT) gene conferred salt tolerance in *Arabidopsis* and also isoflavone accumulation and salt stress tolerance through overexpressed *GmIFS1* in soybean. CIPKs of different types like CIPK1, CIPK3, CIPK16, and CIPK24 (Liu et al. 2000; Kim et al. 2003; D'Angelo et al. 2006) that observed in *Arabidopsis* have correlation with salt stress tolerance. Gene expression of *CIPK6* is correlated with the salt stress response (Chen et al. 2007; Tripathi et al. 2009). It has been observed that CIPK16 from *A. thaliana* reduces sodium ions in leaves.

The proteome is also modulated during salt stresss in plants like *Musa paradisiaca* (Ji et al. 2019). The recent upcoming roles of microRNAs (miRNAs) and also small interfering RNAs (siRNAs) have emerged as important molecular players in stress response of plants (Zhou et al. 2012; Xie et al. 2014). The posttranscriptional gene regulation has an important role in the salt stress response. Role of miRNAs in *A. thaliana* and *O. sativa* has been elucidated in response to abiotic stress in plants. miRNAs profiling under abiotic stress have shown differential expression. In *A. thaliana*, many miRNAs, viz. miR319, miR393, miR394, miR396, miR171, miR169, miR168, and miR167 are upregulated under salt stress, but the miR398 expression is downregulated. This shows the correlation and direct role for miRNAs under salt stress response (Lu et al. 2008). In *Phaseolous vulgaris* (Arenas-Huertero et al. 2009) and *Populus trichocarpa* (Lu et al. 2008) and *Populus cathayana* L. and *Salix matsudana* Koidz (Zhou et al. 2012), expression of few miRNAs is modulated during salinity. A QTL is locus on DNA which correlates with variation in the quantitative trait, i.e. phenotype. Salt-stress-related QTLs in white clover identified at the vegetative stage have shown many QTLs responsible for salinity tolerance (Wang et al. 2010). In the seedling stage of *Solanum pennellii* and *Solanum lycopersicoides*, QTLs responsible for salt tolerance are seen. In *Solanum Pennellii* four major QTLs and in *S. lycopersicoides* six major QTLs for salt tolerance have been detected (Li et al. 2011). There is flooding of high throughput data related to bioinformatics which adds up to the genomic information. These "OMICS" data are integrated to build a signaling network, and the salt stress tolerance is seen as a complex trait (Li et al. 2011). Thus, a complex trait which is governed by all kinds of "OMICS" is a decisive force in salinity tolerance in grain legumes. There are many bioinformatics softwares for studying the OMICS-related matter viz. Chickpea Transcriptome Database (CTDB) in chickpea (Verma et al. 2015), Soybean Knowledge Base in soybean (SKB, http://soykb.org), Cowpea Genespace/Genomics Knowledge Base (CGKB) in cowpea (Chen et al. 2007), and Legume information system (LegumeInfo. org) (Dash et al. 2016). These are very important to study and decipher the signaling pathways using bioinformatics approach.

6.8 Conclusion

Since the world is experiencing continuous and adverse effects of changing climate and various stresses, the pressure is on the farmers to get more produce and on the scientists to make climate resilient crop plants on the decreasing agricultural land. The ever-increasing population further adds to the problem. Thus, an alternative is to grow and cultivate salt stress-tolerant biomass crops as these will help in directly competing with nutritionally important plants. There must be search and identification of endemically growing marginal crops which can grow in low-grade land with high salt concentrations and are also economically useful.

References

Acosta-Motos, J.R., Diaz-Vivancos, P., Álvarez, S. et al. (2015a). NaCl-induced physiological and biochemical adaptative mechanisms in the ornamental *Myrtus communis* L. plants. *J. Plant Physiol.* 183: 41–51.

Acosta-Motos, J.R., Diaz-Vivancos, P., Álvarez, S. et al. (2015b). Physiological and biochemical mechanisms of the ornamental *Eugenia myrtifolia* L. plants for coping with NaCl stress and recovery. *Planta* 242 (4): 829–846.

Acquaah, G. (2009). *Principles of Plant Genetics and Breeding*. Wiley.

Aguirre-Medina, J.F., Acosta Gallegos, J.A., Ruiz Posadas, L.D. et al. (2002). Morphological differences on the leaf epidermis of common bean and their relationship to drought tolerance. *Agric. Techn. En Mex.* 28 (1): 53–64.

Albaladejo, I., Meco, V., Plasencia, F. et al. (2017). Unravelling the strategies used by the wild tomato species *Solanum pennellii* to confront salt stress: from leaf anatomical adaptations to molecular responses. *Environ. Exp. Bot.* 135: 1–12.

Alfocea, F.P., Balibrea, M.E., Alarcón, J.J. et al. (2000). Composition of xylem and phloem exudates in relation to the salt-tolerance of domestic and wild tomato species. *J. Plant Physiol.* 156 (3): 367–374.

Ali, Y., Aslam, Z., Ashraf, M.Y. et al. (2004). Effect of salinity on chlorophyll concentration, leaf area, yield and yield components of rice genotypes grown under saline environment. *Int. J. Environ. Sci. Technol.* 1: 221–225.

An, D., Chen, J.G., Gao, Y.Q. et al. (2017). AtHKT1 drives adaptation of *Arabidopsis thaliana* to salinity by reducing floral sodium content. *PLoS Genet.* 13 (10): e1007086.

Andriolo, J.L., Luz, G.L.D., Witter, M.H. et al. (2005). Growth and yield of lettuce plants under salinity. *Hortic. Bras.* 23 (4): 931–934.

Arenas-Huertero, C., Pérez, B., Rabanal, F. et al. (2009). Conserved and novel miRNAs in the legume *Phaseolus vulgaris* in response to stress. *Plant Mol. Biol.* 70 (4): 385–401.

Ashraf, M. and Akram, N.A. (2009). Improving salinity tolerance of plants through conventional breeding and genetic engineering: an analytical comparison. *Biotechnol. Adv.* 27: 744–752.

Assaha, D.V., Ueda, A., Saneoka, H. et al. (2017). The role of Na+ and K+ transporters in salt stress adaptation in glycophytes. *Front. Physiol.* 8: 509.

Atabayeva, S., Nurmahanova, A., Minocha, S. et al. (2013). The effect of salinity on growth and anatomical attributes of barley seedling (*Hordeum vulgare* L.). *Afr. J. Biotechnol.* 12 (18): 2366–2377.

Bagheri, A. and Sadeghipour, O. (2009). Effects of salt stress on yield, yield components and carbohydrates content in four Hullless barley (*Hordeum vulgare* L.) cultivars. *J. Biol. Sci.* 9: 909–912.

Belamkar, V., Weeks, N.T., Bharti, A.K. et al. (2014). Comprehensive characterization and RNA-Seq profiling of the HD-zip transcription factor family in soybean (*Glycine max*) during dehydration and salt stress. *BMC Genomics* 15 (1): 950.

Breckle, S.W. (1995). How do halophytes overcome salinity. *Biol. Salt Tol. Plant* 23: 199–203.

Buyuk, I., Ilhan, E., Sener, D. et al. (2019). Genome-wide identification of CAMTA gene family members in phaseolus vulgaris L. And their expression profiling during salt stress. *Mol. Biol. Rep.* 46: 2721–2732.

Cachorro, P., Ortiz, A., Barcelo, A.R., and Cerda, A. (1993). Lignin deposition in vascular tissues of *Phaseolus vulgaris* roots in response to salt stress and Ca2+ ions. *Phyton* 33: 33–40.

Cameron, R.W.F., Harrison-Murray, R.S., Atkinson et al. (2006). Regulated deficit irrigation – a means to control growth in woody ornamentals. *J. Hortic. Sci. Biotechnol.* 81 (3): 435–443.

Chakraborty, K., Sairam, R.K., and Bhattacharya, R.C. (2012). Differential expression of salt overly sensitive pathway genes determines salinity stress tolerance in Brassica genotypes. *Plant Physiol. Biochem.* 51: 90–101.

Chauhan, S.V.S. and Singh, V. (2005). Chemical induction of male sterility in *Brassica juncea* and its utilization for hybrid seed production. *Brassica* 7: 117.

Cheeseman, J.M. (2013). The integration of activity in saline environments: problems and perspectives. *Funct. Plant Biol.* 40 (9): 759–774.

Chen, C., Tao, C., Peng, H., and Ding, Y. (2007). Genetic analysis of salt stress responses in asparagus bean (*Vigna unguiculata* [L.] ssp. *sesquipedalis Verdc.*). *J. Heredity.* 98 (7): 655–665.

Chen, J.H., Jiang, H.W., Hsieh, E.J. et al. (2012). Drought and salt stress tolerance of an *Arabidopsis* glutathione S-transferase U17 knockout mutant are attributed to the combined effect of glutathione and abscisic acid. *Plant Physiol.* 158 (1): 340–351.

Christensen, J.H., Bauw, G., Welinder, K.G. et al. (1998). Purification and characterization of peroxidases correlated with lignification in poplar xylem. *Plant Physiol.* 118 (1): 125–135.

Chrominski, A., Bhat, R.B., Weber, D.J., and Smith, B.N. (1988). Osmotic stress-dependent conversion of 1-aminocyclopropane-1-carboxylic acid (ACC) to ethylene in the halophyte, *Allenrolfea occidentalis*. *Environ. Exp. Bot.* 28 (3): 171–174.

D'Angelo, C., Weinl, S., Batistic, O. et al. (2006). Alternative complex formation of the Ca2+-regulated protein kinase CIPK1 controls abscisic acid-dependent and independent stress responses in Arabidopsis. *Plant J.* 48 (6): 857–872.

Dantas, B.F., Ribeiro, L.D.S., and Aragão, C.A. (2005). Physiological response of cowpea seeds to salinity stress. *Revista Bras. de Semen.* 27 (1): 144–148.

Dash, S., Campbell, J.D., Cannon, E.K. et al. (2016). Legume information system (LegumeInfo.org): a key component of a set of federated data resources for the legume family. *Nucleic Acids Res.* 44 (D1): D1181–D1188.

Ding, F., Cui, P., Wang, Z. et al. (2014). Genome-wide analysis of alternative splicing of pre-mRNA under salt stress in *Arabidopsis*. *BMC Genomics* 15 (1): 431.

El Mahi, H., Pérez-Hormaeche, J., De Luca, A. et al. (2019). A critical role of sodium flux via the plasma membrane Na+/H+ exchanger SOS1 in the salt tolerance of rice. *Plant Physiol.* 180 (2): 1046–1065.

Fatehi, F., Hosseinzadeh, A., Alizadeh, H. et al. (2012). The proteome response of salt-resistant and salt-sensitive barley genotypes to long-term salinity stress. *Mol. Biol. Rep.* 39 (5): 6387–6397.

Fernandez, G.C.J. (1992) Effective selection criteria for assessing plant stress tolerance. *Proceedings of the international symposium on adaptation of vegetable and other food crops in temperature and water stress*, pp. 257–270 (Taiwan, 1992).

Fernández-García, N., Olmos, E., Bardisi, E. et al. (2014). Intrinsic water use efficiency controls the adaptation to high salinity in a semi-arid adapted plant, henna (*Lawsonia inermis* L.). *J. Plant Physiol.* 171 (5): 6475.

Flowers, T.J. (1975). Halophytes. In: *Ion Transport in Plant Cells and Tissues* (eds. D.A. Baker and J.L. Hall), 309–334. Amsterdam: Elsevier.

Flowers, T.J. and Colmer, T.D. (2008). Salinity tolerance in halophytes. *New Phytol.*: 945–963.

Flowers, T.J. and Flowers, S.A. (2005). Why does salinity pose such a difficult problem for plant breeders? *Agric. Water. Manag.* 78: 15–24.

Flowers, T.J., Munns, R., and Colmer, T.D. (2015). Sodium chloride toxicity and the cellular basis of salt tolerance in halophytes. *Ann. Bot.* 115 (3): 419–431.

Franco, J.A., BañÒ, S., Ferna Ndez, S., and Leskovar, D.I. (2001). Effect of nursery regimes and establishment irrigation on root development of Lotus creticus seedlings following transplanting. *J. Hortic. Sci. Biotechnol.* 76 (2): 174–179.

Franco, J.A., Bañón, S., Vicente, M.J. et al. (2011). Root development in horticultural plants grown under abiotic stress conditions – a review. *J. Hortic. Sci. Biotechnol.* 86 (6): 543–556.

Godfray, H.C.J., Beddington, J.R., Crute, I.R. et al. (2010). Food security: the challenge of feeding 9 billion people. *Science* 327 (5967): 812–818.

Gómez-Bellot, M.J., Nortes, P.A., Ortuño, M.F. et al. (2015). Influence of arbuscular mycorrhizal fungi and treated wastewater on water relations and leaf structure alterations of *Viburnum tinus* L. plants during both saline and recovery periods. *J. Plant Physiol.* 188: 96–105.

Greenway, H. and Munns, R. (1980). Mechanisms of salt tolerance in nonhalophytes. *Annu. Rev. Plant Physiol.* 31 (1): 149–190.

Gucci, R. and Tattini, M. (1997). Salinity tolerance in olive. *Hortic. Rev.* 21: 177–214.

Gucci, R., Aronne, G., Lombardini, L., and Tattini, M. (1997). Salinity tolerance in Phillyrea species. *New Phytol.* 135 (2): 227–234.

Ha, E., Ikhajiagba, B., Bamidele, J.F., and Ogic-Odia, E. (2008). Salinity effects on young healthy seedling of *kyllingia peruviana* collected from escravos, Delta state. *Global J. Environ. Res.* 2 (2): 74–88.

Hasegawa, P.M., Bressan, R.A., Zhu, J.K., and Bohnert, H.J. (2000). Plant cellular and molecular responses to high salinity. *Annu. Rev. Plant Biol.* 51 (1): 463–499.

Hoang, X.L.T., Du Nhi, N.H., Binh Anh Thu, N. et al. (2017). Transcription factors and their roles in signal transduction in plants under abiotic stresses. *Curr. Genomics* 18 (6): 483–497.

Hsiao, T.C. and Xu, L.K. (2000). Sensitivity of growth of roots versus leaves to water stress: biophysical analysis and relation to water transport. *J. Exp. Bot.* 51 (350): 1595–1616.

Hu, H., Dai, M., Yao, J. et al. (2006). Overexpressing a NAM, ATAF, and CUC (NAC) transcription factor enhances drought resistance and salt tolerance in rice. *Proc. Natl. Acad. Sci.* 103 (35): 12987–12992.

Hu, Y., Chen, L., Wang, H. et al. (2013). *Arabidopsis* transcription factor WRKY 8 functions antagonistically with its interacting partner VQ 9 to modulate salinity stress tolerance. *Plant J.* 74 (5): 730–745.

İbrahimova, U., Kumari, P., Yadav, S. et al. (2021). Progress in understanding salt stress response in plants using biotechnological tools. *J. Biotechnol.* 329: 180–191. https://doi.org/10.1016/j.jbiotec.2021.02.007.

Ji, H., Pardo, J.M., Batelli, G. et al. (2013). The salt overly sensitive (SOS) pathway: established and emerging roles. *Mol. Plant* 6 (2): 275–286.

Ji, F.S., Tang, L., Li, Y.Y. et al. (2019). Differential proteomic analysis reveals the mechanism of *Musa paradisiaca* responding to salt stress. *Mol. Biol. Rep.* 46 (1): 1057–1068.

Johnson, R.R., Wagner, R.L., Verhey, S.D., and Walker-Simmons, M.K. (2002). The abscisic acid-responsive kinase PKABA1 interacts with a seed-specific abscisic acid response element-binding factor, TaABF, and phosphorylates TaABF peptide sequences. *Plant Physiol.* 130 (2): 837–846.

Kamal, A.H.M., Cho, K., Kim, D.E. et al. (2012a). Changes in physiology and protein abundance in salt-stressed wheat chloroplasts. *Mol. Biol. Rep.* 39 (9): 9059–9074.

Kamal, A.H.M., Cho, K., Komatsu, S. et al. (2012b). Towards an understanding of wheat chloroplasts: a methodical investigation of thylakoid proteome. *Mol. Biol. Rep.* 39 (5): 5069–5083.

Khan, M.A. (2008). *Crop and Forage Production Using Saline Waters*. Daya Books.

Kiani-Pouya, A., Roessner, U., Jayasinghe, N.S. et al. (2017). Epidermal bladder cells confer salinity stress tolerance in the halophyte quinoa and Atriplex species. *Plant Cell Environ.* 40 (9): 1900–1915.

Kim, K.N., Cheong, Y.H., Grant, J.J. et al. (2003). CIPK3, a calcium sensor – associated protein kinase that regulates abscisic acid and cold signal transduction in *Arabidopsis*. *Plant Cell* 15 (2): 411–423.

Kim, H.Y., Choi, E.H., Min, M.K. et al. (2015). Differential gene expression of two outward-rectifying shaker-like potassium channels OsSKOR and OsGORK in rice. *J. Plant Biol.* 58 (4): 230–235.

Li, J., Liu, L., Bai, Y. et al. (2011). Identification and mapping of quantitative resistance to late blight (*Phytophthora infestans*) in *Solanum habrochaites* LA1777. *Euphytica* 179 (3): 427–438.

Lin, J., Li, J.P., Yuan, F. et al. (2018). Transcriptome profiling of genes involved in photosynthesis in *Elaeagnus angustifolia* L. under salt stress. *Photosynthetica* 56 (4): 998–1009.

Liu, J., Ishitani, M., Halfter, U. et al. (2000). The *Arabidopsis thaliana* SOS2 gene encodes a protein kinase that is required for salt tolerance. *Proc. Natl. Acad. Sci.* 97 (7): 3730–3734.

Liu, C., Yan, M., Huang, X., and Yuan, Z. (2018). Effects of salt stress on growth and physiological characteristics of pomegranate (*Punica granatum* L.) cuttings. *Pak. J. Bot.* 50: 457–464.

Lohaus, G., Hussmann, M., Pennewiss, K. et al. (2000). Solute balance of a maize (*Zea mays* L.) source leaf as affected by salt treatment with special emphasis on phloem retranslocation and ion leaching. *J. Exp. Bot.* 51 (351): 1721–1732.

Lu, S., Sun, Y.H., and Chiang, V.L. (2008). Stress-responsive microRNAs in Populus. *Plant J.* 55 (1): 131–151.

Maas, E.V. and Hoffman, G.J. (1977). Crop salt tolerance-current assessment. *J. Irrig. Drain. Div.* 103: 115–134.

Mahmoodzadeh, H. and Kabiri, M. (2009). A normal micro-and megagametophyte development in *Brassica napus* induced by high salinity. *Res. J. Environ. Sc.* 3 (5): 514–521.

Marschner, H. (1995). *Marschner's Mineral Nutrition of Higher Plants*. Academic Press.

Matsushita, N. and Matoh, T. (1991). Characterization of Na^+ exclusion mechanisms of salt-tolerant reed plants in comparison with salt-sensitive rice plants. *Physiol. Plant.* 83 (1): 170–176.

McCue, K.F. and Hanson, A.D. (1990). Drought and salt tolerance: towards understanding and application. *Trends Biotechnol.* 8: 358–362.

Memon, S.A., Hou, X., and Wang, L.J. (2010). Morphological analysis of salt stress response of *Pak Choi. Elec. J. Env. Agricult. Food Chem.* 9 (1): 248–254.

Mishra, A.N., Sahu, S.M., Misra, M. et al. (1997). Sodium chloride induced changes in leaf growth, and pigment and protein contents in two rice cultivars. *Biol. Plant.* 39 (2): 257–262.

Muir, C.D. (2015). Making pore choices: repeated regime shifts in stomatal ratio. *Proc. R. Soc. B Biol. Sci.* 282 (1813): 20151498.

Munns, R. and Gilliham, M. (2015). Salinity tolerance of crops – what is the cost? *New Phytol.* 208: 668–673.

Munns, R. and Tester, M. (2008). Mechanisms of salinity tolerance. *Annu. Rev. Plant Biol.* 59: 651–681.

Munns, R., Lorraine Tonnet, M., SHENNAN, C., and Anne Gardner, P. (1988). Effect of high external NaCl concentration on ion transport within the shoot of Lupinus albus. II. Ions in phloem sap. *Plant Cell Environ.* 11 (4): 291–300.

Munns, R., Husain, S., Rivelli, A.R. et al. (2002). Avenues for increasing salt tolerance of crops, and the role of physiologically based selection traits. In: *Progress in Plant Nutrition: Plenary Lectures of the XIV International Plant Nutrition Colloquium* (eds. W.J. Horst, A. Bürkert, N. Claassen, et al.), 93–105. Dordrecht: Springer.

Murillo-Amador, B., Yamada, S., Yamaguch, T. et al. (2007). Salinity toxicity influence of calcium silicate on growth physiological parameters and mineral nutrition in two legume species under salt stress. *J. Agron. Crop Sci.* 193 (6): 413–421.

Muscolo, A., Panuccio, M.R., and Sidari, M. (2003). Effects of salinity on growth, carbohydrate metabolism and nutritive properties of kikuyu grass (*Pennisetum clandestinum Hochst*). *Plant Sc.* 164 (6): 1103–1110.

Namuco, O.S. and O'Toole, J.C. (1986). Reproductive stage water stress and sterility. I. Effect of stress during meiosis 1. *Crop Sci.* 26 (2): 317–321.

Navarro, A., Sanchez-Blanco, M.J., and Bañon, S. (2007). Influence of paclobutrazol on water consumption and plant performance of *Arbutus unedo* seedlings. *Sci. Hortic.* 111 (2): 133–139.

Pardo, J.M. (2010). Biotechnology of water and salinity stress tolerance. *Curr. Opin. Biotechnol.* 21 (2): 185–196.

Parida, A.K. and Das, A.B. (2005). Salt tolerance and salinity effects on plants: a review. *Ecotoxicol. Environ. Saf.* 60 (3): 324–349.

Parida, A.K., Veerabathini, S.K., Kumari, A., and Agarwal, P.K. (2016). Physiological, anatomical and metabolic implications of salt tolerance in the halophyte *Salvadora persica* under hydroponic culture condition. *Front. Plant Sci.* 7: 351.

Pattanagul, W. and Thitisaksakul, M. (2008). Effect of salinity stress on growth and carbohydrate metabolism in three rice (*Oryza sativa* L.) cultivars differing in salinity tolerance. *Ind. J. Exp. Bot.* 46 (10): 736–742.

Rehman, S., Kook, H.S., Lim, J.H. et al. (2004). Varietal responses of pollen development to salt stress in barley. 한국작물학회지 49 (5): 407–409.

Romero-Aranda, R., Moya, J.L., Tadeo, F.R. et al. (1998). Physiological and anatomical disturbances induced by chloride salts in sensitive and tolerant citrus: beneficial and detrimental effects of cations. *Plant Cell Environ.* 21 (12): 1243–1253.

Rouphael, Y., De Micco, V., Arena, C. et al. (2017). Effect of Ecklonia maxima seaweed extract on yield, mineral composition, gas exchange, and leaf anatomy of zucchini squash grown under saline conditions. *J. Appl. Phycol.* 29 (1): 459–470.

Saade, S., Maurer, A., Shahid, M. et al. (2016). Yield-related salinity tolerance traits identified in a nested association mapping (NAM) population of wild barley. *Sci. Rep.* 6: 32586. https://doi.org/10.1038/srep32586.

Sadeghirad, E., Majd, A., Iranbakhsh, A., and Javanshah, A. (2017). The effect of NaCl on anther development in *Pistacia vera* L. *J. Mol. Biol. Res.* 7 (1): p132.

Saini, H.S. (1997). Effects of water stress on male gametophyte development in plants. *Sex. Plant Reprod.* 10 (2): 67–73.

Salter, P.J. (1967). *Crop Responses to Water at Different Stages of Growth.* Farna Royal; Buks; England: Commonwealth Agricultural Bureaux.

Samineni, S., Siddique, K.H., Gaur, P.M., and Colmer, T.D. (2011). Salt sensitivity of the vegetative and reproductive stages in chickpea (*Cicer arietinum* L.): podding is a particularly sensitive stage. *Environ. Exp. Bot.* 71 (2): 260–268.

Schmidt, R., Mieulet, D., Hubberten, H.M. et al. (2013). Salt responsive ERF1 regulates reactive oxygen species – dependent signaling during the initial response to salt stress in rice. *Plant Cell* 25 (6): 2115–2131.

Serra, T.S., Figueiredo, D.D., Cordeiro, A.M. et al. (2013). OsRMC, a negative regulator of salt stress response in rice, is regulated by two AP2/ERF transcription factors. *Plant Mol. Biol.* 82 (4–5): 439–455.

Shabala, S. (2013). Learning from halophytes: physiological basis and strategies to improve abiotic stress tolerance in crops. *Ann. Bot.* 112 (7): 1209–1221.

Shabala, S., Shabala, S., Cuin, T.A. et al. (2010). Xylem ionic relations and salinity tolerance in barley. *Plant J.* 61 (5): 839–853.

Shabala, L., Mackay, A., Tian, Y. et al. (2012). Oxidative stress protection and stomatal patterning as components of salinity tolerance mechanism in quinoa (Chenopodium quinoa). *Physiol. Plant.* 146 (1): 26–38.

Shankar, R., Bhattacharjee, A., and Jain, M. (2016). Transcriptome analysis in different rice cultivars provides novel insights into desiccation and salinity stress responses. *Sci. Rep.* 6: 23719.

Sheoran, I.S. and Saini, H.S. (1996). Drought-induced male sterility in rice: changes in carbohydrate levels and enzyme activities associated with the inhibition of starch accumulation in pollen. *Sex. Plant Reprod.* 9 (3): 161–169.

Shi, H., Lee, B.H., Wu, S.J., and Zhu, J.K. (2003). Overexpression of a plasma membrane Na+/H+ antiporter gene improves salt tolerance in *Arabidopsis thaliana*. *Nat. Biotechnol.* 21 (1): 81–85.

Shinozaki, K. and Yamaguchi-Shinozaki, K. (2007). Gene networks involved in drought stress response and tolerance. *J. Exp. Bot.* 58 (2): 221–227.

Shrivastava, P. and Kumar, R. (2015). Soil salinity: a serious environmental issue and plant growth promoting bacteria as one of the tools for its alleviation. *Saudi J. Biol. Sci.* 22 (2): 123–131.

Smeekens, S. (2000). Sugar-induced signal transduction in plants. *Annu. Rev. Plant Biol.* 51 (1): 49–81.

Sobhanian, H., Aghaei, K., and Komatsu, S. (2011). Changes in the plant proteome resulting from salt stress: toward the creation of salt-tolerant crops. *J. Proteome* 74 (8): 1323–1337.

Song, J. and Wang, B. (2015). Using euhalophytes to understand salt tolerance and to develop saline agriculture: Suaeda salsa as a promising model. *Ann. Bot.* 115 (3): 541–553.

Song, S.Y., Chen, Y., Chen, J. et al. (2011). Physiological mechanisms underlying OsNAC5-dependent tolerance of rice plants to abiotic stress. *Planta* 234 (2): 331–345.

Sulpice, R., Tsukaya, H., Nonaka, H. et al. (2003). Enhanced formation of flowers in salt-stressed Arabidopsis after genetic engineering of the synthesis of glycine betaine. *Plant J.* 36 (2): 165–176.

Sun, K., Hunt, K., and Hauser, B.A. (2004). Ovule abortion in *Arabidopsis* triggered by stress. *Plant Physiol.* 135 (4): 2358–2367.

Taffouo, V.D., Kouamou, J.K., Ngalangue, L.M.T. et al. (2009). Effects of salinity stress on growth, ions partitioning and yield of some cowpea (*Vigna unguiculata* L. Walp.) cultivars. *Int. J. Bot.* 5 (2): 135–143.

Tester, M. and Davenport, R. (2003). Na^+ tolerance and Na^+ transport in higher plants. *Ann. Bot.* 91 (5): 503–527.

Tripathi, S.M. (2008). Abnormal another development and high sporopollenin synthesis in benzotriazole treated male sterile *Helianthus annuus* L. *Indian J. Exp. Biol.* 46: 71–78.

Tripathi, V., Parasuraman, B., Laxmi, A., and Chattopadhyay, D. (2009). CIPK6, a CBL-interacting protein kinase is required for development and salt tolerance in plants. *Plant J.* 58 (5): 778–790.

Verma, M., Kumar, V., Patel, R.K. et al. (2015). CTDB: an integrated chickpea transcriptome database for functional and applied genomics. *PLoS One* 10 (8): e0136880.

Vishal, B., Krishnamurthy, P., Ramamoorthy, R., and Kumar, P.P. (2019). OsTPS8 controls yield-related traits and confers salt stress tolerance in rice by enhancing suberin deposition. *New Phytol.* 221: 1369–1386.

Volkmar, K.M., Hu, Y., and Steppuhn, H. (1998). Physiological responses of plants to salinity: a review. *Canad. J. Plant Sci.* 78: 19–27.

Wang, J., Drayton, M.C., George, J. et al. (2010). Identification of genetic factors influencing salt stress tolerance in white clover (*Trifolium repens* L.) by QTL analysis. *Theor. Appl. Genet.* 120 (3): 607–619.

Wei, S., Wang, L., Zhang, Y., and Huang, D. (2013). Identification of early response genes to salt stress in roots of melon (*Cucumis melo* L.) seedlings. *Mol. Biol. Rep.* 40 (4): 2915–2926.

Xie, F., Stewart, C.N. Jr., Taki, F.A. et al. (2014). High-throughput deep sequencing shows that micro RNA s play important roles in switchgrass responses to drought and salinity stress. *P. Biotechnol. J.* 12 (3): 354–366.

Yang, Z., Wang, Y., Wei, X. et al. (2017). Transcription profiles of genes related to hormonal regulations under salt stress in sweet sorghum. *Plant Mol. Biol. Report.* 35 (6): 586–599.

Yilmaz, H. and Kina, A. (2008). The influence of NaCl salinity on some vegetative and chemical changes of strawberries (*Fragaria x ananssa* L.). *Afr. J. Biotechnol.* 7 (18).

Zhao, Y., Yang, Z., Ding, Y. et al. (2019). Over-expression of an R2R3 MYB gene, GhMYB73, increases tolerance to salt stress in transgenic *Arabidopsis*. *Plant Sci.* 286: 28–36.

Zhou, J., Liu, M., Jiang, J. et al. (2012). Expression profile of miRNAs in *Populus cathayana* L. and *Salix matsudana* Koidz under salt stress. *Mol. Biol. Rep.* 39 (9): 8645–8654.

Zhu, J.K. (2002). Salt and drought stress signal transduction in plants. *Annu. Rev. Plant Biol.* 53 (1): 247–273.

7

Effect of Salt Stress on Soil Chemistry and Plant-Atmosphere Continuum (SPAC)

Gunjan Goyal, Aruna Yadav, and Gunjan Dubey

Department of Botany, University of Rajasthan, Jaipur, Rajasthan, India

7.1 Introduction

World population is increasing at an expeditious rate, and it is envisaged that the total population shall increase by 34% in 2050 reaching about 9.1 billion or more, thus necessitating 70% more food production (FAO 2012). Plants are subjected to a variety of abiotic (salinity, thermal, light, water etc.) and biotic (herbivores) stresses under field conditions. Grime in 1977 defined stress as "any external abiotic or biotic constraint that limits the rate of photosynthesis and reduces a plants' ability to convert energy to biomass". Plants use many strategies in response to such abiotic stresses to maintain their growth rate and productivity in such stressful environments. These strategies include change in morphological, physiological, developmental as well as biochemical processes against several stresses (Tuteja 2007; Saud et al. 2014). Among these abiotic stresses, salinity is one of the most serious factors which:

a) lowers the productivity of agricultural crops (Greenway and Munns 1980) and
b) adversely affects germination of seeds (Xu et al. 2011), rate of photosynthesis (Zhang et al. 2005), yield (Nahar and Hasanuzzaman 2009) etc.

Hence, the overall crop yield is reduced.

Salt stress in soil affects plants in two different phases:

i) **Osmotic effect** – This is the first phase of salinity stress, which is independent of the accumulation of salt in the shoot. Salts dissolved in the soil solution reduce the soil water potential due to higher osmotic pressure and thus thermodynamically hamper water uptake from roots and induce water deficit (Roy et al. 2014). Water deficit signal is rapidly transmitted from roots to shoots causing intracellular turgor reduction and decreased cell expansion (Munns and Tester 2008). This signal also stimulates the biosynthesis of

Physiology of Salt stress in Plants: Perception, Signalling, Omics and Tolerance Mechanism, First Edition.
Edited by Pratibha Singh, Madhulika Singh, Rajiv Kumar Singh, and Sheo Mohan Prasad.

abscisic acid (ABA), which leads to a lower stomatal conductance for less transpiration in water deficit conditions (Munns and Tester 2008; Roy et al. 2014) which ultimately leads to lower carbon assimilation, biomass production, and decline in yield.

ii) **Ionic effect** – This is the second phase of salinity stress. It is caused due to the accumulation of higher concentration of sodium (Na^+) and/or chloride (Cl^-) ions, particularly in older leaves, causing ion toxicity which induces tissue necrosis and early leaf senescence (Roy et al. 2014). For most of the plant species, Na^+ is reported to reach a toxic concentration earlier than Cl^- (Tester and Davenport 2003). For example, Tsai et al. (2004) have reported that Na^+ is the primary toxic ion for rice.

Both osmotic and ionic effects disturb the aerobic metabolism in plants and stimulate the production and/or accumulation of reactive oxygen species (ROS) beyond the detoxifying capacity of the plant. This in turn negatively affects cellular structures and metabolism (Chaves et al. 2009).

7.1.1 Soil Salinization – Types and Causes

Soil salinization occurs due to different causes that have mainly been categorized into two different categories viz., Primary and Secondary salinity.

Primary/Natural salinity occurs due to the accumulation of salts in soil or groundwater over long periods of time through natural processes like:

- Weathering of rocks containing various soluble salts leading to their release mainly as chlorides (of Na, Ca and Mg), sulfates, and carbonates to a lesser extent.
- Deposition of “Cyclic salts” carried by wind and rain. “Cyclic salts” are ocean salts (mainly NaCl) which are carried inland by wind and deposited by rainfall.

Secondary/Human-induced salinity results from anthropogenic activities that change the water balance of soil between water applied (as irrigation/rainfall) and water used by crops in the form of transpiration (Garg and Manchanda 2008). The most common causes include:

- Land clearing and replacing perennial vegetation with annual crops.
- Irrigation schemes which use salt-rich irrigation water or have insufficient drainage.

According to the reports of FAO Land and Plant Nutrition Management Service, over 6% of world’s total agriculturable land area is affected by salinity. The gravity of salinity stress factor can be understood from the fact that around 19.5% of all the irrigated land and approximately 2% of dry land agriculture are affected by salinity (Ghassemi et al. 1995). Thus, secondary or human-induced salinity has proved to be a cause of major concern.

7.2 Effect of Salt Stress on Soil Component

Salt-affected soils occur almost under all climatic conditions. They are characterized by high concentration of soluble salt, together with low concentration of organic matter and nitrogen content. Ions which contribute towards soil salinity are Na^+, Cl^-, Ca^{++} and Mg^{++}.

Deposition of high salt concentration (Na^+ in particular) in soil changes its basic texture resulting in decreased soil porosity ultimately leading to reduced soil aeration and water conductance which ultimately adversely affects the growth rate of plant. High salt depositions in soil generate a low water potential zone in the soil. Consequently, it becomes increasingly difficult for the plant to acquire both water and nutrients (Porcel et al. 2011) from the soil. Based upon the nature, characteristics, and plant growth relationships in salt-affected soils, Szabolcs (1974) coined two main types of soils viz:

Saline soils – soluble salts are chiefly NaCl and Na_2SO_4 with appreciable quantity of $CaCl_2$, $CaSO_4$, $MgCl_2$, and $MgSO_4$. Such soils contain sufficient neutral soluble salts to pose negative effect on growth of most crop plants.

Sodic soils – soil-containing Na^+ salts capable of alkaline hydrolysis, mainly Na_2CO_3. This soil contains a huge amount of exchangeable sodium and low levels of soluble salt, and this excess exchangeable sodium exerts adverse effects on plant growth and soil structure which leads to the reduction of crop yields.

7.2.1 Effect of Salt Stress on Abiotic Component of Soil and Soil Health

Within a certain limit, soil salinity is beneficial for soil aeration, root penetration, and root growth causing fine soil particles to bind together into lumps. This process is called flocculation. Flocculation is important because water and oxygen move mostly in large pores between aggregates. Also, plant roots grow mainly between aggregates. But at high levels, salinity can have negative effects which prove to be lethal for plants. High concentrations of sodium in soil cause reduced infiltration, reduced hydraulic conductivity, surface crusting, soil dispersion, and aggregate swelling as described below:

- **Dispersion** – In the process of dispersion, soil saturated with sodium ions breaks down into fine particles. This is because large-sized sodium ions come in between the soil particles in quantitatively large numbers making the soil vulnerable to erosion. Na^+ particularly has such effects, while other salts such as Ca and Mg do not show these effects due to their smaller size and tendency to cluster closer to the clay particles. The process of soil dispersion hardens soil and blocks water infiltration, making it difficult for plants to establish and grow.
- **Infiltration** – Soil infiltration refers to the ability of soil to allow water to move into or through the soil. It allows the soil to temporarily store water, making it available for plants and soil organisms.

Both dispersion and infiltration are negatively correlated, for example, sodium-induced dispersion causes low infiltration.

- **Hydraulic Conductivity** – Hydraulic conductivity is the rate at which water flows through soil. Soil normally contains large number of pores, cracks, and fissures which allow rapid flow of water through the soil. When sodium-induced soil dispersion causes loss of soil structure, the hydraulic conductivity is also reduced. If water is unable to pass through the soil, soil's upper layer becomes swollen and water-logged. This leads to the development of anaerobic condition in soils which have low organic matter decomposition rates making them infertile, black alkali soils. Such soils reduce/prevent plant growth.

- **Surface crusting** – Surface crusting is the characteristic of sodium-affected soils. The primary causes of surface crusting are

 i) physical dispersion – caused by collision effect of raindrops or irrigation water, and
 ii) chemical dispersion – which depends on the ratio of salinity and sodicity of the applied water.

Surface crusting due to rainfall is highly amplified by clay dispersion induced by sodium. When clay particles disperse within soil water, they seal the macro pores in surface soil via two means. First, they block the path for movement of water and roots through soil. Second, they form hard cement like surface layer when the soil dries. This hardened upper layer/surface crust limits water infiltration and plant emergence.

- **Swelling** – Soils swell in volume when subjected to moisture. Such swollen soils typically contain clay minerals that attract and absorb water. Upon introduction of water to these soils, the water molecules are pulled into gaps between the soil plates. As more water is absorbed, the plates are forced further apart, resulting in an increase in soil pore pressure (Handy 1995). Such swelling reduces the cross-sectional area of soil pores. This process of swelling is more pronounced in soil water in the presence of high sodium or low salt concentrations, or both.

7.2.2 Effect of Salt Stress on Biotic Component of Soil and Soil Health

Apart from physical and chemical properties of soil, little attention has been paid to the microbiological aspects/biotic soil components of saline environments. Soil salinity is a threat to food production worldwide as it reduces plant growth and microbial functioning (Yan et al. 2015). Soil micro-organisms constitute <0.5 (w/w) of soil mass, but they play a key role in determining soil properties and processes. Soil microbes include bacteria, archaea, fungi, protozoa, and viruses. All these micro-organisms participate in biological processes like oxidation, nitrogen-fixation, ammonification, nitrification etc. leading to the degradation of soil organic matter and transformation of nutrients in soil. They also store carbon and other nutrients in their biomass which are mineralized after their death by living microbes present in the soil. Salinity leads to poor plant growth or low microbial activity in soil due to osmotic stress and toxic ions. Soluble salts in soil increase the osmotic potential of soil water which kills microbes as it draws water out of their cells and from plant roots through plasmolysis. Low osmotic potential also makes it more difficult for roots and microbes to absorb water from soil (Oren 1999). Thus, plants and microbes cope up with low osmotic potential by accumulation of osmolytes. This allows them to maintain their cell turgor and metabolism. However, synthesis of osmolytes requires large amount of energy, and it results in reduced activity and growth of microbes (Oren 1999; Wichren et al. 2006). Proline and glycine are the main organic osmolytes; among inorganic osmolytes, K^+ ions are most commonly accumulated by salinity tolerant microbes (Csonka 1989) to help them to overcome salinity stress. Differences in salinity tolerance amongst microbes result in changes in community structure compared to non-saline soils (Gros et al. 2003).

7.3 Soil Chemistry Affecting Factors in Agricultural Land

Soil chemistry is the branch of soil science that deals with the chemical composition, properties, and reactions of soils. The ingredients of soil are minerals, organic matter, living organisms, gas, and water. The factors affecting the soil chemistry and quality are chemical fertilizers, salinization, heavy metal accumulation, acidification, nutrient depletion, leaching, desertification etc.

Inorganic fertilizers contain phosphate, nitrate, ammonium, and potassium salts; they are considered as a source of heavy metals because they contain Hg, Cd, As, Pb, Cu, Ni, and natural radioactive elements like ^{238}U, ^{232}Th, and ^{210}Po (Sonmez et al. 2007). Thus, excessive use of fertilizers causes environmental health problems viz. soil salinity, accumulation of heavy metal in soil, eutrophication of water, and accumulation of nitrates. Plants absorb these fertilizers from the soil which ultimately enters the food chain. Thus, fertilization leads to water, soil, and air pollution. Application of ammonium-based fertilizers also causes *soil acidification* in which concentration of H^+ increases in soil. This contributes to soil acidification if nitrogen is leached rather than being taken up by plants. Leached ammonium nitrogen in soil is converted to nitrate and H^+ by bacteria already present as soil micro fauna. As a result, amount of H^+ increases in soil adding to acidification.

7.4 Soil Salinity Effect on Crop Plants

Soil salinity may inhibit plant growth for two reasons.

i) The presence of salt in the soil solution reduces the ability of plants to absorb water, and this leads to growth rate reduction. This is referred as osmotic or water deficit effect of salinity.
ii) Excessive amount of salt present in soil water enters in transpiration stream of plants thus causing injury to cells in transpiring leaves causing further reduction in growth. This is called salt-specific or ion-excess effect of salinity (Greenway and Munns 1980).

Both these effects exert certain repercussions on plants like, reduction in water potential causing ion imbalance, disturbance in ion homeostasis toxicity, and altered water status causing initial growth reduction and limitation of plant productivity. The detrimental effect is observed at the level of the whole plant as death of the plant or as decline in productivity. Salt stress affects all major processes and metabolisms like germination, growth, photosynthesis, photosynthetic pigment, water relation, nutrient imbalance, oxidative stress, and ultimately yield which have been described in detail (Figure 7.1).

7.4.1 Germination

Seed germination is defined as the rupture of seed coat and the emergence of radical through it (Copeland and Mcdonald 1995). It is one of the most fundamental and vital phases in the growth cycle of a plant which ultimately determines the yield. It is well established that salinity adversely affects the process of germination. Investigations regarding seed germination have proved that seeds of most plants show maximum

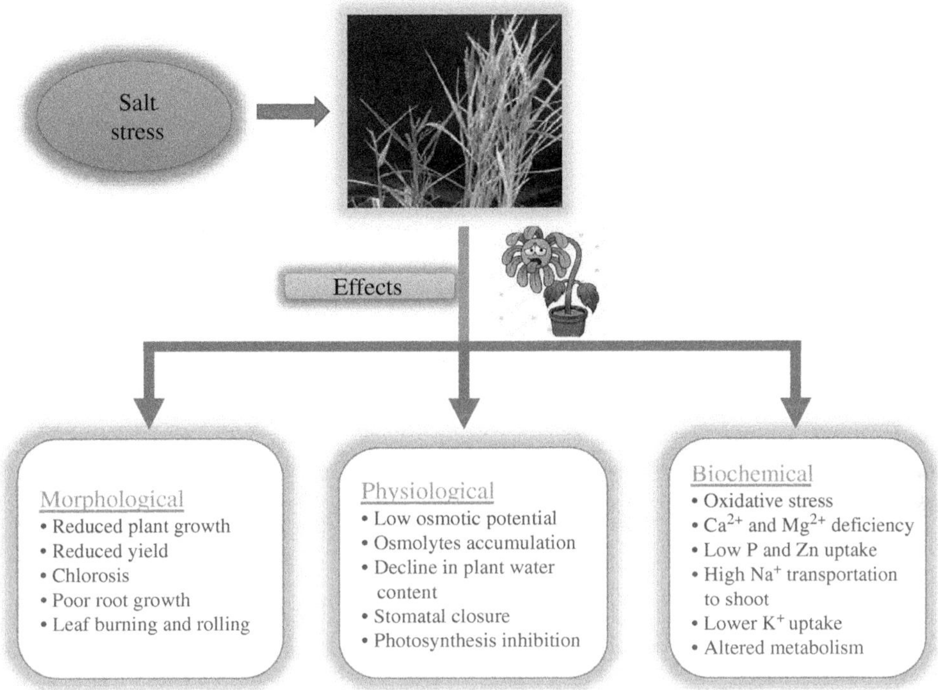

Figure 7.1 Salinity stress induced major responses in plants.

germination in conditions of distilled water. Contrarily, seeds are very sensitive to high salinity conditions during their germination and seedling phase of development (Gulzar et al. 2003). Seeds of most halophytes attain maximum germination in the absence of NaCl. They are very sensitive to high salinity conditions during germination and early establishment phases (Khan et al. 2000). Othman et al. (2006) reported that seed germination can be initiated by water imbibition, and shortage in water supply at this stage leads to stress conditions for the seed. It is reported that salt stress negatively affected seed germination, either osmotically through reduced water absorption or ionically through the accumulation of Na^+ and Cl^- ions causing imbalance in nutrient uptake and exerting toxic effect.

Lauchli and Grattan (2007) proposed a generalized relationship between percent germination and time after adding different concentrations of saline water (Figure 7.2). Kaveh et al. (2011) found a significant negative correlation between salinity and germination percentage manifested as delayed germination and reduced germination percentage in plants.

7.4.2 Growth

Salt stress initially reduces the growth rate either by osmotic/water deficit effect or by salt-specific/ion-excess effect. Osmotic effect reduces the plant's ability to take up water, while the ion-excess effect injures the cells in transpiring leaves further reducing growth. These effects give rise to a two-phase growth response to salinity as shown in Figure 7.3 given by Munns (2005).

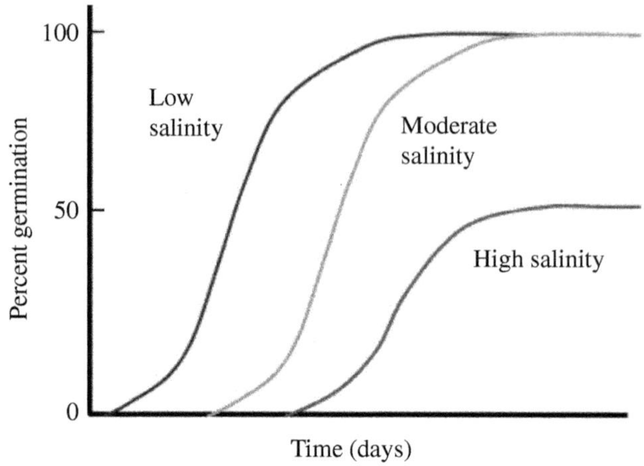

Figure 7.2 Relationship between rate of germination and time duration after sowing at different salinity levels. *Source:* Lauchli and Grattan (2007).

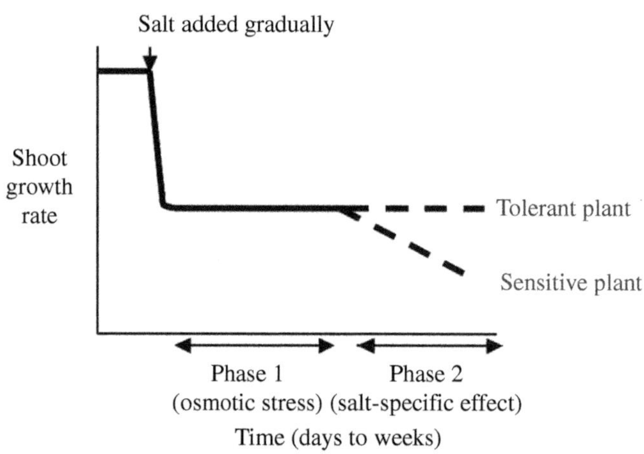

Figure 7.3 Overview of the two-phase growth response to salt stress for plants differing in salt sensitivity. *Source:* Munns (2005).

Phase 1: The first phase of growth reduction occurs soon after exposure to salt stress. **This phase results from the effect of salt present in soil water outside the plant**. The changes due to salt outside the root cause changes in cell water relations or osmotic effect. The osmotic effect initially reduces the water absorbance capability of the plant. The cellular and metabolic processes involved in these are quite similar to those of drought-affected plants. Later on, after the initial decrease in leaf growth, there is a gradual recovery of growth rate until a new steady state is reached, dependent upon the salt concentration outside the root (Munns 2005).

Phase 2: The second phase results from the toxic effect of salt present inside the plants. It is a much slower effect, taking days or months. Primarily, salt is taken up by the older leaves where Na^+ and Cl^- build up in transpiring leaves gradually over time resulting in salt toxicity and death of leaves. Thus, the total photosynthetic leaf area reduces; as a result, the supply of total photosynthate to the plant reduces, affecting the overall carbon balance required for sustained growth (Munns 2005). Exceedingly high salt concentrations in leaf (beyond the capacity of its compartmentation in vacuoles) accumulate toxic levels of salt in the cytoplasm causing injury and death of leaf (Munns 2005; Munns et al. 2006). The leaves die and thus reduce the total photosynthetic leaf area. The rate at which the leaves die determines the survival of the plant. If new leaves are produced at a rate greater than the rate of death of older leaves, the plant survives. However, if old leaves die faster than the rate of development of new leaves, the plant is unable to supply sufficient photosynthate to the reproductive organs to produce viable seeds. Based on this two-phase concept, the initial reduction in growth is caused by the osmotic effect of salts in the medium outside the roots in both salt-sensitive and salt-tolerant plants. In contrast, in the second phase, a salt-sensitive species or genotype differs from a more salt-tolerant species by its inability to prevent salt from accumulating in transpiring leaves to toxic levels (Munns et al. 2006).

7.4.3 Photosynthesis and Photosynthetic Pigments

Cell growth and photosynthesis are among the primary processes which are affected by salinity (Munns et al. 2006). The effects can be either direct or indirect. Direct effects include decreased CO_2 availability due to its limited diffusion through the stomata and the mesophyll (Flexas et al. 2004), while indirect effects arise due to altered photosynthetic metabolism (Lawlor and Cornic 2002). Secondary effects such as oxidative stress also arise due to salinity stress (described in detail below). Oxidative stress is a complex chemical and physiological phenomenon that goes hand in hand with virtually all biotic and abiotic stresses in higher plants and develops because of overproduction and/or accumulation of ROS. This stress arises because of single or multiple stress conditions (Chaves and Oliveria 2004) under both biotic and abiotic stressors. It affects the leaf photosynthetic machinery severely (Ort 2001). Munns and Tester (2008) reported reduction in photosynthetic rates due to inhibited leaf development/expansion and early leaf abscission during Phase I of growth reduction. Under conditions of prolonged salt stress, ion toxicity, membrane disruption, and complete stomatal closure are the primary causes responsible for photosynthetic inhibition. Munns and Tester (2008) identified that curtailment of stomatal aperture was the most perceptible and easily quantifiable whole-plant response to salinity stress. They also concluded that the salt present outside the roots induces stomatal responses due to the osmotic effect. Chlorophyll is an important ingredient in the process of photosynthesis. It directly correlates with the productivity of the plant. Inhibition of photosynthesis takes place when high concentration of Na^+ and Cl^- is accumulated in chloroplasts (Zhang et al. 2005). High salt stress conditions lead to decrease in chlorophyll content. Saha et al. (2010) observed a linear decline in total chlorophyll, chlorophyll *a*, chlorophyll *b*, carotenoids, and xanthophylls content in mung bean. Photosystem II (PS II) is a relatively sensitive component of the photosynthetic system with respect to salt stress (Allakhverdiev et al. 2000). A considerable decline in PSII efficiency, electron transport chain activity, and assimilation rate of CO_2 has been observed under the influence of salinity (Piotr and Grazyna 2005; Mittal et al. 2012). It is reported that salinity stress affects growth by altering chlorophyll florescence and function of oxygen-evolving complex.

To summarize, stomatal as well as non-stomatal limitations are associated with salt stress-induced reductions in photosynthesis. Stomatal limitations cause decrease in intercellular CO_2, whereas major non-stomatal constraints include salinity-induced oxidative damages causing reduction in canopy size, photosynthetic pigments (Saha et al. 2010), and activity of certain photosynthetic enzymes.

7.4.4 Mineral Uptake and Assimilation

Soil is the major source of nutrients both macro and trace nutrients, absorbed by plants for growth starting from their emergence ensuing through their development, productivity, and metabolism reaching up to the promotion and protection of plants. The bioavailability of these soil nutrients is controlled by various factors like soil pH, salinity, nutrient biogeochemical cycles, and environmental and climatic changes. Soil salinization induces nutrient disorders in plants. These disorders result from effects of salinity on availability, absorption, and transport of nutrients within the plant (Munns and Tester 2008). Salt stress causes deficiencies or imbalances in plant nutrients, due to competition of Na^+ and Cl^- with various other plant nutrients like Ca^{2+}, K^+, and NO_3^- (Forni et al. 2017). High concentration of NaCl has been reported to induce increase in Na^+ and Cl^- and decrease in N, P, K, and Mg level in many studied crop plants (Naeem et al. 2017). Reports indicate toward the reduced accumulation of nitrogen in plants due to salinity (Rozeff 1995). Under saline conditions, decreased nitrogen uptake occurs because of interactions between Na^+ and NH_4^+ or between Cl^- and NO_3^- that ultimately reduces the growth and crop's yield (Rozeff 1995). This reduced uptake of NO_3^- is associated with Cl^- antagonism (Bar et al. 1997).

For many crops, Na^+ is the most toxic ion (Sümer et al. 2004), as it interferes with K^+ uptake and transport resulting in disturbance in stomatal conductance and consequently leading to water loss and necrosis. Competition between K^+ and Na^+ under salinity stress greatly decreases K^+ content in both leaves and roots of crops (Kaya et al. 2010). Salt stress also hinders the translocation of K^+ ions from roots to shoots (Shahzad et al. 2012). Net absorption of K^+ is dependent on its concentration in the root medium and on root K^+ status. Inhibition in K^+ translocation is usually higher at low K^+ contents under salt stress (Botella et al. 1997).

Along with above nutrients, the availability of phosphorous is also reduced in saline soils because:

a) ionic strength effects which decrease the activity of PO_4^{3-},
b) concentration of phosphate in soil solution is rigorously controlled by sorption process, and
c) low solubility of Ca and P minerals.

Hence, due to the above factors phosphate concentration in crops also decreases as salinity increases (Qadir and Schubert 2002).

Besides potassium, Ca^{2+} uptake and translocation are also severely inhibited under salt stress. Particularly, reports suggest that increased accumulation of sodium leads to reduced calcium content in crops' tissues (Hu et al. 2007; Shahzad et al. 2012). During the first phase of salt stress, calcium transport to the tissue of youngest leaves is impaired. In saline environments, lesser calcium reaches the expanding shoot tissues which leads to reduced leaf expansion (Hu et al. 2007). The high values for Na/K, Na/Ca, and Na/Mg ratios in the total plant, apoplast, and symplast of expanding tissues in crops confirmed that impaired transport of potassium, calcium, and magnesium by sodium might upset plant metabolism leading to reduced growth under saline conditions (Shahzad et al. 2012). According to Hussain et al. (2014), $Ca^{2+}$and Mg^{2+} concentrations of all plant organs transiently declined in response to external high NaCl concentration in saline environment. In conclusion, increasing concentration of NaCl levels in rhizosphere reduces the nutrient assimilation of K and Ca resulting in ion imbalance of K, Ca, and Mg (Keutgen and Pawelzik 2009).

7.4.5 Oxidative Stress

Salt stress leads to oxidative stress through the increase in ROS such as superoxide ($O_2^{\cdot-}$), hydrogen peroxide (H_2O_2), and hydroxyl radicals ($OH^{\cdot}$) which are highly reactive and causative agents for damage to lipid membranes, essential molecules, including nucleic acids (RNA and DNA), pigments, fatty acids, and proteins that subsequently result in cell damage (Kumar et al. 2009; Khare et al. 2015). One of the chief sites for the production of ROS is the chloroplasts because in existing stress condition, there is enough formation of superoxide radicals and singlet oxygen during the unbalanced redox status of the electron transport machinery (Zimmermann and Zentgraf 2005). To overcome this stress, plants possess a complex antioxidant system, composed of low molecular mass non-enzymatic antioxidants (glutathione, carotenoids, and ascorbate) and ROS-scavenging enzymatic antioxidants such as superoxide dismutase, catalase, ascorbic peroxidase, guaiacol peroxidase, and glutathione reductase. These antioxidant enzymes either directly react with ROS, keeping their concentration low or generate oxidized forms of antioxidants to maintain the redox balance.

Since, non-enzymatic antioxidants like ascorbic acid, phenolic compounds, reduced glutathiones, flavonoids, and tocopherols can quench all types of ROS (Kumar et al. 2009; Khare et al. 2015); hence, all these mechanisms can resist the negative effect of ROS and are collectively considered as oxidative stress-induced plant defenses.

7.4.6 Yield

The ultimate effect of salt stress on crops leads to yield reduction in general which is the most countable effect in agriculture. Grain weight and number of grains are important parameters for evaluating final grain yield of crops (Farooq et al. 2014). Salinity stress in crops during the reproductive phase decreases grain weight and number resulting in substantial reduction in grain yield (Abdullah et al. 2001; Schubert et al. 2009; Kaya et al. 2013).

7.5 An Introduction to Soil, Plant-Atmosphere Continuum (SPAC)

Water is the link for soil, plant, and atmosphere continuum. Soil, plant-atmosphere continuum (SPAC) is defined as the movement of water from soil through the plants to the atmosphere along an interconnected film of water (Lambers et al. 2008) as depicted in Figure 7.4.

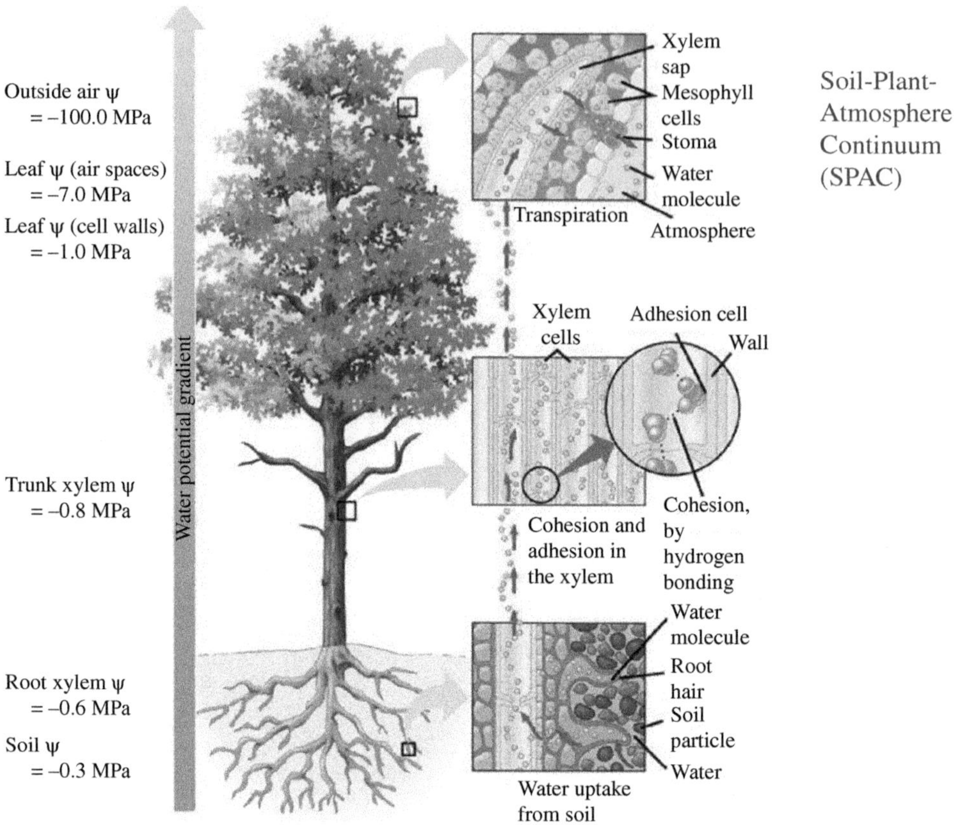

Figure 7.4 Diagrammatic representation of the SPAC (soil, plant-atmosphere continuum). *Source:* Soo-Hyung (2011).

Movement of water occurs from higher to lower water potential due to the formation of an energy/diffusion gradient. The diffusion gradient across the stomatal pores of leaves is responsible for moving water out of leaves to the external environment (in the form of vapor). This process is driven by low water potential of external environment as compared to that of leaves (as shown in the above figure). As water transpires out from the leaves (in the form of vapor) through the stomata, more water molecules evaporate off from the surface of mesophyll cells to replace the lost water molecules. This is because water in the air inside the leaves is maintained at saturated vapor pressure. Water lost at the surface of cells as water vapor is replaced by water from the xylem. The water present in the xylem of plants (due to its cohesion-tension properties) generates a drawing force to pull additional water molecules through the xylem from the roots towards the leaf.

7.6 Salt Absorption by Root Tissues and Their Effect on Plant-Atmosphere Continuum

Mineral salts present in the soil are absorbed in the form of ions chiefly through the meristematic regions of root. Plasma membrane of the root cells does not allow equal absorption of all the ions because of its semi permeable nature. It is selectively permeable. This means that all the ions of the same salt are not absorbed at an equal rate. The movement of minerals by the roots and their transfer to the aerial parts involves two main steps:

i) absorption of ions from soil by epidermal cells, cortical, and endodermal cells and
ii) their secretion inside the vessels at the level of xylem parenchyma cells.

Plant–water relations help to explain the behavior of plants in terms of how they control the hydration of their cells (Passioura 2010). Plants generally survive in two different environments, the soil and the atmosphere, and the conditions in these two media do not necessarily always behave in coordination. The evaporative demand of the atmosphere is continuous, whereas the water supply by rain or irrigation can be sufficient/insufficient. Most of the water absorbed by the roots of the plants from the soil is lost to the atmosphere by transpiration. In such a situation, a hydraulic gradient is established in the xylem, causing water flow from the roots to the leaves. This water movement is proportional to a driving force and is inversely proportional to the resistance encountered through the soil–plant system. Transpiration creates the internal conditions under which water is absorbed by the root system and transported to the canopy (Chavarria and Pessoa dos Santos 2012). As water is lost by transpiration, leaf water potential and the pressure in the xylem are reduced. Water transpiration from leaves is primarily controlled by stomata, but sustained transpiration, without sufficient replenishment, leads to progressive dehydration (loss of leaf turgidity). The factors on which degree of dehydration of plants depends are the climate and environment, the characteristics of the canopy, the growth stage of the plant, and water quality.

Transpiration is the most effective means of leaf cooling in those plants which get sufficient water supply (Passioura 2010). Under water deficit condition, the stomata are closed and transpiration occurs only through the cuticle and the plant canopy. As a result, temperature tends to rise, which causes an increase in stomatal resistance to transpiration and an increase in the rate of tissue respiration. These conditions lead to limited photosynthetic capacity, and the plant uses its own photo-assimilates, resulting in decreased growth.

Across the soil–plant–atmosphere continuum, water flow is preferentially helped by the gradient formed between the soil water potential and leaf water potential. The main factors which control the movement of water through the soil–plant system are hydraulic conductivity of the whole plant and individual conductance components like plant transpiration and other physiological processes related to transpiration (Stiller et al. 2003; Tyree 2003). Differences in root hydraulic conductance cause differences in water transport from soil to plant, affecting the leaf water status and thus the growth and other physiological responses of the plant (Lo Gullo et al. 2007). Plants under salt stress conditions exhibit dehydration, as indicated by lower water potential, due to greater difficulty in taking up water from the roots or less availability of water to the roots (Álvarez et al. 2012). Most negative values for leaf and stem water potential have been found in plants exposed to both water and salinity stress because in both cases passive dehydration and salt accumulation contribute to the lowering of leaf water potential (Slama et al. 2008). Salinity disturbs plants' water relations due to decreased availability of water from the soil solution which causes lowering of osmotic potential and is activated by the toxic effects of the sodium and chloride ions (Munns 2005). Accordingly, an increase in resistance to water flow from soil to plant under salinity has been observed in many species (Navarro et al. 2007; Álvarez et al. 2012). For example, root hydraulic resistance has been found to increase in salt-stressed plants such as *Euonymus japonica*, *Phlomis purpurea*, and *Rosmarinus officinalis*, in which the lowest leaf water potential values have been recorded (Alarcón et al. 2006; Álvarez et al. 2012; Gómez-Bellot et al. 2013). Salinity also causes a decrease in stomatal conductance and evaporation, and, as a result lower rate of transpiration.

It is reported that accumulation of salts in the root zone causes a decline in osmotic potential and subsequently lesser water is available to plants (Sánchez-Blanco et al. 2004; Franco et al. 2011). The reduced water potential induces stomatal closure, which causes decline in rate of transpiration. Under salt stress, plant balances its water potential by losing water, which induces a decrease in osmotic potential and turgor. This creates chemical signals that triggers adaptive responses (Hasegawa et al. 2000; Abdul-Jaleel et al. 2007). Kaldenhoff et al. (2008) proposed that during the process of acclimation, the number of water channels/aquaporins is reduced. This consequently causes a reduction in the hydraulic conductivity of the membranes. Simultaneously, an accumulation of organic and inorganic solutes occurs which reduces cellular osmotic potential (Munns 2002).

7.7 Translocation of Salt Ions in the Vascular System of Crop Plants

Like most multicellular organisms and animals, plants consist of certain tube-like structures called xylem and phloem that enable the mass flow of solvents and solutes like ions, sugars, hormones etc. Under salt stress, salt accumulation in soil decreases the osmotic potential and water potential of soil solution, diminishing the total water availability in root zone for plant uptake. This disturbs the thermodynamic balance for uptake of water by roots and their availability for plant cells (Roy et al. 2014). The increased concentration of Na^+ and Cl^- in plants induces specific toxicities as well as alterations in water relations.

Absorption of salts from soil solution occurs in two ways:

- **Passive absorption** occurs without expense of metabolic energy, along with the concentration gradient that is when the concentration of mineral salts is higher in the outer solution than in the cell sap of root cells.
- **Active absorption** occurs against the concentration gradient at the cost of metabolic energy expenditure through respiration.

Under saline conditions, Na^+ is transported into plants through the pathways that are generally functioning for K^+ influx due to having similar ionic radii of both, thus making discrimination between these two ions difficult. As a result, plants suffer from Na^+ toxicity and deficiency of K^+. The toxic levels of Na^+ present in the cytoplasm must be lowered to maintain a low cytoplasmic Na^+ concentration and a high K^+/Na^+ ratio by mechanisms that function to:

- reduce Na^+ influx into root cells,
- Na^+ compartmentalization into vacuoles, and
- increase Na^+ efflux from root cells through SOS pathway (Pardo and Rubio 2011).

These processes of Na^+ detoxification and cellular osmotic adjustment are important for plants to tolerate salt stress (Blumwald et al. 2000).

7.7.1 Mechanism of Sodium Influx Into Cytosol

Under normal physiological conditions, plants generally maintain a high cytosolic K^+/Na^+ ratio and a negative electrical membrane potential difference across the plasma membrane. Contrarily, under salinity conditions, increased Na^+ concentration in the soil establishes an electrochemical gradient favoring the passive transport of Na^+ from the soil into the cytosol (Blumwald et al. 2000). Different channels and transporters help in this process of Na^+ influx into the roots. The channels which are involved in Na^+ influx are non-selective cation channels (NSCC) which cannot differentiate between Na^+ and K^+ (having same radii), including the cyclic nucleotide-gated channels (CNGCs) and glutamate receptors (GLRs). The transporters involved in Na^+ influx are high-affinity potassium transporters (HKTs and HAKs) (Horie et al. 2009). Through the roots, Na^+ is transported into the xylem by other channels and transporters and then delivered to the shoot. Voltage-insensitive NSCCs (VI-NSCCs) are an important route for influx of Na^+ into the roots and hence have been studied regarding salt stress tolerance. There are two subgroups of VI-NSCCs categorized based on the blocking effect of cations, such as Ca^{2+}, Zn^{2+}, Mg^{2+}, and Ba^{2+}. The cations can freely pass through the channels of one group, whereas the other group is partly blocked by these cations. The permeability of Na^+ through VI-NSCCs has been demonstrated in epidermal cells of *Pisum sativum* leaves (Elzenga and Volkenburgh 1994), protoplasts of rye roots, protoplasts of wheat root cortex (Tyerman et al. 1997), and root cells of *Arabidopsis* (Maathuis and Sanders 2001). The studies showed that Na^+ influx into intact tissues via NSCCS is partially inhibited by Ca^{2+} and it is sensitive to its blockers such as quinine. Additionally, two subclasses of NSCCs have been proposed to be involved in Na^+ influx, namely, CNGCs and GLRs.

7.7.2 Mechanism of Na^+ Compartmentalization in Vacuoles

Compartmentalization of Na^+ in plants is necessary to reduce the toxic concentration of Na^+ in the cytosol. Vacuolar Na^+/H^+ exchangers (NHXs) are important regulators of the above process. *Arabidopsis* NHX1 (AtNHX1), the first NHX to be identified in plants, is localized in tonoplast (Apse et al. 1999), and constitutive overexpression of *AtNHX1* in *Arabidopsis* improves salt stress tolerance (Rodríguez-Rosales et al. 2009).

In *Arabidopsis*, NHX1 and NHX2 are the most abundant NHXs and reported to be involved not only in Na^+ compartmentalization but also in uptake of K^+ into the vacuoles and regulation of vacuole pH (Yokoi et al. 2002). Induction of NHX1 and NHX2 in response to salt stress is ABA-dependent (Shi and Zhu 2002). NHX2 and NHX5 have been reported to be the determinants of salt stress tolerance, but their direct role in Na^+ sequestration is not clear till date (Li et al. 2009). It is widely accepted that, during salt stress, the activity of NHXs increases, which promotes salt stress tolerance in many plants (Silva and Geros 2009).

7.7.3 The SOS Pathway

Under salt stress, this pathway plays an important role in regulating ionic (Na^+/K^+) homeostasis (Figure 7.5). In this, the functions of three genes have been studied by analyzing *Arabidopsis* salt overly sensitive (*sos*) mutants which includes:

a) SOS1, a plasma membrane-localized Na^+/H^+ antiporter that mediates Na^+ efflux from the roots and loading of Na^+ ions in the xylem (Munns 2002; Olìas et al. 2009) SOS2, a Ser/Thr protein kinase
b) SOS3, calcium-binding protein, which is a myristoylated, EF-hand-containing protein. The EF hand is a helix-loop-helix structural domain/motif found in a large family of calcium-binding proteins. They bind calcium ion with high affinity and selectivity, and

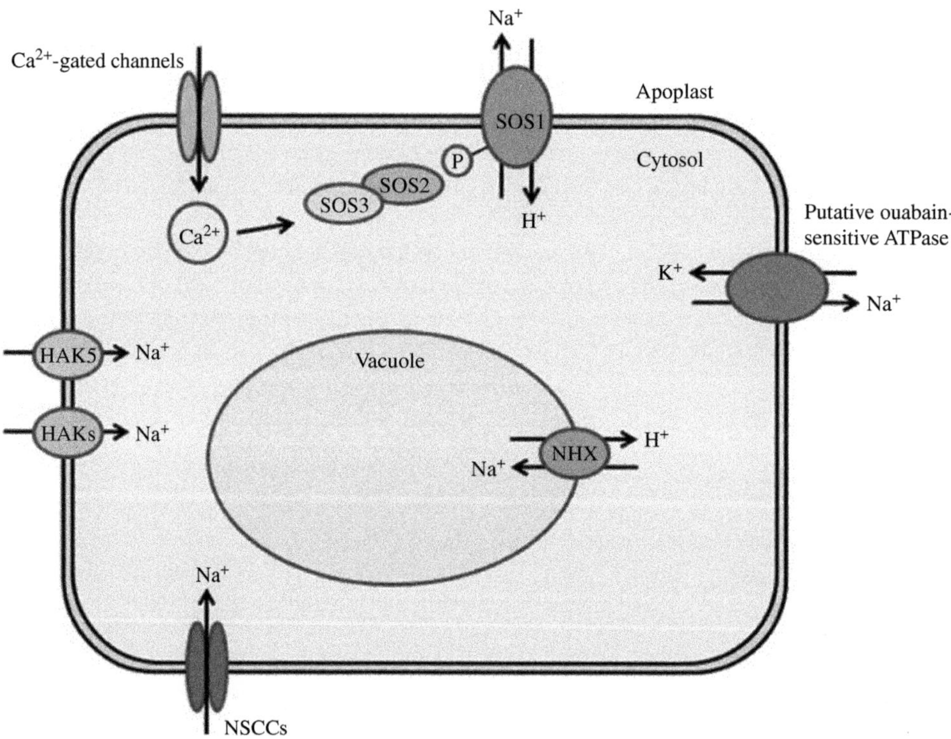

Figure 7.5 Pictorial representation of Na^+ entering the cell via channels such as NSCCs and transporters such as HAK5 and HKTs. Na^+ is sensed by an unidentified sensory mechanism, and Ca^{2+} signaling cascade is activated. The salt stress-induced increase in cytosolic concentration of Ca^{2+} is sensed by SOS3 (a myristoylated Ca^{2+} sensor protein). SOS3 interacts with SOS2 and activates its kinase activity. The SOS2–SOS3 complex becomes localized to the plasma membrane. SOS2 then phosphorylates SOS1 and activates its NA^+/H^+ antiporter activity facilitating Na^+ efflux from the cell. Na^+ is also compartmentalized inside vacuoles by NHX (a Na^+/H^+ exchanger) in response to salt stress. (Abbreviations used: NSCCs, non-selective cation channels; HAK5, high-affinity potassium transporter 5; HKTs, high-affinity potassium transporters; SOS, salt overly sensitive; NHX, Na^+/H^+ exchanger). *Source:* Keisham et al. (2018).

this property enables EF hand proteins to sense changes in intracellular calcium. Ca^{2+} which acts as a secondary messenger mediates plant responses to hormones, developmental cues, and external stimuli (Zielinski 1998). Cytoplasmic Ca^{2+} concentration is maintained in the nanomolar range (approximately 100–200 nM), whereas the concentration in organelles and cell wall is in the millimolar range (Zielinski 1998; White 2000). Various signals (hormonal, abiotic, and biotic) have been shown to cause transient elevation of cytoplasmic Ca^{2+} concentration (White 2000; Reddy and Reddy 2001) which is sensed by Ca^{2+}-binding proteins (Zielinski 1998; Snedden and Fromm 1998). The conformation of the Ca^{2+}-binding protein changes on binding Ca^{2+}, resulting in modulation of its activity or its ability to interact with other proteins or nucleic acids and modulate their function or activity.

The SOS pathway invokes a salt stress-generated Ca^{2+} signal that is perceived by SOS3, enabling the formation of the SOS3–SOS2 complex, which phosphorylates and activates the transport activity of SOS1. The activated SOS3–SOS2 complex also stimulates, among other genes, the transcriptional activation of SOS1 and stabilizes cellular levels of SOS1 mRNA (Ji et al. 2013).

The Ca^{2+}-activated SOS pathway affects K^+ homeostasis. sos1, sos2, and sos3 mutants are hypersensitive to K^+ deficiency. However, activated SOS1 does not transport K^+ (Quintero et al. 2002). K^+ uptake in *sos* mutants is inhibited by even mild NaCl stress. The activity of *Arabidiopsis* K Transporter (AKT1), a major highly selective potassium channel mediating potassium uptake under low K^+ conditions, is inhibited by high cytoplasmic NaCl concentration (Pyo et al. 2010), suggesting that the SOS pathway is affected by the activity of AKT1. Some mutations and deletions of the Na^+-transporter HKT1 suppress the salt-hypersensitive phenotype of *sos* mutants by inhibiting the accumulation of Na^+. Hypersensitivity to K^+ deficiency of the *sos* mutants was suppressed by the expression of mutated HKT1 (Rus et al. 2004). The SOS pathway is also suggested to mediate salt tolerance by modulating the activity of vacuolar H^+-ATPase (V-ATPase), which is both tonoplast and endosome localized. SOS2 binds to regulatory subunits of VATPase and stimulates its activity (Batelli et al. 2007). An endosomal V-ATPase mediates salt tolerance (Krebs et al. 2010). Finally, the SOS pathway controls endocytosis, vacuolar shape, and function, and intracellular pH (Oh et al. 2010). The given figure shows mechanism of Na^+ influx, Na^+ compartmentalization in vacuoles, and Na^+ sensing in response to salt stress.

7.7.4 Effect of Salt Stress on Xylem Transport

High salinity in soil causes osmotic stress which reduces the capacity of root systems to absorb water and together with ion toxicity causes accumulation of toxic concentrations of Na^+ and Cl^- in the cells. Under osmotic conditions created due to water uptake reduction by plants, there are gradual changes in plant internal structure which reduces the xylem and the phloem area (Hameed et al. 2010). Cortex area of root and stem is also significantly decreased. High salinity causes cell injury. These cell injuries reduced the growth and area of cortex tissue in roots and stem. These results are supported by the findings of Casenave et al. (1999), in which they observed smaller cortex in cotton seedlings at high salinity level.

On the other hand, Walsh (1990) suggested that the thickness of cortical area increases under salinity stress only up to a certain level; subsequently, it decreases under severe saline conditions. The reduction in cortical cell area at higher salinity levels has also been reported by Akram et al. (2002), but increased cortical area under salt stress in salt tolerant species may be critical under physiological drought for its better storage capacity (Baloch et al. 1998).

7.7.5 Effect of Salt Stress on Phloem Loading

Under salinity conditions, sodium contained in xylem would be absorbed and stored into leaf cells during its movement, and therefore, there would also be transportation of sodium from xylem to phloem in the leaves; due to this, the concentration of sodium increases in phloem sap (Shabala 2013). Rate of phloem loading is also strongly influenced by the frequency of plasmodesmata between cells. The fact is that higher the number of plasmodesmata at a given interface, higher is the potential of symplastic transfer across the interface.

7.8 Current Status and Conclusion

Salinity stress severely affects crop productivity due to its impact on photosynthesis, respiration, ionic imbalance, hormonal imbalance, nutrient assimilation, etc. Production of ROS is an indirect adverse effect in salinity-stressed plants which subsequently causes damage to macromolecules like lipids, proteins, and nucleic acids which ultimately limit crop productivity. Therefore, to engineer more salt tolerant plants, it is necessary to pick up the key components of plant tolerance network. However, the development of salt-tolerant crops has been delayed too much. Plants have evolved remarkable mechanisms to control K^+ and Na^+ tissue and cellular homeostasis under salinity. Many of these mechanisms depend on H^+, K^+, and Na^+ transporters. In the past years, various molecular studies established involvement of SOS1, HKTs, and NHXs transporters in K^+ and Na^+ homeostasis. Recently, genomics, transcriptomics, proteomics, and metabolomics have been successfully applied to yield exciting outcomes to investigate different components of salt tolerance. These components include various genes, transcription factors, proteins, and metabolites like osmolytes, phytohormones, lipids, etc. which may be used to modify plants for their increased salt tolerance. This multidisciplinary approach will be helpful in increasing crop productivity to satisfy the increasing food demand for the rapidly increasing population.

Acknowledgements

Gunjan Goyal and Aruna Yadav gratefully acknowledge the financial support they have received from CSIR as JRF-Fellowship. Authors also acknowledge UGC-DRS-II, and RUSA I, Department of Botany, University of Rajasthan, Jaipur, for providing necessary facilities.

References

Abdul-Jaleel, C., Gopi, R., Manivannan, P., and Panneerselvam, R. (2007). Responses of antioxidant defense system of *Catharanthus roseus* (L.) G. Don. to paclobutrazol treatment under salinity. *Acta Physiol. Plant.* 29: 205–209.

Abdullah, Z., Khan, M.A., and Flowers, T.J. (2001). Causes of sterility in seed set of rice under salinity stress. *J. Agron. Crop Sci.* 167: 25–32.

Akram, M., Akhtar, S., Javed, I.H. et al. (2002). Anatomical attributes of different wheat (*Triticum aestivum*) accessions/varieties to NaCl salinity. *Int. J. Agric. Biol.* 4: 166–168.

Alarcón, J.J., Morales, M.A., Ferrández, T., and Sánchez-Blanco, M.J. (2006). Effects of water and salt stresses on growth, water relations and gas exchange in *Rosmarinus officinalis*. *J. Hortic. Sci. Biotechnol.* 81: 845–853.

Allakhverdiev, S.I., Sakamoto, A., Nishiyama, Y. et al. (2000). Ionic and osmotic effects of NaCl-induced inactivation of photosystem I and II in *Synechococcus sp. Plant Physiol.* 123: 1047–1056.

Álvarez, S., Gómez-Bellot, M. J, Castillo, M., Bañón, S., and Sánchez-Blanco, M. J. (2012). Osmotic and saline effect on growth, water relations, and ion uptake and translocation in *Phlomis purpurea* plants. *Environ. Exp. Bot.* 78:138–145.

Apse, M., Aharon, G.S., Snedden, W.A., and Blumwald, E. (1999). Salt tolerance conferred by over expression of vacuolar Na^+/H^+ antiporter in *Arabidopsis*. *Science* 285: 1256–1258.

Baloch, A.H., Gates, P.J., and Baloch, V. (1998). Anatomical changes brought about by salinity in stem, leaf and root of *Arabidopsis thaliana* (L.) Heynh (Thale cress). *Sarhad J. Agri.* 14: 131–142.

Bar, Y., Apelbaum, A., Kafkafi, U., and Goren, R. (1997). Relatioship between chloride and nitrate and its effect on growth and mineral composition of avocado and citrus plants. *J. Plant Nutr.* 20: 715–731.

Batelli, G., Verslues, P.E., Agius, F. et al. (2007). *SOS2* promotes salt tolerance in part by interacting with the vacuolar H^+-ATPase and up regulating its transport activity. *Mol. Cell. Biol.* 27: 7781–7790.

Blumwald, E., Aharon, G.S., and Apse, M.P. (2000). Sodium transport in plant cells. *Biochem. Biophys. Acta* 1465 (1-2): 140–151.

Botella, M.A., Martinez, V., Pardines, J., and Cerda, A. (1997). Salinity induced potassium deficiency in maize plants. *J. Plant Physiol.* 150 (1–2): 200–205.

Casenave, E.C., Degano, C.A.M., Toselli, M.E., and Catan, E.A. (1999). Statistical studies on anatomical modifications in the radicle and hypocotyl of cotton induced by NaCl. *Biol. Res.* 32: 1–8.

Chavarria, G. and Pessoa dos Santos, H. (2012). Plant water relations: absorption, transport and control mechanisms. In: *Advances in Selected Plant Physiology Aspects*, vol. 5 (ed. G. Montanaro), 105–132. InTech.

Chaves, M.M., Flexas, J., and Pinheiro, C. (2009). Photosynthesis under drought and salt stress: regulation mechanisms from whole plant to cell. *Ann. Bot.* 103: 551–560.

Chaves, M.M. and Oliveria, M.M. (2004). Mechanisms underlying plant resilience to water deficit prospects for water saving agriculture. *J. Exp. Bot.* 55 (407): 2365–2384.

Copeland, L.O. and McDonald, M.B. (1995). *Principles of Seed Science and Technology*, 3e, 409. USA: Chapman and Hall.

Csonka, L.N. (1989). Physiological and genetic responses of bacteria to osmotic stress. *Microbiol. Rev.* 53: 121–147.

Elzenga, J.T.M. and Volkenburgh, E.V. (1994). Characterization of ion channels in the plasma membrane of epidermal cells of expanding pea (*Pisum sativum* arg) leaves. *J. Membr. Biol.* 137: 227–235.

FAO (Food and Agriculture Organization) (2012). *High Level Expert Forum – How to Feed the World in 2050*. Rome: Economic and Social Development Department. Food and Agricultural Organization of the United Nations.

Farooq, M., Hussain, M., and Siddique, K.H.M. (2014). Drought stress in wheat during flowering and grain-filling periods. *Crit. Rev. Plant Sci.* 33: 331–349.

Flexas, J., Bota, J., Loreto, F. et al. (2004). Diffusive and metabolic limitations to photosynthesis under drought and salinity in C3 plants. *Plant Biol.* 6: 269–279.

Forni, C., Daiana, D., and Bernard, R.G. (2017). Mechanisms of plant response to salt and drought stress and their alteration by rhizobacteria. *Plant Soil* 410: 335–356.

Franco, J.A., Bañón, S., Vicente, M.J. et al. (2011). Root development in horticultural plants grown under abiotic stress conditions – A review. *J. Hortic. Sci. Biotechnol.* 86: 543–556.

Garg, N. and Manchanda, G. (2008). Salinity and its effect on the functional biology of legumes. *Acta Physiol. Plant.* 30: 595–618.

Ghassemi, F., Jakeman, A.J., and Nix, H.A. (1995). *Salinisation of Land and Water Resources: Human Causes, Extent, Management and Case Studies*. Sydney, Australia, and Wallingford, Oxon, UK: UNSW Press, CAB International.

Gómez-Bellot, M.J., Álvarez, S., Castillo, M. et al. (2013). Water relations, nutrient content and developmental responses of *Euonymus* plants irrigated with water of different degrees of salinity and quality. *J. Plant Res.* 126: 567–576.

Greenway, H. and Munns, R. (1980). Mechanisms of salt tolerance in non-halophytes. *Annu. Rev. Plant Physiol.* 31: 149–190.

Grime, J.P. (1977). Evidence for the existence of three primary strategies in plants and its relevance to ecological and revolutionary theory. *Am. Nat.* 111: 1169–1194.

Gros, R., Poly, F., Monrozier, L.J., and Faivre, P. (2003). Plant and soil microbial community responses to solid waste leachates diffusion on grassland. *Plant Soil* 255: 445–455.

Gulzar, S., Khan, M., and Ungar, I. (2003). Effects of salinity on growth, ionic content, and plant–water status of *Aeluropus lagopoides*. *Commun. Soil Sci. Plant Anal.* 34: 1653.

Hameed, M., Basra, S., Aqeel, S., and Naz, N. (2010). Structural and functional adaptations in plants for salinity tolerance. In: *Plant Adaptation and Phytoremediation* (eds. M. Ashraf, M.S.A. Ahmad and M. Ozturk), 151–170. Dordrecht: Springer.

Handy, R.L. (1995). A stress path model for collapsible loess. In: *Genesis and Properties of Collapsible Soils*, NATO ASI Series (Series C: Mathematical and Physical Sciences), vol. 468 (eds. E. Derbyshire, T. Dijkstra and I. Smalley), 33–47. Dordrecht: Springer.

Hasegawa, P.M., Bressan, R.A., Zhu, J.K., and Bohnert, H.J. (2000). Plant cellular and molecular responses to high salinity. *Annu. Rev. Plant Physiol. Plant Mol. Biol.* 51: 463–499.

Horie, T., Hauser, F., and Schroeder, J.I. (2009). HKT transporter-mediated salinity resistance mechanisms in *Arabidopsis* and monocot crop plants. *Trends Plant Sci.* 14: 660–668.

Hu, Y., Burucs, Z., Tucher, S.V., and Schmidhalter, U. (2007). Short-term effects of drought and salinity on mineral nutrient distribution along growing leaves of maize seedlings. *Environ. Exp. Bot.* 60: 268–275.

Hussain, S., Peng, S., Fahad, S. et al. (2014). Rice management interventions to mitigate greenhouse gas emissions: a review. *Environ. Sci. Pollut. Res.* https://doi.org/10.1007/s11356-014-3760-4.

Ji, H., Pardo, J.M., Batelli, G. et al. (2013). The salt overly sensitive (SOS) pathway: established and emerging roles. *Mol. Plant* 6: 275–286.

Kaldenhoff, R., Ribas-Carbo, M., Flexas, J. et al. (2008). Aquaporins and plant water balance. *Plant Cell Environ.* 31: 658–666.

Kaveh, H., Nemati, H., Farsi, M., and Jartoodeh, S.V. (2011). How salinity affect germination and emergence of tomato lines. *J. Biol. Environ. Sci.* 5: 159–163.

Kaya, C., Ashraf, M., Dikilitas, M., and Tuna, A.L. (2013). Alleviation of salt stress induced adverse effects on maize plants by exogenous application of indole acetic acid (IAA) and inorganic nutrients- a field trial. *Aust. J. Crop. Sci.* 7: 249–254.

Kaya, C., Tuna, A.L., and Okant, A.M. (2010). Effect of foliar applied kinetin and indole acetic acid on maize plants grown under saline conditions. *Turk. J. Agri. For.* 34 (6): 529–538.

Keisham, M., Mukherjee, S., and Bhatla, S.C. (2018). Mechanisms of sodium transport in plants – progresses and challenges. *Int. J. Mol. Sci.* 19: 647.

Keutgen, A.J. and Pawelzik, E. (2009). Impacts of NaCl stress on plant growth and mineral nutrient assimilation in two cultivars of strawberry. *Environ. Exp. Bot.* 65: 170–176.

Khan, M.A., Ungar, I.A., and Showalter, A.M. (2000). Effects of salinity on growth, water relation and ion accumulation of the subtropical perennial halophyte, *Atriplex griffithii var. stocksii*. *Ann. Bot.* 85: 225–232.

Khare, T., Kumar, V., and Kishor, P.B.K. (2015). Na^+ and Cl^- ions show additive effects under NaCl stress on induction of oxidative stress and the responsive antioxidative defense in rice. *Protoplasma* 252: 1149–1165.

Krebs, M., Beyhl, D., Görlich, E. et al. (2010). *Arabidopsis* V-ATPase activity at the tonoplast is required for efficient nutrient storage but not for sodium accumulation. *Proc. Natl. Acad. Sci.* 107: 3251–3256.

Kumar, G., Purty, R.S., Sharma, M.P. et al. (2009). Physiological responses among *Brassica* species under salinity stress responses show strong correlation with transcript abundance for SOS pathway related genes. *J. Plant Physiol.* 166: 507–520.

Lambers, H., Raven, J., Shaver, G., and Smith, S. (2008). Plant nutrient-acquisition strategies change with soil age. *Trends Ecol. Evol.* 23: 95–103.

Lauchli, A. and Grattan, S.R. (2007). Plant growth and development under salinity stress. In: *Advances in Molecular Breeding Towards Drought and Salt Tolerant Crops* (eds. M.A. Jenks, P.M. Hasegawa and J.S. Mohan), 1–32. Berlin: Springer.

Lawlor, D. and Cornic, G. (2002). Photosynthetic carbon assimilation and associated metabolism in relation to water deficits in higher plants. *Plant Cell Environ.* 25: 275–294.

Li, H., Liu, H., Gao, X., and Zhang, H. (2009). Knock-out of *Arabidopsis* AtNHX4 gene enhances tolerance to salt stress. *Biochem. Biophys. Res. Commun.* 382: 637–641.

Lo Gullo, M.A., Trifilo, P., and Raimondo, F. (2007). Hydraulic characteristics and water relations in pigment-less mutant shoots of an orange tree. *Tree Physiol.* 27: 209–217.

Maathuis, F.J.M. and Sanders, D. (2001). Sodium uptake in *A. thaliana* is regulated by cyclic nucleotides. *Plant Physiol.* 127: 1617–1625.

Mittal, S., Kumari, N., and Sharma, V. (2012). Differential response of salt stress on *Brassica juncea*: photosynthetic performance, pigment, proline, D1 and antioxidant enzymes. *Plant Physiol. Biochem.* 54: 17–26.

Munns, R. (2002). Comparative physiology of salt and water stress. *Plant Cell Environ.* 25: 239–250.

Munns, R. (2005). Genes and salt tolerance: bringing them together. *New Phytol.* 167: 645–663.

Munns, R., James, R.A., and Läuchli, A. (2006). Approaches to increasing the salt tolerance of wheat and other cereals. *J. Exp. Bot.* 57: 1025–1043.

Munns, R. and Tester, M. (2008). Mechanisms of salinity tolerance. *Ann. Rev. Plant Biol.* 59: 651–681.

Naeem, M., Ansari, A.A., Gill, S.S. et al. (2017). Regulatory role of mineral nutrients in nurturing of medicinal legumes under salt stress. In: *Essential Plant Nutrients: Uptake, use Efficiency, and Management* (eds. M. Naeem, A.A. Ansari and S.S. Gill), 309–334. Cham: Springer International Publishing.

Nahar, K. and Hasanuzzaman, M. (2009). Germination, growth, nodulation and yield performance of three mung bean varieties under different levels of salinity stress. *Green Farming* 2: 825–829.

Navarro, A., Bañón, S., Olmos, E., and Sánchez-Blanco, M.J. (2007). Effects of sodium chloride on water potential components, hydraulic conductivity, gas exchange and leaf ultrastructure of *Arbutus unedo* plants. *Plant Sci.* 172: 473–480.

Oh, D.H., Lee, S.Y., Bressan, R.A. et al. (2010). Intracellular consequences of SOS1 deficiency during salt stress. *J. Exp. Bot.* 61: 1205–1213.

Olías, R., Eljakaoui, Z., Li, J., De Morales, A., Carmen, M. M. M, Pardo, J. M., and Belver, A. (2009). The plasma membrane Na^+/H^+ antiporter SOS1 is essential for salt tolerance in tomato and affects the partitioning of Na^+ between plant organs. *Plant Cell Environ.* 32:904–916.

Oren, A. (1999). Bioenergetic aspects of halophilism. *Microbiol. Mol. Biol. Rev.* 63: 334–340.

Ort, D.R. (2001). When there is too much light. *Plant Physiol.* 125: 29–32.

Othman, Y., Al-Karaki, G., Al-Tawaha, A.R., and Al-Horani, A. (2006). Variations in germination and ion uptake in barley genotypes under salnity. *World J. Agric. Sci.* 2: 11–15.

Pardo, J. and Rubio, F. (2011). Na^+ and K^+ transporters in plant signaling. In: *Transporters and Pumps in Plant Signalling, Signalling and Communication in Plants* (eds. M. Geisler and K. Venema), 7. Berlin, Heidelberg: Springer.

Passioura, J.B. (2010). Plant–water relations. In: *Encyclopedia of Life Sciences (ELS)* (ed. N.J. Hoboken), 1–7. Chichester, UK: Wiley.

Piotr, S. and Grazyna, K. (2005). Antioxidant defence in the leaves of C3 and C4 plants under salinity stress. *Physiol. Plant.* 125: 31–40.

Porcel, R., Aroca, R., and Ruiz-Lozano, J.M. (2011). Salinity stress alleviation using arbuscular mycorrhizal fungi. A review. *Agron. Sustain. Dev.* 32: 181–200.

Pyo, Y.J., Gierth, M., Schroeder, J.I., and Cho, M.H. (2010). High-affinity K^+ transport in *Arabidopsis*: AtHAK5 and AKT1 are vital for seedling establishment and post germination growth under low-potassium conditions. *Plant Physiol.* 153: 863–875.

Qadir, M. and Schubert, S. (2002). Degradation processes and nutrient constraints in sodic soils. *Land Degrad. Dev.* 13: 275–294.

Quintero, F., Ohta, M., Shi, H. et al. (2002). Reconstitution in yeast of the *Arabidopsis* SOS signaling pathway for Na^+ homeostasis. *Proc. Natl. Acad. Sci. U. S. A.* 99: 9061–9066.

Reddy, A.S.N. and Reddy, V. (2001). Calcium as a messenger in stress signal transduction. In: *Handbook of Plant and Crop Physiology* (ed. M. Pessarakali), 697–732. New York: Mercel Dekker.

Rodríguez-Rosales, M.P., Galvez, F.J., Huertas, R. et al. (2009). Plant NHX cation/proton antiporters. *Plant Signal. Behav.* 4: 265–276.

Roy, S.J., Negrão, S., and Tester, M. (2014). Salt resistant crop plants. *Curr. Opin. Biotechnol.* 26: 115–124.

Rozeff, N. (1995). Sugarcane and salinity- a review paper. *Sugarcane* 5: 8–19.

Rus, A., Lee, B., Muñoz-Mayor, A. et al. (2004). AtHKT1 facilitates Na^+ homeostasis and K^+ nutrition in plants. *Plant Physiol.* 136: 2500–2511.

Saha, P., Chatterjee, P., and Biswas, A.K. (2010). NaCl pretreatment alleviates salt stress by enhancement of antioxidant defence system and osmolyte accumulation in mungbean (*Vigna radiata* L. Wilczek). *Indian J. Exp. Biol.* 48: 593–600.

Sánchez-Blanco, M.J., Rodríguez, P., Morales, M.A. et al. (2004). Comparative growth and water relations of *Cistus albidus* and *Cistus monspeliensis* plants during water deficit conditions and recovery. *Plant Sci.* 162: 107–113.

Saud, S., Li, X., Chen, Y. et al. (2014). Silicon application increases drought tolerance of Kentucky Bluegrass by improving Plant Water Relations and Morphophysiological functions. *Sci. World J.*: 368694. https://doi.org/10.1155/2014/368694.

Schubert, S., Neubert, A., Schierholt, A. et al. (2009). Development of salt resistant maize hybrids: the combination of physiological strategies using conventional breeding methods. *Plant Sci.* 177: 196–202.

Shabala, S. (2013). Learning from halophytes: physiological basis and strategies to improve abiotic stress tolerance in crops. *Ann. Bot.* 112 (7): 1209–1221.

Shahzad, M., Witzel, K., Zorb, C., and Muhling, K.H. (2012). Growth-related changes in subcellular ion patterns in maize leaves (*Zea mays* L.) under salt stress. *J. Agron. Crop Sci.* 198: 46–56.

Shi, H.Z. and Zhu, J.K. (2002). Regulation of expression of the vacuolar Na^+/H^+ antiporter gene AtNHX1 by salt stress and abscisic acid. *Plant Mol. Biol.* 50: 543–550.

Silva, P. and Gerós, H. (2009). Regulation by salt of vacuolar H^+-ATPase and H^+-pyrophosphatase activities and Na^+/H^+ exchange. *Plant Signal. Behav.* 4: 718–726.

Slama, I., Ghnaya, T., Savouré, A., and Abdelly, C. (2008). Combined effects of long-term salinity and soil drying on growth, water relations, nutrient status and proline accumulation of *Sesuvium portulacastrum*. *C.R. Biol.* 331: 442–451.

Snedden, W.A. and Fromm, H. (1998). Calmodulin, calmodulin-related proteins and plant responses to the environment. *Trends Plant Sci.* 3: 299–304.

Sönmez, İ., Kaplan, M., and Sonmez, S. (2007). Investigation of seasonal changes in nitrate contents of soils and irrigation waters in greenhouses located in Antalya-Demre region. *Chem. Asian J.* 19: 5639–5646.

Soo-Hyung, K. (2011). Soil, water plant interaction. *IPM Workshop-Water is the link for Soil, Plant and Atmosphere Continuum.*

Stiller, V., Lafitte, H.R., and Sperry, J.S. (2003). Hydraulic properties of rice and the response of gas exchange to water stress. *Plant Physiol.* 132: 1698–1706.

Sümer, A., Zörb, C., Yan, F., and Schubert, S. (2004). Evidence of sodium toxicity for the vegetative growth of maize (*Zea mays* L.) during the first phase of salt stress. *J. Appl. Bot.* 78: 135–139.

Szabolcs, I. (1974). *Salt Affected Soils in Europe*, vol. 7 (ed. J. Fink), 57. Netherlands: Springer.

Tester, M. and Davenport, R. (2003). Na^+ tolerance and Na^+ transport in higher plants. *Ann. Bot.* 91: 503–527.

Tsai, Y.C., Hong, C.Y., Liu, L.F., and Kao, C.H. (2004). Relative importance of Na^+ and Cl^- in NaCl-induced antioxidant systems in roots of rice seedlings. *Physiol. Plant.* 122: 86–94.

Tuteja, N. (2007). Mechanisms of high salinity tolerance in plants. *Methods Enzymol.* 428: 419–438.

Tyerman, S.D., Skerrett, M., Garrill, A. et al. (1997). Pathways for the permeation of Na^+ and Cl^- into protoplast derived from the cortex of wheat roots. *J. Exp. Bot.* 48: 459–480.

Tyree, M.T. (2003). Hydraulic limits on tree performance: transpiration, carbon gain and growth of trees. *Trees Struct. Funct.* 17: 95–100.

Walsh, G.E. (1990). Anatomy of the seed and seedling of *Spartina alterniflora* Lois. (Poaceae). *Aquat. Bot.* 38: 177–193.

White, P.J. (2000). Calcium channels in higher plants. *Biochem. Biophys. Acta* 1465: 171–189.

Wichern, J., Wichern, F., and Joergensen, R.G. (2006). Impact of salinity on soil microbial communities and decomposition of maize in acidic soils. *Geoderma* 137: 100–108.

Xu, S., Hu, B., He, Z. et al. (2011). Enhancement of salinity tolerance during rice seed germination by pre-soaking with haemoglobin. *Int. J. Mol. Sci.* 12: 2488–2501.

Yan, N., Marschner, P., Cao, W. et al. (2015). Influence of salinity and water content on soil microorganisms. *Int. Soil Water Conserv. Res.* 3: 316–323.

Yokoi, S., Bressan, R.A., and Hasegawa, P.M. (2002). Differential expression and function of *A. thaliana* NHX Na^+/H^+ antiporters in the salt stress response. *J. Plant.* 30: 529–539.

Zhang, M.H., Qin, Z.H., and Liu, X. (2005). Remote sensed spectral imagery to detect the late blight in field tomatoes. *Precis. Agric.* 6: 489–508.

Zielinski, R.E. (1998). Calmodulin and calmodulin-binding proteins in plants. *Annu. Rev. Plant Physiol. Plant Mol. Biol.* 49: 697–725.

Zimmermann, P. and Zentgraf, U. (2005). The correlation between oxidative stress and leaf. *Cell. Mol. Biol. Lett.* 10: 515–534.

8

Effects of Salt Stress on Nutrient Cycle and Uptake of Crop Plants

Lav Kumar Jaiswal[1], Prabhakar Singh[2], Rakesh Kumar Singh[1], Tanamyee Nayak[1], Yashoda Nandan Tripathi[2], Ram Sanmukh Upadhyay[2], and Ankush Gupta[1]

[1] *Department of Biochemistry, Institute of Science, Banaras Hindu University, Varanasi, Uttar Pradesh, India*
[2] *Center of Advance Study, Department of Botany, Institute of Science, Banaras Hindu University, Varanasi, Uttar Pradesh, India*

8.1 Introduction

The biogeochemical cycling of elements involves the interactions between the biosphere and the surroundings in the form of cycling of elements throughout the different spheres viz., atmosphere, hydrosphere, and lithosphere. However, the elements interact with each other, and the cycling of one element is largely influenced by other elements.

Salinity among abiotic stresses is the second major impediment to crop productivity all over the world after drought. It is caused due to irrigation with highly saline underground water, disintegration and release of seawater in coastal areas, and accumulation of salts in arid or semiarid area due to insufficient leaching of ions (Chinnusamy and Zhu 2003). The extent of agricultural land that is affected by high salinity is continuously increasing worldwide, due to both natural and human agricultural practices such as irrigation systems (Munns and Tester 2008). Salinity has two major threats to plant growth: osmotic stress and ionic stress (Flowers and Colmer 2008) in addition to oxidative stress. Salinity has adverse effects on plant growth and productivity and is one of the major factors responsible for crop losses throughout the world. The noxious effects of salinity affect different physiological and metabolic processes of plants. The reactance to these changes is often inclusive of a variety of symptoms, such as the decrease in leaf area, increase of leaf thickness, and abscission of leaves, necrosis of root & shoot, and decrease of internode lengths (Kozlowski 1997; Parida and Das 2005). Salt stress influences all the major vital processes of plants such as photosynthesis, protein synthesis, energy, and lipid metabolism (Parida and Das 2005). The high salinity leads to large accumulation of toxic ions (Na^+, Cl^-) and

Physiology of Salt stress in Plants: Perception, Signalling, Omics and Tolerance Mechanism, First Edition.
Edited by Pratibha Singh, Madhulika Singh, Rajiv Kumar Singh, and Sheo Mohan Prasad.

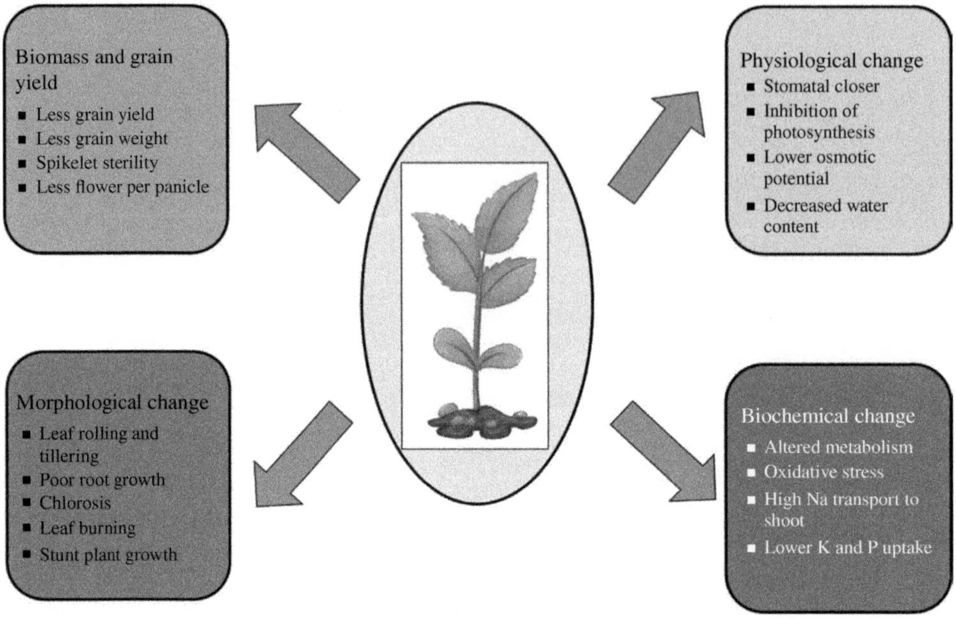

Figure 8.1 Adverse effect of salinity on plant growth.

nutritional imbalance that affects plant enzymes and causes cell swelling, resulting in reduced energy production and other physiological changes. Salinity stress has a negative impact on plant nutrition and mineral homeostasis, particularly for Ca^{2+} and K^{+}. Figure 8.1 briefs the adverse effects of salinity on plants. How plants face with salinity and how successful their strategies are vary widely from species to species (Flowers and Colmer 2008). The uptake of high amounts of salt by the plant leads to the increase of the osmotic pressure in the cytosol of the cell. Under these circumstances, cell homeostasis is regulated by an osmotic adjustment mechanism which consists of the accumulation of large amounts of salt ions in the vacuole or synthesis of organic osmolytes under salt stress (Munns 2002). The compatible osmoprotectants such as proline and soluble sugars are produced to protect the cells against the adverse effects from salt stress in many species of plants. High accumulation of proline is associated with tolerance to stress (Sairam and Tyagi 2004; Hokmabadi et al. 2005). Salinity affects the soil aeration, enzyme activity, soil microbial biomass, and the soil microorganisms' growth rate influencing nutrient cycling (Tripathi et al. 2006; Rousk et al. 2011). The study of the soil microbial community and its function in saline soil represents a crucial target in ecology because it will be helpful to overcome the salinity problem. Soil microorganisms play an important role in maintaining ecosystem function (Perroni et al. 2014; Qi and Yang 2017). The bacterial composition of saline soil was largely affected by salinity than by any other factors such as acidity and nutrient availability (Lozupone and Knight 2007). By affecting the soil organic matter, soil aggregation, porosity, and nutrient availability, soil microorganisms could drive different nutrient cycles (Xu et al. 2016). Hence, any effects of salt on soil microorganism will cause implications for the soil organic matter and nutrient availability (Setia et al. 2012). Understanding the effects of soil salinity on soil microorganism and its soil ecological function is of great importance to find proper methods to enhance soil fertility of saline land (Zhang et al. 2019). Thus, in this chapter, we comprehensively elucidate the effects of salinity stress on biogeochemical cycle and soil microorganisms (nitrogen fixing bacteria, phosphorous bacteria, and sulfur bacteria) that directly or indirectly affect the crop production.

8.2 Limitation of Nutrient Cycle and Uptake of Nutrients

Ecosystem productivity and nutrient retention capacity are largely influenced by plant species, functional groups, or diversity level (Tilman et al. 2001). Studies demonstrated that components of global change such as warming and elevated CO_2 level alter nutrient cycle directly or indirectly (Hobbie 1995). The previous hypothetical model for element inputs and outputs from terrestrial ecosystems shown that the most inputs are controlled by process occurring outside of ecosystems and controlled by biological processes that cause short supply of nutrients within the terrestrial ecosystem (Hedin et al. 1995). This process suggested that nutrient limitation could be caused by sustained or losses of nutrients in forms that are not available to organisms and so cannot be retained by them (Vitousek 2004). Nutrient limitation is defined as by how primary production of ecosystem and other biological processes respond to added nutrients, as in the studies summarized by Elser et al. 2007 and LeBauer and Treseder 2008.

Nutrient uptake, primary productivity, and other biological processes are widely affected by nutrient limitation in terrestrial ecosystems. Nitrogen (N_2) and Phosphorus (P) are the most common limiting elements individually as well as in combination. There are six mechanisms identified that could drive P limitation in terrestrial ecosystems. The well-known of these is depletion-driven limitation, in which accumulated P losses during long-term from old soil. Walker and Syers 1976 termed this mechanism "terminal steady state" of profound P depletion and limitation. The other mechanisms are soil barriers that prevent access to P, transactional limitation, in which weathering of P-containing minerals does not keep pace with the supply of other resources, low-P parent materials, P sinks, and anthropogenic changes that increase the supply of other resources (often N_2) relative to P. There are two types of nutrient limitations occurring in ecosystems. One is proximate in which addition of a nutrient stimulates biological process; other is ultimate where addition of a nutrient can transform ecosystems. The mechanisms that drive P limitation suggest that depletion, soil barriers, and low-P parent material in soil often cause ultimate limitation because they control the mass balance of P in the ecosystem (Vitousek 2004). Similarly, constraints and demand-independent losses of N_2 fixation can control the mass balance of nitrogen level in ecosystem and cause it to be an ultimate limiting element. Nutrient limitation occurs in ecosystems when significant additions of an essential element in biologically available forms lead to increase in the rate of a biological process such as primary productivity or in biomass. Nutrient limitation may be calculated by indirect measurements in which the most common is nutrient availability in soil (Powers 1980), plant investments in acquiring particular nutrients (Harrison and Helliwell 1979), and concentrations of nutrients in plant cells (Koerselman and Meuleman 1996).

Table 8.1 Pathways, mechanisms, and timescales of P limitation to primary production in terrestrial ecosystems.

Pathway	Mechanism	Timescale
Depletion driven	Loss of inorganic and dissolved organic P via leaching: exhaustion of primary minerals in soil	Millions of years
Soil barrier	Formation of soil layers that physically prevent/ inhibit access by roots to potentially available P	Hundreds to tens of thousands of years
Transactional	Slow release of P from mineral forms, relative to the supply of other resources	Decades to centuries
Low-P parent material	Low inputs of P via weathering due to low concentrations of P in rock	All develops quickly and persists
Sink driven	Sequestration of available P in an accumulating pool within ecosystems	Decades to millenia
Anthropogenic	Enhanced supply of other resources (especially N_2) causes P limitation	Years to decades

Source: Vitousek et al. (2010).

8.2.1 Phosphorus Limitation

Phosphorus (P) is primarily derived from weathering of rocks. The ecosystems start their existence with a fixed complement of P from which even very small losses cannot readily be replenished. Hence, P depletion can be seen in ecosystems with very old soils, and pathway or mechanisms followed are shown in Table 8.1. Walker and Syers (1976) measured total soil P and its chemical forms in many places of New Zealand chronosequences (substrate age gradients) and found that a pattern of lower total P and an increased fraction of recalcitrant P in older soil. They suggested that these old soil sites represent a "terminal steady state" of natural P depletion and biological limitation. Phosphorus is one of an ultimate limiting nutrient here because additions of P cause inputs of biological N_2 fixation and C (by enhancing CO_2 diffusion into lakes) together, due to which N_2 and C transform oligotrophic lakes into eutrophic ones. In contrast, nitrogen and carbon are not ultimate limiting nutrients because adding them does not increase the input of P and lakes remain oligotrophic.

8.2.2 Nitrogen Limitation

Most terrestrial ecosystems are largely deprived of nitrogen in initial stage of their development (Walker and Syers 1976). Unlike P, N_2 is accumulated from the atmosphere rapidly by the help of biological nitrogen fixation or more slowly via atmospheric deposition and dispersed sources of biological fixation. The symbiotic N_2 fixers are dominated by an early stage of ecosystem development. A detail of N_2 limitation to primary productivity in terrestrial ecosystems is shown in Table 8.2. The biological processes in many ecosystems on young soils may be limited by low supplies of N_2 (Cleveland et al. 1999).

Table 8.2 Pathways, mechanisms, and timescales of nitrogen limitation to primary production in terrestrial ecosystems.

Pathway	Mechanism	Timescale
Demand-independent losses	Losses of combined N_2 that organisms cannot prevent, including leaching of DON, post-disturbance losses, some gaseous pathways	Decades to centuries, depends on loss pathway
Constraints to biological N_2 fixation	Biological N_2 fixation is slow or absent even when N_2 is limiting, could be due to energetic costs, differential grazing, demands for P, Mo, or other essential elements	Decades to centuries
Transactional	Slow release of N_2 from complex organic into soluble forms, relative to the supply of other resources	Years to centuries
Sink driven	Sequestration of available N_2 in an accumulating pool within ecosystems	Years to centuries

Sources: Adopted from Vitousek and Field (1999), Vitousek and Howarth (1991), and Vitousek (2004).

8.3 Nutrient Cycle or Biogeochemical Cycle

Nutrient cycling is one of the most important phenomena that occur in an ecosystem. The nutrient cycle represents the use, movement, and recycling of nutrients in the environment. The elements such as carbon, oxygen, hydrogen, phosphorus, and nitrogen are essential for life and must be recycled (Stevenson and Cole 1999). The flow of these elements through the ecosystem must be cyclic with matter being consistently reused. Nutrient cycle is the flux of nutrients within and between the various biotic and abiotic environmental factors in which nutrients occur in the soil (Yan et al. 2015).

On the basis of the replacement period, a nutrient cycle is classified as perfect and imperfect cycles. A perfect nutrient cycle is one in which nutrients are replaced as fast as they are utilized. Most gaseous cycles are generally considered as perfect cycle (Jean-claude Bertrand et al. 2015). Water cycle, carbon cycle, nitrogen cycle, and oxygen cycle come under perfect cycle. While sedimentary cycles are considered as relatively imperfect because some amount of nutrients are continuously lost from cycle and get locked into natural sediments and so become inaccessible for plants. Sedimentary cycle includes phosphorus cycle, sulfur cycle, and calcium cycle.

8.3.1 Water Cycle or Hydrological Cycle

The water from the different water bodies evaporates, cools, condenses, and falls back to the Earth as rain. This biogeochemical cycle is responsible for maintaining weather conditions (Figure 8.2). This cycle is numerous and imperishable and changes with the change of place, time, and water in their various forms. The influence of latitude, the amount of solar energy, the capacity of the earth for holding water, geological, and natural are the factors that affect hydrological cycle. The hydrological cycle works more actively in the equatorial region that in the polar region. With the help of hydrological cycle, different kinds of

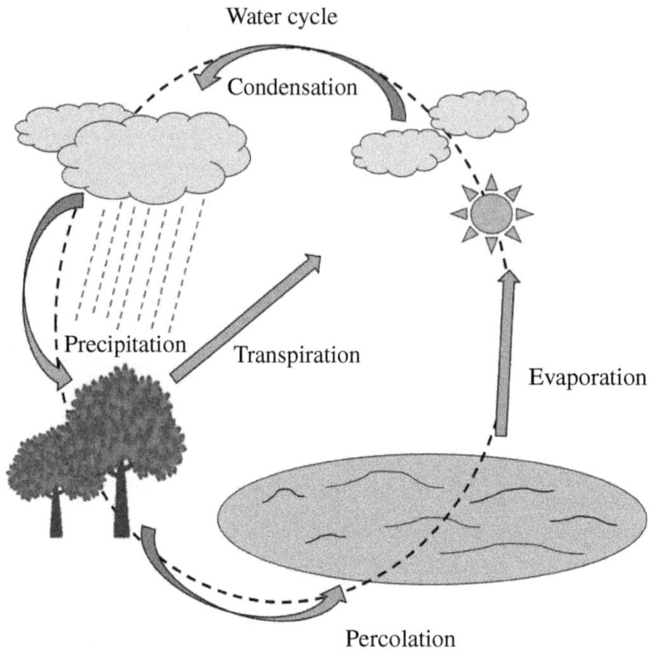

Figure 8.2 Terrestrial water cycle and its components.

hydrological information, hydrological budget of any country, rains, information about climate, evaporation etc. can be reckon and analyzed (Chahine 1992).

8.3.2 Carbon Cycle

Carbon cycle (Figure 8.3) can be classified into two types based on their duration of the process:

Short term – It is occurs within a relatively short period of time and takes just days, months, or years for carbon to flow across the various carbon reservoirs.
Long term – This type of carbon cycle takes relatively thousands of years to occur. The excess carbon from the short-term cycle is stored for a long time before they are released.

Even though carbon dioxide is found in small traces in the atmosphere, it plays a vital role in balancing the energy and traps the long-wave radiations from the sun. Therefore, it acts like a protecting shield over the planet. Carbon is an integral component of every life form on earth, from proteins and lipids to even our DNA.

Furthermore, all known life on earth is based on carbon (Schidlowski 2001). Hence, the carbon cycle along with other biogeochemical cycles like nitrogen, sulfur, calcium, and oxygen cycle etc. plays a vital role in the existence of life on earth.

8.3.3 Nitrogen Cycle

Nitrogen (N_2) is one of the important elements among six (CHNOPS) major essential elements of life and forms the building blocks and intact molecules for metabolism (e.g.

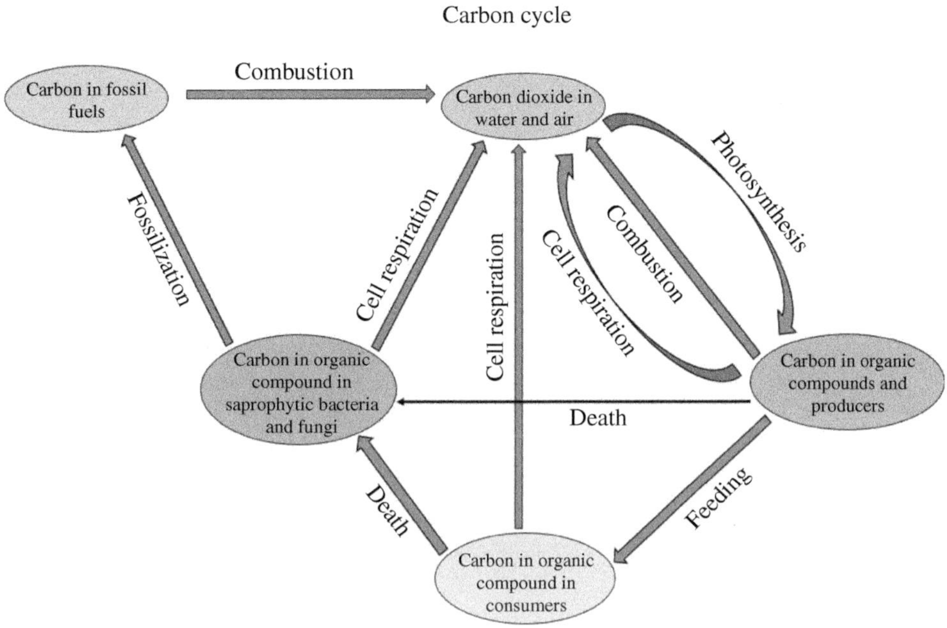

Figure 8.3 Illustration of global carbon cycle.

amino acids and proteins), hereditary materials (nucleotides and nucleic acids), and other important biological functions (Takai 2019). Among the essential elements, circulating volumes of biologically available nitrogen and phosphorus are slightly limited in many biospheres of the planet and thus as a whole, the biological demand for nitrogen and phosphorus and their circulation speeds are markedly greater and faster than those for carbon (Gruber and Galloway 2008).

Nitrogen fixation is a unique ability possessed by microorganisms called diazotrophs and involves the conversion of very inert N_2 to reactive nitrogen (including NH_3) (Dixon and Kahn 2004). The conversion of N_2 to reactive nitrogen (such as NO) always occurs with lightning which provides reactive nitrogen to the land and ocean (Navarro-González et al. 2001).

As essential compound of the DNA, protcins, or enzymes, nitrogen is a key component of life and a fundamental nutrient. Its most abundant form atmospheric dinitrogen (N_2) is unavailable for most organisms, making nitrogen a limiting nutrient (Galloway et al. 2004). Therefore, the conversion of N_2 into reactive nitrogen (Nr) directly or indirectly supports biomass growth.

There are two forms of nitrogen present on Earth: one is atmospheric molecular nitrogen (N_2) and second is biologically reactive nitrogen (e.g. NO_3, NH_4, and organic nitrogen) (Gruber and Galloway 2008). The biological nitrogen cycle mainly consists of internal interactions within the various forms of nitrogen pool and in- and out-flow between reactive nitrogen and atmospheric N_2. The cycle between the two large nitrogen pools is mainly controlled by only two microbial biological processes, nitrogen fixation, and denitrification (Takai 2019). The global nitrogen biochemical cycle is divided into several successive stages as shown in Figure 8.4.

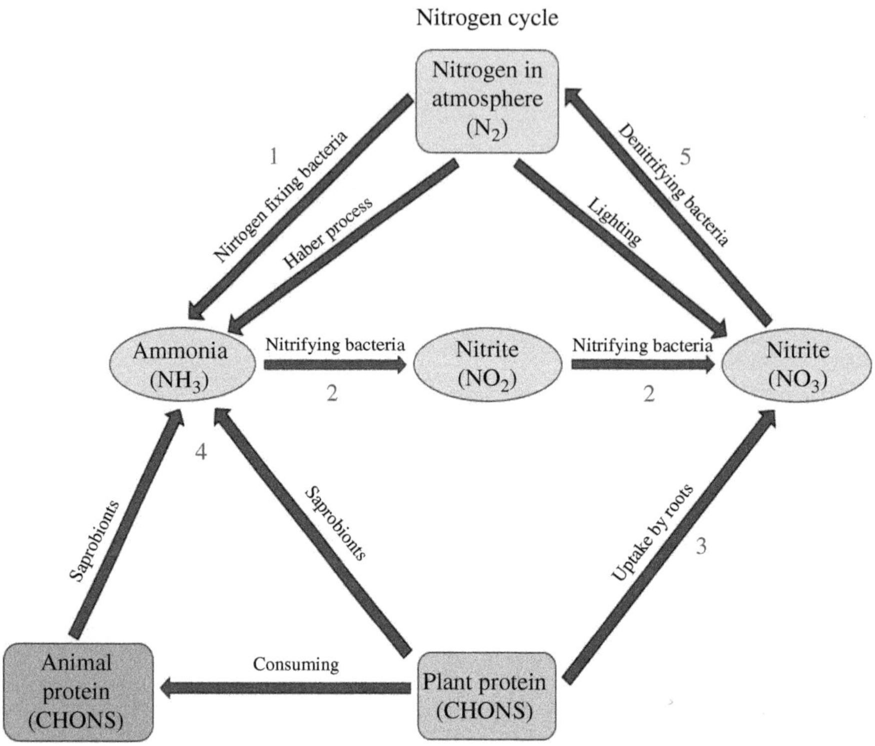

Figure 8.4 Showing different stages of nitrogen cycle.

By the process of nitrogen cycle, plants synthesize chlorophyll from the nitrogen compounds and convert inert nitrogen into a usable form for the plants through the biochemical process. In the ammonification process, the bacteria help in decomposing the organic matter animal and plant after death, which indirectly helps to clean up the nature. The released nitrates and nitrites in soil help in enriching the soil with necessary nutrients required for plant nutrition. Nitrogen forms many crucial compounds and important biomolecules and an essential component of the cell. Nitrogen is also recycled by human activities such as combustion of fuels and the use of nitrogen fertilizers. These processes increase the levels of nitrogen-containing compounds in the atmosphere (Vitousek et al. 2002).

8.3.4 Oxygen Cycle

Oxygen (O_2) is the most abundant element in the earth's crust and second most abundant element in the earth's atmosphere; it accounts for 89% mass of the ocean (Keeling 1995). Most of the oxygen is bound tightly in silicate minerals which make a large fraction of Earth's mantle and crust (Kasting and Canfield 2012). Oxygen is also a stable and important component of many gases, liquids, and ices, including molecular oxygen (O_2), carbon dioxide (CO_2), water (H_2O), and ozone (O_3), and it is an important component of the atmospheres, cryospheres, and hydrospheres of Earth (Farquhar and Johnston 2008). In environment, it is seen that even relatively small changes in oxygen production and

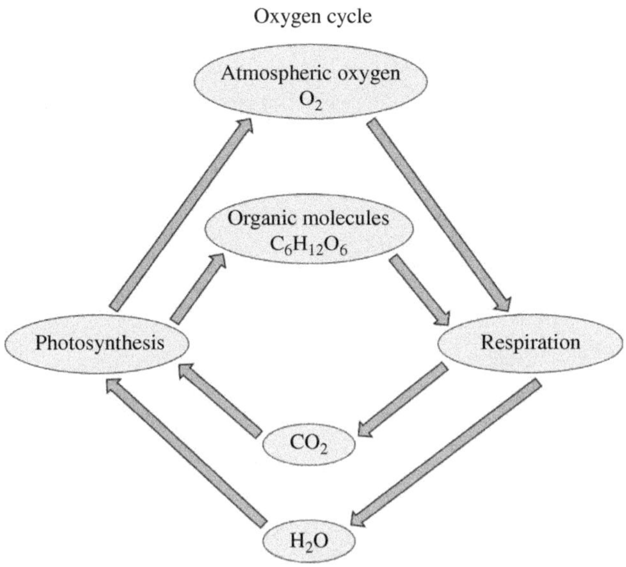

Figure 8.5 Illustration of cyclic oxygen cycle.

consumption have significant impact on atmospheric O_2 concentration within geologically short periods of time. Till now, all available vindication supports the conclusion that stasis in O_2 variation is a significant feature of the Earth's atmosphere over wide spans of the geologic past. Study of the oxygen cycle is therefore important because while an equable O_2 atmosphere is central to life as we know it, we do not have much information about the process by which oxygen concentrations remain constant over large spans of geologic time scale (Petsch 2003).

This nutrient cycle moves throughout the various components of environment that is atmosphere, lithosphere, and the biosphere (Figure 8.5). Plants utilize carbon dioxide in process of photosynthesis and produce oxygen in the presence of sun light. Oxygen is an important element required for life (biochemical reactions) however; sometimes, it may be toxic and life threatening to some obligate anaerobic bacteria present in the nature. The level of oxygen in the Earth atmosphere is maintained through oxygen cycle. In short, we can say that oxygen cycle starts with the process of photosynthesis in the presence of sunlight in which oxygen releases into the atmosphere after that humans and animals inspire the oxygen in lungs and exhale the carbon dioxide and this carbon dioxide reuse by the plants, in this way cycle continues (Lane 2003). This also proves that both the oxygen and carbon cycle occur independently and are interconnected to each other.

8.3.5 Phosphorus Cycle

Phosphorus (P) is an essential element of life and is a basic component of the modern agricultural practice. The presence of this element in affecting ecosystem comes not only from its predominance in cells but also from its being relatively plenty in the environment (Blank 2012). P is considered a potentially critical resource because of its limited reserves.

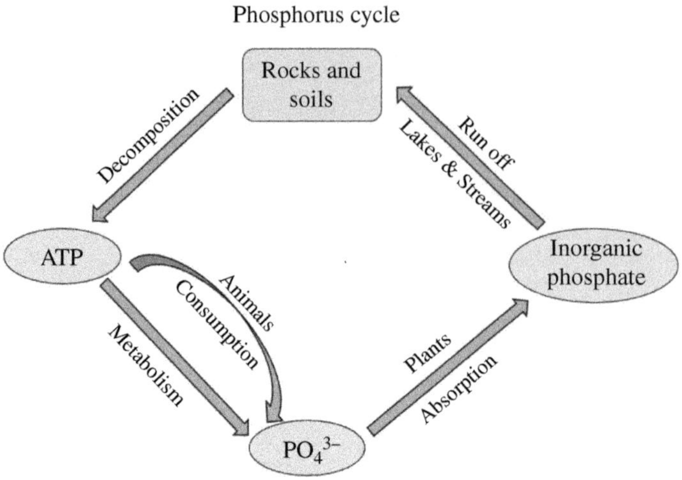

Figure 8.6 Processes and different stages of phosphorus cycle.

The free P present in soil is actually unavailable to most plant life. It plays a prime role in aquatic ecosystems and water quality. Unlike C and N element which come primarily from the atmosphere, P found in large amounts as a mineral in phosphate rocks and mountains and enters in the P cycle (Figure 8.6) via erosion and mining activities. Most of the phosphate in the biosphere originates from calcium phosphate minerals, e.g. apatite ($Ca_5(PO_4)_3OH$) (Brunner and Bachofen 1997). In soil, P is present in esterified form as well as organic phosphate or as insoluble iron or aluminum phosphate (Cross and Schlesinger 1995).

Currently, P compounds are mostly used for both agricultural purpose and animal husbandry. From a total global production of 24.87 Mt of P, fertilizers accounted for 82% of the total phosphate rock consumption (Scholz et al. 2013). The main reserves of phosphate rock are found in relatively few countries (China, United States, and Marocco) (Jasinski 2012). A critical issue in the P cycle is the low efficiency and losses associated with the production and use of phosphate fertilizers. According to Cordell et al. 2009, only about 10% of mined P is currently consumed. The two largest causes of P loss globally have been identified as erosion and runoff of mined P (estimated at 46%) and animal waste (40%). P is one of a limiting nutrient for terrestrial ecosystem and productivity that commonly plays a key role in net carbon uptake (Tiessen et al. 1984). Because of the limitations of P availability in ecosystems, it is restricted by the rate of release of this element during soil weathering; P is generally recycled to various extents in ecosystems depending on climate, soil, and ecosystem type. The release of P from apatite dissolution is a key control on ecosystem productivity (Filippelli 2002). The average amount of phosphorus in the air is estimated to be $1–8 \times 10^{-8}$ mg/l as phosphate which is lower than the concentrations of some important trace gases (Fitter and Hay 1987). The P concentrations in rain are much higher in the range of $1–10 \times 10^{-3}$ mg/l (Brunner and Bachofen 1997). Condensation nuclei and droplets in fog contain both soluble P and particulate P. The average P content in biomass is given as 0.2% based on dry weight (Salisbury and Ross 1992); the amount of P present in the hydrosphere, in the lithosphere, and in the air is 0.003, 0.08, and 10^{-9}%, respectively (Yanai 1992).

8.3.6 Sulfur Cycle

The sulfur cycle illustrated in Figure 8.7 affects the global carbon cycle and different ecosystems primarily through the remineralization of organic carbon. Over geological timescales, recycling of sulfur element is closely linked to the redox state of earth's exosphere through the burial of oxidized (sulfate) and reduced (sulfide) sulfur species in marine sediments. This biogeochemical cycle moves through the rocks, water bodies, and living systems.

This cycle plays a crucial role in regulating Earth's ecosystem via the activity of a variety of metabolic processes occurs within the microorganisms that change the oxidation state of sulfur by natural process. These metabolic activities are tightly linked to the global carbon cycle. Sulfur is an essential element for all living organisms and main constituent of biomolecules (e.g. cysteine and methionine), certain coenzymes, and cosubstrates inorganic biomolecules (e.g. the Fe–S complexes in ETS). However, sulfur makes up $<1\%$ by weight of typical biomass (Canfield 2001), and organic sulfur is generally thought to play only a minor role in the biogeochemical sulfur cycle, although it can provide insights into the fate of organics during diagenesis (Amrani et al. 2006; Raven et al. 2015). The different microbial processes transform the oxidation state of sulfur for energy generation, coupled to

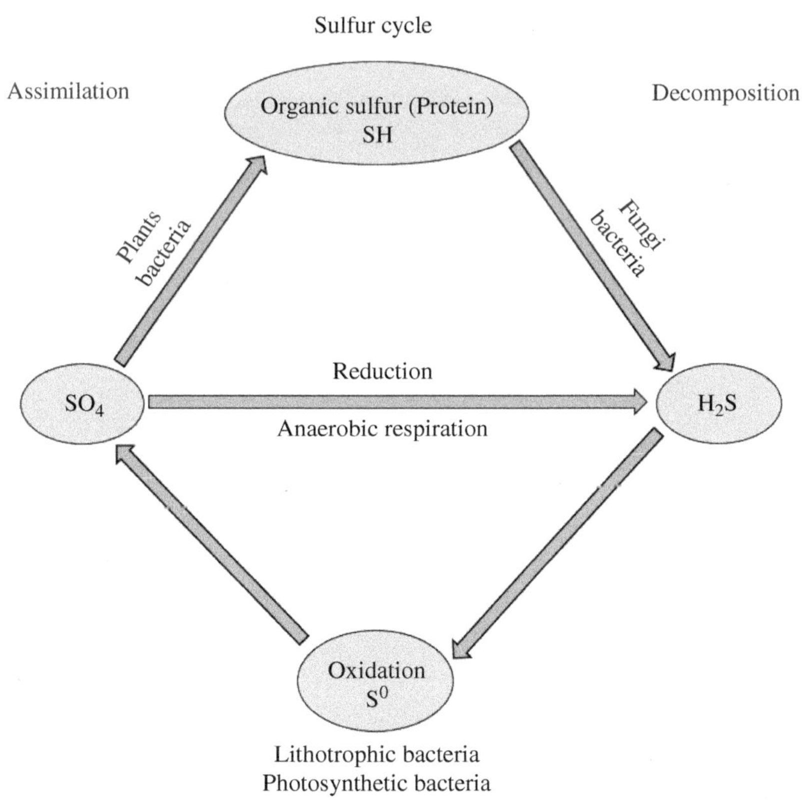

Figure 8.7 The diagram is indicative of the oxidation and reduction that bring about the SO_4 pool and the reservoir hydrogen-sulfide pool extended in soil and sediments.

metabolism. A specific taxonomic group of microbes can perform these transformations on a selective organic and inorganic sulfur compounds. Three dissimilatory metabolic processes are known that are significant for sulfur cycle: sulfate reduction, sulfide oxidation, and sulfur disproportionation reactions. The common example of reduced sulfur compounds is hydrogen sulfide, sulfur, and thiosulfate that can also serve as electron donors for photosynthesis and thus for carbon fixation (Bryant and Frigaard 2006). The sulfur is the important component of the sedimentary cycle except two of its compounds hydrogen sulfide (H_2S) and sulfur dioxide (SO_2), which add a gaseous component (Dick et al. 2008). Sulfur enters in the atmosphere from several natural disasters like volcanic eruptions and combustion of fossil fuels (coal, diesel etc.), from the surface of the ocean and gases released by decomposition. Atmospheric hydrogen sulfide also gets oxidized into sulfur dioxide. Atmospheric sulfur dioxide reacts with atmospheric vapors and falls back to the earth surface in the form of sulfuric acid (acid rain). Plants absorb sulfur from soil in the form of sulfates and incorporated this through a series of metabolic processes into sulfur bearing amino acid in the form of proteins; it then passes through the grazing food chain of ecosystems.

8.3.7 Calcium Cycle

Calcium (Ca) is the fourth most abundant element in mineral nutrition and makes up 3.4% of the total earth's crust mass and plays key physiological and structural roles in plant metabolism. The natural reservoir of calcium is calcium silicates and in sedimentary and metamorphic rocks as calcium carbonates (Alloway 2008). The calcium cycle (Figure 8.8) is one of the key metal cycles coupled to the carbon cycle because the dominant source of calcium is weathering igneous rocks and the dominant sink is through sedimentary

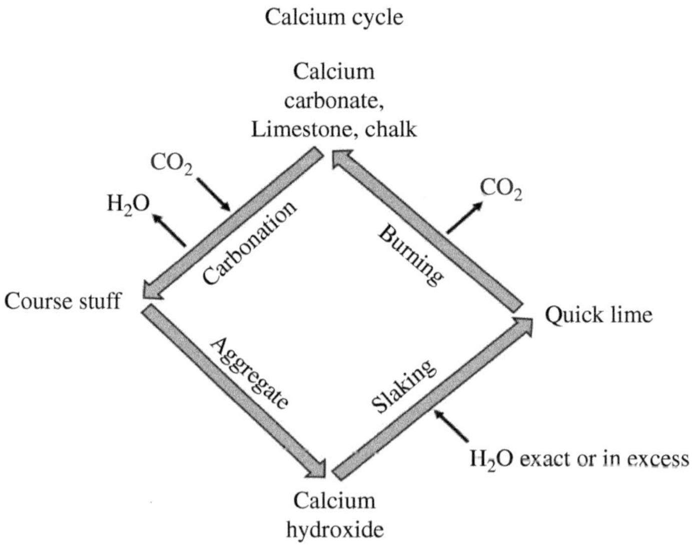

Figure 8.8 Illustration of cyclic calcium cycle in terrestrial ecosystem.

carbonate deposition, similar to carbon cycle (Gadd 2013). Calcium (Ca) comes under a critical element at a variety of scales: it is soluble in natural waters, a biological nutrient, and a major constituent of the dominant mineral sink for carbon in the ocean (as calcium carbonate). There is 4% variation in the Ca isotopic composition ($^{44}Ca/^{40}Ca$) of various Ca reservoirs on Earth surface. In the soil microenvironment, most of the calcium is present as insoluble form until it is "weathered off" of minerals or when organic matter is broken down by microbes into soluble calcium form (Oates 2000). When animals or plants dead decay, their bodies decompose by microorganisms and the calcium is mineralized and released back into the soil. Plant roots also regularly leak minerals, sugars, and other compounds back into the soil including calcium (Ulrich and Reinhard 1997; De-La Rocha et al. 2008; Matthew et al. 2014).

There is a continuous supply of calcium ions into river and ocean by various ways from rocks, organisms, and soils (Walker 1981; Berner 2004). Calcium ions are consumed and removed from aqueous environments as they react to form insoluble structures such as calcium carbonate and calcium silicate (Walker 1981; Ridgwell and Zeebe 2005) which can accumulate to form sediments or the shells of the marine organisms (Raisman and Daniel 2014). The calcium cycle is a common thread between terrestrial, marine, geological, and biological processes (Fantle and Tipper 2014).

8.4 Effect of Salt Stress on Carbon Cycle

Stomatal closure is one of the significant responses of salinity-sensitive plants toward salt stress, subsequently a reduction in carbon dioxide fixation. As CO_2 diffuses through stomata to internal portion of leaves, closure of stomata causes photosynthesis to halt (Heuer 2005). Salt stress also has adverse effect on chlorophyll contents of plants by affecting enzyme involved in chlorophyll biosynthesis (Yang et al. 2020). Cholorophyll content decreases in higher salt-sensitive plants like tomato, potato, pea, and *Phaseolus vulgaris*, while salt stress induce increase in chlorophyll content in salt tolerant plants like pearl millet, mustard, and wheat. Excess NaCl stress inside the chloroplasts of higher plants or in blue green algae cytoplasm causes declined growth rate, which is attributed to decline electron transport activities associated with photosynthesis (Seemann and Critchley 1985; Sudhir and Murthy 2004).

Salt stress is known to inhibit soil bacterial growth hence inhibiting decomposition of organic material and their mineralization. The negative effects exerted by salt stress are decreased carbon dioxide evolution, lower enzymatic activity, and reduction in microbial biomass (León-Lorenzana et al. 2018). Also in drought condition, microbes residing in saline soil face osmotic stress causing cell lysis (Andrei et al. 2012). Salinity is one of the crucial factors of tidal wetlands. Increased global warming, climate change, and rise in sea level have impact on salt content of tidal wetlands. In one study, it has been shown that elevated soil salinity had a negative impact on decomposition rate of organic carbon (Qu et al. 2019). However, the mechanism how salinity affects soil microbes still remains unknown (Chowdhury et al. 2011).

We can conclude that salt stress or salinization is one of the major abiotic stress factors which affect various components of ecosystem as well as different spheres of earth. Salt stress has an overall negative effect on carbon cycle which is implemented by its effect on various components of carbon cycle viz. photosynthesis, microbial decomposers etc. Salt stress causes rate of photosynthesis to decrease causing lower fixation of atmospheric carbon, in case of soil microbes, salt stress causes lowering of microbial biomass hence lowering their benefits in decomposition and carbon mineralization process.

8.5 Effect of Salt Stress on Oxygen and Water Cycle

Oxygen and water cycle or hydrological cycle are two biogeochemical cycles through which oxygen and water moves through biotic and abiotic components of earth and are essential for survival of life (Chahine 1992). Salt stress or salinity has tremendous effect on soil structure, soil microbes, and physiology of plants (Martinez-Beltran 2005). It is estimated that approximately 7% of the globe's land area is affected by salinity (Rhoades and Loveday 1990).

8.5.1 Effect of Salt Stress on Oxygen Cycle

The oxygen cycle involves transition of oxygen through various spheres of earth, which are also known as source of oxygen. The oxygen molecule transits in its different oxidation states in various forms such as ions, oxides, and molecular form (Knoll et al. 2012). Apart from the various abiotic sources of oxygen contributing to oxygen cycle, the abundance of free oxygen on earth's atmosphere and ocean is accredited to the process of photosynthesis performed by plants and phytoplanktons.

Photosynthesis is one of the major parts of oxygen cycle which is known to be affected by salt stress. The effect of salt stress is demonstrated in a number of plants. It has been observed that salt stress can affect various components of photosynthesis such as pigments like chlorophyll, photosynthetic electron transport reactions, fixation etc. In salt-sensitive higher plants like tomato, potato, and pea, chlorophyll content is known to decrease in salt-stressed conditions (Sudhir and Murthy 2004). But in saline resistant plants, chlorophyll contents are known to increase in case of salt stress (Reddy and Vora 1986). Salt stress causes to decrease in photosynthetic electron transport chain activities, which is observed in cyanobacterial cells in which growth rate was decreased when NaCl was accumulated in cytoplasm (Sudhir and Murthy 2004). In another experiment involving *Corylus heterophylla* and *Corylus avellan* Seedlings, it was observed that net photosynthetic rate decreased by 20.5 and 43.2% when plants were subjected to moderate and severe salt stress treatments, respectively. From the above various experimental studies on plants subjected to salt stress, we can conclude that salt stress has overall negative influence on photosynthetic rate. Reduced photosynthetic rate will lead to reduced evolution of oxygen to atmosphere.

8.5.2 Effect of Salt Stress on Water Cycle

In water cycle, there are two ways through which water goes into atmosphere viz., evaporation and transpiration. Water from its sources on earth surface viz. ocean, rivers, lakes, ponds etc., evaporates into the atmosphere with the help of sunlight (Sung 2017). In the process of transpiration, water vapors are exhaled through stomata (Huntington 2006).

More than 90% of water from plants is lost through stomata. Salt stress has major impact on the process of transpiration. In saline condition, water uptake of plants declines as a result of lower osmotic potential of soil water, resulting in a decreased transpiration rate (Nishida et al. 2009). In a study involving effect of salt stress in *Amaranthus tricolor*, a decrease in transpiration rate was observed when the plants were grown in the growth medium with 300 mM NaCl and gradual reversal of transpiration rate was observed when salt stress was removed (Wang and Nii 2000). In another study involving *C. heterophylla* and *C. avellan* seedlings, it has been observed that the rate of transpiration decreased by 2.0, 16.3, 32.0% in the mild, moderate, and severe salt stress treatments, respectively, as compared to those of the control (Luo et al. 2019). Residual transpiration rate also decreases in saline condition where water availability is low. In this case, reduction in residual transpiration rate takes place as a result of improvement of plants performance under above stressed condition, which leads to low water loss to atmosphere as compared to that of non-stressed condition (Hasanuzzaman et al. 2017). As transpiration plays a cardinal role in water cycle, providing 10% of the total water to the atmosphere, we can conclude that the reduction of transpiration rate due to salt stress definitely affects the water cycle in a negative way (Schlesinger and Jasechko 2014).

8.6 Effect of Salt Stress on Nitrogen Fixing Bacteria and Biogeochemical Cycle of Nitrogen

Various environmental stresses negatively affect crop productivity (Battisti and Naylor 2009). Soil salinity, an environmental stress has been considered as the most devastating abiotic stresses (Al-Karaki 2006; Abd-Alla et al. 2019), and it is regarded as the most widespread soil degradation process and a major cause of desertification (Pla and Cobos-Porras 2015). The United Nations Food and Agriculture Organization (FAO) estimated that globally, 19.5% of irrigated agricultural lands and 2.1% of dry-land agriculture were salt-affected soil (FAO 2018; Abd-Alla et al. 2019).

Deforestation has been recognized as one of the major causes of salinization and alkalinization of soils, and thus, soil salinity is a great concern of arid and semiarid regions, where soil salt content is often high and precipitation is insufficient for leaching. The crop yield and productivity of an arable land are severely affected due to salinity in its soil. According to Flowers 2004, the crop yield of over 20% of the cultivated and half of the irrigated lands that are encountering salinity stress of different magnitudes is reduced at par with their genetic potentials. The global change prediction models expect projection of salinity in not very distant future. But, some plants have given relief by showing development of tolerance mechanisms such as changes in gene expression patterns and in metabolic responses related to adaptation/tolerance (ion homeostasis, osmoprotectants, and antioxidant system) in order to face the challenges posed by salt stress like water deficit, ion toxicity, and oxidative stress.

Soil salinization is of two different types: Primary and Secondary. Primary salinization occurs naturally when the soil parent material is rich in soluble salts, or in the presence of shallow saline groundwater table. This may occur either due to weathering of parent material containing soluble salts of various types, mainly chlorides of sodium, calcium, and magnesium, and, to a lesser extent, sulfate and carbonates or because of deposition of microscopic salt particles carried by the wind and rain to oceans (Munns 2009; Pla and Cobos-Porras, 2015). A high concentration of salt gets accumulated in those arid and semi-arid regions, where rainfall is insufficient to wash out soluble salts from the soil or where drainage is restricted (USDA-ARS 2008). In case of secondary salinization, certain anthropogenic activities create changes in the hydrologic balance in the soil between precipitation and water used by crops and by using poor-quality water for irrigation.

Kennedy and Islam (2001) found that free-living and rhizosphere N_2-fixing bacteria associated with wheat could produce 10–30 kg N ha^{-1} $crop^{-1}$. Similarly, Roper et al. 1989 found that free-living bacteria in field conditions could fix 1–12 kg N ha^{-1} during short periods of two to four weeks, but where warm, moist conditions coincide with fresh stubbles, annual potentials of up to 38 kg N ha^{-1} $year^{-1}$ have been calculated (Gupta et al. 2006). High salinity affects not only plant growth but also symbiotic relations in legumes mainly by inhibiting specific nitrogenase activity (Tejera et al. 2005; Lopez et al. 2008). The nitrogenase enzyme complex which catalyses biological nitrogen fixation in legume-*Rhizobium* association is extremely sensitive to oxygen, and a protein named *leghemoglobin* is responsible for regulating O_2 level in the nodule/infected plant cells. Delgado et al. 1993 reported that inhibition of nitrogenase activity was due to reduction of leghemoglobin, by oxidative stress and by carbohydrate contents of both the cytosol and the bacteroides. The sucrose synthase and phosphoenolpyruvate carboxylase activity is inhibited under salt stress which lowers carbon supply to bacteroides, mainly in the form of malate and thus interfering with the nitrogenase activity and plant growth (Delgado et al. 1994; Soussi et al. 1999). Bolanos et al. 2003 found that salinity adversely affects nodulation and nitrogen fixation in legume-*Rhizobium* association leading to prevention of establishment of legumes and hence, crop yield. Unsuccessful symbiosis has been observed under saline conditions probably due to the effect of salinity on the establishment of rhizobia. Also, the legumes and the nodule formation are more susceptible to salt stress than the rhizobia.

The legume genotypes that are resistant to hypersaline conditions are now selected as a measure to increase legume production (Kucuk and Kivanc 2008). Lauter et al. (1981) and Singleton et al. (1982) reported that the rhizobia grows and exists better under salinity and high osmotic stress than their hosts. The endophytic bacteria *Stenotrophomonas maltophilia* and *Stenotrophomonas rhizophila* are salt resistant and able of nitrogen-fixing (Park et al. 2005; Berg et al. 2013).

8.7 Effect of Salt Stress on Phosphorous Bacteria and Biogeochemical Cycle of Phosphorus

Morrissey et al. (2014) studied the harmful effect of salinity and concluded that salinity could decrease soil organic matters essential for plant growth (by enhancing the activity of carbon-degrading extracellular enzymes) and affecting soil microorganism community.

P bacteria play a central role in soil P cycle and significant role in mediating the transfer of P between different organic and inorganic soil fractions, subsequently releasing available P for plant requirement (McLaughlin et al. 1988; Oberson et al. 2001). There are two forms of microbial P solubilization: first, inorganic phosphate (*Pi*) released by the solubilization processes (Rodriguez and Fraga 1999) and second, inorganic phosphate released from accumulated P from biomass of microorganisms (Oehl et al. 2001). Inorganic phosphate-solubilizing microorganisms constitute portions of the soil microbial population and vary from soil to soil (Kucey et al. 1989). Srinivasan et al. (2012) performed an experiment to find out the effect of salt on phosphorus bacteria. They isolated phosphate-solubilizing bacteria (PSB) from rhizosphere soils of different crops and selected 12 PSB that solubilize tricalcium phosphate (TCP) in the presence of different concentrations of NaCl. First, the primary culture of bacteria grown overnight and then next day were inoculated into 50 ml Luria broth containing different concentrations of NaCl (0.09 (control), 0.4, 0.8, 1.2, 1.6 and 2.0 M), and the growth was observed at an interval of 24 hours up to eight days by determining the viable cell counts by serial dilution plate count method using Pikovskaya's agar medium. Among 12 PSB, *Aerococcus sp.* strain $PSBCRG_1$-1 recorded the highest (12.15) log viable cell count at 0.4 M NaCl concentration after seven days after incubation (DAI) and the lowest log cell count (1.39) was recorded by *Pseudomonas aeruginosa* strain $PSBI_3$-1 at 2.0 M NaCl concentration after 24 hours of incubation. In general, the population counts (log viable cell counts) of majority of the isolates were found to be higher than that of the reference strain (*Poecilotheria striata*), particularly at higher NaCl concentrations. At 15 DAI, strain *Aerococcus sp.* strain $PSBCRG_1$-1 irrespective of NaCl concentrations showed the maximum P-solubilization (12.12%) which was significantly superior over all other isolates. Among different salt concentrations tested, the maximum solubilization was found at 0.8 M NaCl (6.93%) which was significantly superior over all other concentrations including control. The population counts of a majority of the isolates were found to be higher than that of the reference strain (*P. striata*) particularly at higher NaCl concentrations. This could be due to less stress adaptability of the reference strain where as the isolates could tolerate more stress as they were isolated from salt-affected soils.

8.8 Effect of Salt Stress on Sulfur Bacteria and Biogeochemical Cycle of Sulfur

Sulfur is an essential element for plant growth, involves in many biochemical processes, and constituents of major metabolic compounds such as antioxidant (e.g. glutathione), amino acids (e.g. methionine and cysteine), proteins, and sulfo-lipids (Abdallah et al. 2010). Sulfur is a ubiquitous and essential element for plants as well as all living organisms. Sulfur cycling, which is primarily driven by sulfate reduction, is interconnected with other important biogeochemical cycles such as carbon, nitrogen, oxygen, calcium and therefore has profound implications for both cellular and ecosystem level processes. Sulfur transforming microorganisms have evolved diverse genetic, metabolic, and peculiar phenotypic characteristics to fill an array of ecological niches in terrestrial as well as marine sediments. The Chiprana Lagoon located in the Ebro depression (Arago, Spain) is the only permanent hypersaline endorheic lake of Western Europe. The water has an athalassohaline composition with magnesium sulfate and sodium chloride as the main salts and an overall salinity from 53 to 84%. The development of anoxygenic photosynthetic bacteria in the hypolimnion had been documented previously during stratification periods. The community of anoxygenic phototrophs was mainly dominated by dense populations of green sulfur bacteria (GSB) (Guerrero et al. 1991; Vila et al. 2002) resembling the formerly named *Chlorobium vibrioforme*. Most of the *Prosthecochloris* strains isolated were able to grow at high salt concentrations (up to 7% NaCl or even 9–10% NaCl) and were regarded as halophilic bacteria. The phylotypes clustered into the genus *Prosthecochloris* were exclusively found at salinity concentrations over 1%, and they were the most common phylotypes in hypersaline water as determined by the denaturing gradient gel electrophoresis (DGGE) technique. *Vibrioformis DSM* 260 T were found to grow in a wide range of salinities. These results contribute to confirm that all members of the *Prosthecochloris* group are homogeneous from both an ecological and physiological point of view, since all of them have been exclusively isolated from saline environments (Imhoff 2001; Vila et al. 2002; Alexander and Imhoff 2006).

Thus, adenosine-5- phosphosulfate (APS) reductase α (aprA) and methyl coenzyme M reductase α (mcrA) gene enzymes are specific for both sulfur-oxidizing prokaryotes (SOP) and sulfur-reducing prokaryotes (SRP), respectively (Montoya et al. 2011). The important sulfate-reducing prokaryotes are: acetoclastic sulfate belonging to *Desulfobacteraceae* family, *Desulfobacter halotolerans* (grows up to 13% NaCl concentration) (Foti et al. 2007) and *Desulfohalobiaceae* members have higher salt tolerances (up to 25% NaCl concentration). Several studies have also reported that a high genomic CG content (often upwards of 60%) and a GC bias at the codon usage level are common adaptations to hypersaline environments.

In the natural habitat of environment, the microbial sulfur cycle is most active in soda lakes. There are two parts of global sulfur cycle: the first oxidative part of the sulfur cycle is driven by chemolithoautotrophic haloalkaliphilic sulfur-oxidizing bacteria (SOB), which are unique for soda lakes and second the halo alkaliphilic SOB are present in high numbers at the upper surface sediment layer of various soda lakes. The reductive part of the sulfur cycle is active in the anoxic layers of the sediments of soda lakes. The high sulfate reduction rates only are impeded by high salt (salt-saturated) conditions. This was proved by experiments by the statistical measurements of *in situ* of sulfate reduction rates and with sediment slurries using sulfate, thiosulfate, or elemental sulfur as electron acceptors. The major haloalkaliphilic autotrophic sulfate-reducing bacteria belong to the three major groups: *Desulfonatronovibrio, Desulfonatronum and Desulfonatronospira* with a clear propensity to grow at high salt-saturating conditions. Soda lakes form a specific type of salt lakes where sodium carbonate/bicarbonate is among the dominant salts. At low-to-moderate salinity, the long sulfur cycle is possible with complete oxidation of sulfide/thiosulfate to sulfate by moderately natronophilic SOB and sulfate/thiosulfate reduction by natronophilic SRB. In saturated soda brines, only a limited number of players can operate such as the extremely natronotolerant SOB *Thioalkalivibrio* and the extremely natronophilic sulfur reducers, such as very specialized representatives of the *Halanaerobiales* and of the *natronoarchaea*.

8.9 Future Prospective and Conclusion

Nutrients are chemical elements that are essential for plant metabolism, microorganisms, and ecosystems. The deficiency of these elements largely affects the ecological productivity. The excess presence of these nutrients in soil can also cause environmental damage and plant life cycle, caused by toxicity and other problems. These nutrients continuously recycle among inorganic and organic forms within ecosystems. The key components of nutrient cycles are illustrated by the carbon, nitrogen, water, phosphorus, oxygen, calcium, and sulfur cycles. Salinity is one of the dangerous threats worldwide to agriculture production and ecosystem because it affects nutrient uptake and growth of plants as well as soil microbial community beneficial for plant and soil fertility. Effects of salinity on plants nutrient uptake and soil microbes have been studied extensively, but usually separately. The comparative study of interaction between soil microbes, salinity, and nutrient cycle is needed for better understanding of salinity effect on these biotic and abiotic environmental components. These components and parameters are discussed here in single platform. Soil salinization could change soil microbial community, which is well known for driving soil nutrient cycle in various land ecosystems. Addition of organic materials such as plant residues or manures as nutrient sources for microbes may be as important strategy to ameliorate saline soils. It is important to carry out detailed studies of internal and external factors that affect the physiological and biochemical processes of microorganisms under salt stress conditions. In this chapter, we conclude that soil salinity largely affects the plant nutrition uptake, productivity, soil microorganisms, and nutrient cycle. Future strategies shall focus on the regulation and expression of microorganism specific genes and pathways under salinity stress.

References

Cross, A.F. and Schlesinger, W.H. (1995). A literature review and evaluation of the Hedley fractionation: Applications to the biogeochemical cycle of soil phosphorus in natural ecosystems. *Geoderma* 64: 197–214.

Fitter, A.H. and Hay, R.K.M. (1987). *Environmental Physiology of Plants*. Academic Press: London.

Abd-Alla, M.H., Nafady, N.A., Bashandy, S.R., and Hassan, A.A. (2019). Mitigation of effect of salt stress on the nodulation, nitrogen fixation and growth of chickpea (*Cicer arietinum L.*) by triple microbial inoculation. *Rhizosphere* 10: 100148.

Abdallah, M., Dubousset, L., Meuriot, F. et al. (2010). Effect of mineral sulfur availability on nitrogen and sulfur uptake and remobilization during the vegetative growth of Brassica napus L. *J. Exp. Bot.* 61: 2335–2346.

Al-Karaki, G.N. (2006). Nursery inoculation of tomato with arbuscular mycorrhizal fungi and subsequent performance under irrigation with saline water. *Sci. Hortic.* 109: 1–7.

Alexander, B. and Imhoff, J.F. (2006). Communities of green sulfur bacteria in marine and saline habitats analyzed by gene sequences of 16S rRNA and Fenna-Matthews-Olson protein. *Int. Microbiol.* 9: 259–266.

Amrani, A., Said-Ahamed, W., Lewan, M.D., and Aizenshtat, Z. (2006). Experiments on δ34S mixing between organic and inorganic sulfur species during thermal maturation. *Geochim. Cosmochim. Acta Theriol.* 70: 5146–5161.

Andrei, A.Ş., Banciu, H.L., and Oren, A. (2012). Living with salt: metabolic and phylogenetic diversity of archaea inhabiting saline ecosystems. *FEMS Microbiol. Lett.* 330 (1): 1–9.

Battisti, D.S. and Naylor, R.L. (2009). Historical warnings of future food insecurity with unprecedented seasonal heat. *Science* 323: 240–244.

Berg, G., Alavi, M., Schmidt, C.S. et al. (2013). Biocontrol and osmoprotection for plants under salinated conditions. In: *Molecular Microbial Ecology of the Rhizosphere*, 1e, vol. 2 (ed. F.J. de Bruijn), 587–592. Wiley.

Blank, L.M. (2012). The cell and P: from cellular function to biotechnological application. *Curr. Opin. Biotechnol.* 23: 846–851.

Bolanos, L., El-Hamdaoui, A., and Bonilla, I. (2003). Recovery of development and functionality of nodules and plant growth in salt-stressed Pisum sativum-Rhizobium leguminosarum symbiosis by boron and calcium. *J. Plant Physiol.* 160 (12): 1493–1497.

Bryant, D.A. and Frigaard, N.U. (2006). Prokaryotic photosynthesis and phototrophy illuminated. *Trends Microbiol.* 14: 488–496.

Canfield, D.E. (2001). Biogeochemistry of sulfur isotopes. *Rev. Mineral. Geochem.* 43: 607–636.

Chahine, M.T. (1992). The hydrological cycle and its influence on climate. *Nature* 359 (6394): 373–380.

Chinnusamy, V. and Zhu, J.K. (2003). Plant salt tolerance. In: *Plant Responses to Abiotic Stress*, 241–270. Berlin, Heidelberg: Springer.

Chowdhury, N., Marschner, P., and Burns, R.G. (2011). Soil microbial activity and community composition: impact of changes in matric and osmotic potential. *Soil Biol. Biochem.* 43 (6): 1229–1236.

Cleveland, C.C., Townsend, A.R., Schimel, D.S. et al. (1999). Global patterns of terrestrial biological nitrogen (N2) fixation in natural ecosystems. *Global Biogeochem. Cycles* 13 (2): 623–645.

Cordell, D., Drangert, J.O., and White, S. (2009). The story of phosphorus: global food security and food for thought. *Glob. Environ. Chang.* 19: 292–305.

De-La Rocha, C.L., Hoff, C., and Bryce, J. (2008). Calcium cycle. In: *Encyclopedia of Ecology* (ed. S.E.J.D. Fath), 507–513. Oxford: Academic Press.

Delgado, M., Garrido, J., Ligero, F., and Lluch, C. (1993). Nitrogen fixation and carbon metabolism by nodules and bacteroids of pea plants under sodium chloride stress. *Physiol. Plant.* 89 (4): 824–829.

Delgado, M.J., Ligero, F., and Lluch, C. (1994). Effects of salt stress on growth and nitrogen fixation by pea, faba-bean, common bean and soybean plants. *Soil Biol. Biochem.* 26 (3): 371–376.

Dick, W.A., Kost, D., and Chen, L. (2008). Availability of sulfur to crops from soil and other sources. *Sulf. A Miss. Link Betw. Soils Crops Nutr.* 50: 59–82.

Dixon, R. and Kahn, D. (2004). Genetic regulation of biological nitrogen fixation. *Nat. Rev. Microbiol.* 2: 621–631.

Elser, J.J., Bracken, M.E.S., Cleland, E.E. et al. (2007). Global analysis of nitrogen and phosphorus limitation of primary producers in freshwater, marine, and terrestrial ecosystems. *Ecol. Lett.* 10: 1135–1142.

Matthew, F., Tipper, S., and Edward, T. (2014). Calcium isotopes in the global biogeochemical Ca cycle: implications for development of a Ca isotope proxy. *Earth Sci. Rev.* 131: 148–177.

FAO (2018). http://www.fao.org/soils-portal/soil-management/management-of-some-problem-soils/salt-affected-soils/more-information-on-salt-affected-soils/en.

Farquhar, J. and Johnston, D.T. (2008). The oxygen cycle of the terrestrial planets: insights into the processing and history of oxygen in surface environments. *Rev. Mineral. Geochem.* 68 (1): 463–492.

Flowers, T.J. (2004). Improving crop salt tolerance. *J. Exp. Bot.* 55 (396): 307–319.
Flowers, T.J. and Colmer, T.D. (2008). Salinity tolerance in halophytes. *New Phytol.*: 945–963.
Filippelli, G.M. (2002). The global phosphorus cycle. *Rev. Mineral. Geochem.* 48 (1): 391–425.
Galloway, J.N., Dentener, F.J., Capone, D.G. et al. (2004). Nitrogen cycles: past, present, and future. *Biogeochemistry* 70 (2): 153–226.
Gruber, N. and Galloway, J.N. (2008). An Earth-system perspective of the global nitrogen cycle. *Nature* 451: 293–296.
Guerrero, M.C., Balsa, J., Pascale, M. et al. (1991). Caracterizacio´n limnolo´gica de la laguna Salada de Chiprana (Zaragoza, Espan˜a) y sus comunidades de bacterias fototro´ficas. *Limnetica* 7: 83–96.
Gupta, V.V.S.R., Roper, M.M., and Roget, D.K. (2006). Potential for non-symbiotic N_2-fixation in different agroecological zones of southern Australia. *Aust. J. Soil Res.* 44 (4): 343–354.
Harrison, A.F. and Helliwell, D.R. (1979). A bioassay for comparing phosphorus availability in soils. *J. Appl. Ecol.* 16: 497–505.
Hedin, L.O., Armesto, J.J., and Johnson, A.H. (1995). Patterns of nutrient loss from unpolluted, old growth temperate forests: evaluation of biogeochemical theory. *Ecology* 76 (2): 493–509.
Hasanuzzaman, M., Davies, N.W., Shabala, L. et al. (2017). Residual transpiration as a component of salinity stress tolerance mechanism: a case study for barley. *BMC Plant Biol.* 17 (1): 107.
Huntington, T.G. (2006). Evidence for intensification of the global water cycle: review and synthesis. *J. Hydrol.* 319 (1-4): 83–95.
Hobbie, S.E. (1995). Direct and indirect effects of plant species on biogeochemical processes in arctic ecosystems. In: *Arctic and Alpine Biodiversity: Patterns, Causes and Ecosystem Consequences*, 213–224. Berlin, Heidelberg: Springer.
Hokmabadi, H., Arzani, K., and Grierson, P.F. (2005). Growth, chemical composition, and carbon isotope discrimination of pistachio (*Pistacia vera L.*) rootstock seedlings in response to salinity. *Aust. J. Agric. Res.* 56: 135–144.
Imhoff, J.F. (2001). True marine and halophilic anoxygenic phototrophic bacteria. *Arch. Microbiol.* 176: 243–254.
Jasinski, S.M. (2012). Phosphate rock. *Mineral Commodity Summaries*, US Geological Survey, pp. 118–119.
Kasting, J.F. and Canfield, D.E. (2012). The global oxygen cycle. In: *Fundamentals of Geobiology*, 93–104. Blackwell Publishing.
Keeling, R.F. (1995). The atmospheric oxygen cycle: The oxygen isotopes of atmospheric CO_2 and O_2 and the O_2/N_2 ratio. *Rev. Geophys.* 33 (S2): 1253–1262.
Kennedy, I. and Islam, N. (2001). The current and potential contribution of asymbiotic nitrogen fixation to nitrogen requirements on farms: a review. *Aust. J. Exp. Agric.* 41 (3): 447–457.
Koerselman, W. and Meuleman, A.F.M. (1996). The vegetation N: P ratio: a new tool to detect the nature of nutrient limitation. *J. Appl. Ecol.* 33: 1441–1450.
Kozlowski, T.T. (1997). Responses of woody plants to flooding and salinity. In: *Tree Physiology Monograph*, vol. 1 (ed. J. Cooke), 1–29. Victoria, Canada: Heron Publishing.
Kucey, R.M.N., Janzen, H.H.A.F.M., and Leggett, M.E. (1989). Microbiologically mediated increases in plant-available-phosphorus. *Adv. Agron.* 42: 199–228.
Knoll, A.H., Canfield, D.E., and Konhauser, K.O. (eds.) (2012). *Fundamentals of Geobiology*. Wiley.

Kucuk, C. and Kivanc, M. (2008). Preliminary characterization of Rhizobium strains isolated from chickpea nodules. *Afr. J. Biotechnol.* 7: 6.

Lane, N. (2003). *Oxygen: The Molecule that Made the World*. Oxford: Oxford University Press.

Lauter, D.J., Munns, D.N., and Clarkin, K.L. (1981). Salt response of chickpea as influenced by N supply 1. *Agron. J.* 73: 961–966.

LeBauer, D.S. and Treseder, K.K. (2008). Nitrogen limitation of net primary productivity in terrestrial ecosystems is globally distributed. *Ecology* 89: 371–379.

León-Lorenzana, D., Arit, S., Delgado-Balbuena, L. et al. (2018). Soil salinity controls relative abundance of specific bacterial groups involved in the decomposition of maize plant residues. *Front. Ecol. Evol.* 6: 51.

Lozupone, C.A. and Knight, R. (2007). Global patterns in bacterial diversity. *Proc. Natl. Acad. Sci. U. S. A.* 104: 11436–11440.

Luo, D., Hou, X., Zhang, Y. et al. (2019). CaDHN5, a dehydrin gene from pepper, plays an important role in salt and osmotic stress responses. *Int. J. Mol. Sci.* 20 (8): 1989.

Foti, M., Sorokin, D.Y., Lomans, B. et al. (2007). Diversity, activity, and abundance of sulfate-reducing bacteria in saline and hypersaline soda lakes. *Appl. Environ. Microbiol.* 73: 2093–2100.

Montoya, L., Lozada-Chávez, I., Amils, R. et al. (2011). The sulfate-rich and extreme saline sediment of the ephemeral tirez lagoon: a biotope for acetoclastic sulfate-reducing bacteria and hydrogenotrophic methanogenic archaea. *Int. J. Microbiol.* 2011: 753–758.

Morrissey, E.M., Gillespie, J.L., Morina, J.C., and Franklin, R.B. (2014). Salinity affects microbial activity and soil organic matter content in tidal wetlands. *Glob. Chang. Biol.* 20: 1351–1362.

Munns, R. (2009). Strategies for crop improvement in saline soils. In: *Salinity and Water Stress*, vol. 44 (eds. M. Ashraf, M. Ozturk and H.R. Athar), 99–110.

Martinez-Beltran, J., (2005). Overview of salinity problems in the world and FAO strategies to address the problem. Managing saline soils and water: science, technology and social issues. *Proceedings of the International Salinity Forum*, Riverside, California.

Munns, R. and Tester, M. (2008). Mechanisms of salinity tolerance. *Annu. Rev. Plant Biol.* 59: 651–681.

Munns, R. (2002). Comparative physiology of salt and water stress. *Plant Cell Environ.* 25: 239–250.

Navarro-González, R., Villagran-Muniz, M., Sobral, H. et al. (2001). The physical mechanism of nitric oxide formation in simulated lightning. *Geophys. Res. Lett.* 28: 3867–3870.

Nishida, K., Khan, N.M., and Shiozawa, S. (2009). Effects of salt accumulation on the leaf water potential and transpiration rate of pot-grown wheat with a controlled saline groundwater table. *Soil Sci. Plant Nutr.* 55 (3): 375–384.

Oates, T. (2000). Lime and limestone. In: *Kirk-Othmer Encyclopedia of Chemical Technology*.

Oberson, A., Friesen, D.K., Rao, I.M. et al. (2001). Phosphorus transformations in an oxisol under contrasting landuse system: The role of the microbial biomass. *Plant Soil* 237: 197–210.

Oehl, F., Oberson, M., Probst, A. et al. (2001). Kinetics of microbial phosphorus uptake in cultivated soils. *Biol. Fertil. Soils* 34: 31–41.

Parida, A.K. and Das, A.B. (2005). Salt tolerance and salinity effects on plant: a review. *Ecotoxicol. Environ. Saf.* 60: 324–349.

Park, M., Kim, C., Yang, J. et al. (2005). Isolation and characterization of diazotrophic growth promoting bacteria from rhizosphere of agricultural crops of Korea. *Microbiol. Res.* 160: 127–133.

Perroni, Y., Garcia-Oliva, F., Tapia-Torres, Y., and Souza, V. (2014). Relationship between soil P fractions and microbial biomass in an oligotrophic grassland-desert scrub system. *Ecol. Res.* 29: 463–472.

Petsch, S.T. (2003). The global oxygen cycle. *Treat. Geochem.* 8: 682.

Pla, C.L. and Cobos-Porras, L. (2015). Salinity: physiological impacts on legume nitrogen fixation. In: *Legume Nitrogen Fixation in a Changing Environment* (eds. S. Sulieman and L.-S.P. Tran), 35–65. Cham: Springer.

Powers, R.F. (1980). Mineralizable soil nitrogen as an index of nitrogen availability to forest trees. *Soil Sci. Soc. Am. J.* 44: 1314–1320.

Qu, W., Li, J., Han, G. et al. (2019). Effect of salinity on the decomposition of soil organic carbon in a tidal wetland. *J. Soils Sediments* 19 (2): 609–617.

Qi, L. and Yang, J. (2017). Microbial community composition regulates SOC decomposition response to forest conversion in a Chinese temperate forest. *Ecol. Res.* 32: 163–172.

Yanai, R.D. (1992). Phosphorus budget of a 70-year-old northern hardwood forest. *Biogeochemistry* 17: 1–22.

Raisman, S.M. and Daniel, T. (2014). *Ocean Acidification: Elements and Considerations.* Hauppauge, New York: Nova Science.

Raven, M.R., Adkins, J.F., Werne, J.P. et al. (2015). Sulfur isotopic composition of individual organic compounds from Cariaco Basin sediments. *Org. Geochem.* 80: 53–59.

Ridgwell, A. and Zeebe, R.E. (2005). The role of the global carbonate cycle in the regulation and evolution of the Earth system. *Earth Planet. Sci. Lett.* 234 (3-4): 299–315.

Rodriguez, H. and Fraga, R. (1999). Phosphate solubilizing bacteria and their role in plant growth promotion. *Biotechnol. Adv.* 17: 319–339.

Roper, M., Marschke, G., and Smith, N. (1989). Nitrogenase activity (C_2H_2 reduction) in soils following wheat straw retention: effects of straw management. *Aust. J. Agric. Res.* 40 (2): 241–253.

Rousk, J., Elyaagubi, F.K., Jones, D.L., and Godbold, D.L. (2011). Bacterial salt tolerance is unrelated to soil salinity across an arid agroecosystem salinity gradient. *Soil Biol. Biochem.* 43: 1881–1887.

Reddy, M.P. and Vora, A.B. (1986). Salinity induced changes in pigment composition and chlorophyllase activity of wheat. *Indian J. Plant Physiol.* 29 (4): 331–334.

Rhoades, J.D. and Loveday, J. (1990). Salinity in irrigated agriculture. *Agronomy* 30: 1089–1142.

Sairam, R.K. and Tyagi, A. (2004). Physiology and molecular biology of salinity stress tolerance in plants. *Curr. Sci.* 86 (3): 407–421.

Scholz, R.W., Ulrich, A.E., Eilitta, M., and Roy, A. (2013). Sustainable use of phosphorus: a finite resource. *Sci. Total Environ.* 461: 799–803.

Salisbury, F.B. and Ross, C.W. (1992). *Plant Physio*. Belmont, California-USA: Wadsworth Pub. Com., Inc.

Schlesinger, W.H. and Jasechko, S. (2014). Transpiration in the global water cycle. *Agric. For. Meteorol.* 189: 115–117.

Sudhir, P. and Murthy, S.D.S. (2004). Effects of salt stress on basic processes of photosynthesis. *Photosynthetica* 42 (2): 481–486.

Sung, L. (2017). *Effect of Hydrological Cycle on Climate*. Seagull Publications.

Setia, R., Smith, P., Marschner, P., Gottschalk, P., Baldock, J., Verma, V., Setia, D and, Smith J. (2012). Simulation of salinity effects on past, present, and future soil organic carbon stocks. *Environ. Sci. Technol.*, (46): 1624-1631.

Singleton, P.W., El Swaify, S.A., and Bohlool, B.B. (1982). Effect of salinity on Rhizobium growth and survival. *Appl. Environ. Microbiol.* 44: 884–890.

Soussi, M., Lluch, C., and Ocana, A. (1999). Comparative study of nitrogen fi xation and carbon metabolism in two chick-pea (*Cicer arietinum* L.) cultivars under salt stress. *J. Exp. Bot.* 50 (340): 1701–1708.

Seemann, J.R. and Critchley, C. (1985). Effects of salt stress on the growth, ion content, stomatal behaviour and photosynthetic capacity of a salt-sensitive species *Phaseolus vulgaris* L. *Planta* 164 (2): 151–162.

Srinivasan, R., Yandigeri, M.S., Kashyap, S., and Alagawadi, A.R. (2012). Effect of salt on survival and P-solubilization potential of phosphate solubilizing microorganisms from salt affected soils. *Saudi J. Biol. Sci.* 19 (4): 427–434.

Stevenson, F.J. and Cole, M.A. (1999). *Cycles of Soils: Carbon, Nitrogen, Phosphorus, Sulfur, Micronutrients*. Wiley.

Takai, K. (2019). The nitrogen cycle: a large, fast, and mystifying cycle. *Microbes Environ.* 34 (3): 223–225.

Tejera, N.A., Campos, R., Sanjuan, J., and Lluch, C. (2005). Effect of sodium chloride on growth, nutrient accumulation, and nitrogen fixation of common bean plants in symbiosis with isogenic strains. *J. Plant Nutr.* 28 (11): 1907–1921.

Tiessen, H., Stewart, J.W.B., and Cole, C.V. (1984). Pathways of phosphorus transformations in soils of differing pedogenesis. *Soil Sci. Soc. Am. J.* 48: 853–858.

Tilman, D., Reich, P.B., Knops, J. et al. (2001). Diversity and productivity in a long-term grassland experiment. *Science* 294 (5543): 843–845.

Tripathi, S., Kumari, S., Chakraborty, A. et al. (2006). Microbial biomass and its activities in salt-affected coastal soils. *Biol. Fertil. Soils* 42: 273–277.

Ulrich, B. and Reinhard, B. (1997). The Biogeochemical cycles of Phosphorus: A Review of local and global consequences of the atmospheric input. *Toxicol. Environ. Chem.* 67: 171–188.

USDA-ARS. (2008). Research databases. http: //www.ars.usda.gov/Services/docs.htm/ docid=8908.

Vila, X., Guyoneaud, R., Cristina, X.P. et al. (2002). Green sulfur bacteria from hypersaline Chiprana Lake (Monegros, Spain): habitat description and phylogenetic relationship of isolated strains. *Photosynth. Res.* 71: 165–172.

Vitousek, P.M. (2004). *Nutrient Cycling and Limitation: Hawai'i as a Model System*. Princeton University Press.

Vitousek, P.M. and Howarth, R.W. (1991). Nitrogen limitation on landand in the sea: how can it occur? *Biogeochemistry* 13: 87–115.

Vitousek, P.M. and Field, C.B. (1999). Ecosystem constraints to symbiotic nitrogen fixers: a simple model and its implications. *Biogeochemistry* 46:179–202.

Vitousek, P.M., Cassman, K.E.N., Cleveland, C. et al. (2002). Towards an ecological understanding of biological nitrogen fixation. In: *The Nitrogen Cycle at Regional to Global Scales*, 1–45. Dordrecht: Springer.

Vitousek, P.M., Porder, S., Houlton, B.Z., and Chadwick, O.A. (2010). Terrestrial phosphorus limitation: mechanisms, implications, and nitrogen–phosphorus interactions. *Ecol. Appl.* 20 (1): 5–15.

Walker, T.W. and Syers, J.K. (1976). The fate of phosphorus during pedogenesis. *Geoderma* 15: 1–19.

Wang, Y. and Nii, N. (2000). Changes in chlorophyll, ribulose bisphosphate carboxylase-oxygenase, glycine betaine content, photosynthesis and transpiration in Amaranthus tricolor leaves during salt stress. *J. Hort. Sci. Biotechnol.* 75 (6): 623–627.

Xu, J., Feng, Y., Wang, Y. et al. (2016). The foliar spray of Rhodopseudomonas palustris grown under Stevia residue extract promotes plant growth via changing soil microbial community. *J. Soils Sediments* 16: 916–923.

Yan, N., Marschner, P., Cao, W. et al. (2015). Influence of salinity and water content on soil microorganisms. *Int. Soil Water Coner. Res.* 3 (4): 316–323.

Yang, Z., Li, J.L., Liu, L.N. et al. (2020). Photosynthetic regulation under salt stress and salt-tolerance mechanism of sweet sorghum. *Front. Plant Sci.* 10: 1722.

Zhang, W.W., Chong, W.A.N.G., Rui, X.U.E., and Wang, L.J. (2019). Effects of salinity on the soil microbial community and soil fertility. *J. Integr. Agric.* 18 (6): 1360–1368.

9

Salt-Induced Effects on Crop Plants and Counteract Mitigating Strategy by Antioxidants System

Indrajeet Kumar, Umesh Kumar, Prince Kumar Singh, and Rajesh Kumar Sharma

Department of Botany, Banaras Hindu University, Varanasi, Uttar Pradesh, India

9.1 Introduction

Plants suffer with numbers of abiotic and biotic stresses resulting in the reduction of growth and developmental attributes such as height, number of leaves, leaf area, number of branches, root volume, and physiological processes which ultimately leads to the reduction of biomass production. Salinity is one of the important menaces to crop plants (Kumar and Sharma 2018). Flowers et al. (1986) divided all the plants into glycophytes and halophytes based on their abilities to grow on various salt concentrations. Plants which grow and complete their life cycle in saline environment are halophytes. Most of the plants including agricultural crops are glycophytic and are unable to complete their life cycle in high concentration of salt. Therefore, presence of salt in soil is one of the most significant abiotic stresses in agriculture sector (Liang et al. 2018). Currently, about one-third of agricultural land is affected by salinity worldwide.

Globally, about 1128 m ha area is affected by salinity and sodicity stresses. The regions with emphasis of salt-affected soils are Middle East (189 m ha) followed by Australia (169 m ha) and North Africa (144 m ha). South Asia, including India, has about 52 m ha of salt-affected area. In India, about 6.73 m ha area is salt-affected out of which 3.77 m ha are sodic and 2.96 being saline, in which with Gujarat (2.23 m ha), Uttar Pradesh (1.37 m ha), Maharashtra (0.61 m ha), West Bengal (0.44 m ha), and Rajasthan (0.38 m ha) together cover almost 75% of saline and sodic soils of the country (Singh et al. 2018). Hyper-saline soils with soluble salts inhibit plant growth at different stages causing to yield differences between crops and also the differences in their ion compositions at maturity. The loss of farmable land due to salinization directly affects the food requirement of the world population (Liang et al. 2018). Different levels of salinity and their effects on crop plants are shown in Table 9.1.

Physiology of Salt stress in Plants: Perception, Signalling, Omics and Tolerance Mechanism, First Edition.
Edited by Pratibha Singh, Madhulika Singh, Rajiv Kumar Singh, and Sheo Mohan Prasad.

Table 9.1 Different levels of salinity and their effects on crops.

S. No.	Electrical conductivity (dS/m)	Salinity level	Consequences on crops
1	Less than 2.0	Nonsaline	No effect on crop plants
2	2.0–4.0	Slightly saline	Salt-sensitive crop plants affected
3	4.1–8.0	Moderately saline	Many crop plants are restricted to appear or yield will be highly affected
4	8.1–16.0	Highly saline	Salt-tolerant crops can appear
5	Above 16	Extremely saline	Only highly salt-resistant crops can appear

In crop plants, salinity may cause three major types of stress namely ionic, osmotic, and oxidative (Zhu 2002). Under the imbalance of ion homeostasis, plants suffer with ionic stress. In the case of NaCl salinity, Na^+ competes with K^+ for its uptake and results in the deficiency of K^+ in the cytosol, and balance of ion concentration in cytosol get disturbed. Osmotic stress is the results of excess of Na^+ and Cl^- in the environment which decrease the osmotic potential of the soil solution and hence poor water and other nutrient uptake by the plant roots. There are numbers of low molecular mass compounds such as compatible solutes or osmolytes like proline, protein, mannitol, sorbitol, glycine betaine, etc., are accumulated by the plants to cope with osmotic stress. Salt-induced osmotic stress is responsible for the oxidative stress caused by reactive oxygen species (ROS). Although ROS are harmful to the crop plants, they act as important signaling molecules. ROS, along with other messengers like Ca^{2+}, induces a phosphorylation cascade which activates major stress-responsive genes leading to plant adaptation and helping them to cope under stress conditions (Mahajan and Tuteja 2005). Detrimental effect of ROS is mainly because of their ability to induce a variety of autoxidative chain reactions on unsaturated fatty acids (Rasool et al. 2013) and produce malondialdehyde (MDA). Oxidative attack on proteins results in site-specific amino acid modifications, fragmentation of the peptide chain, aggregation of cross-linked reaction products, and increased susceptibility to proteolysis (Anjum et al. 2015). ROS can also induce numerous lesions in DNA that cause deletions, mutations, and other lethal genetic effects (Srivalli and Khanna-Chopra 2008). The present chapter will summarize the consequences of salinity on crop plants, oxidative biomarkers, and their effects in plants under salt stress and also discuss the mitigation strategies of the plants to cope with salt stress by both the enzymatic and nonenzymatic defense system.

9.2 Formation of Salt-Induced Indirect Products (Oxidative Biomarkers) in Crops

Salinity stress accounts for formation and hyper-accumulation of ROS, which disturb the redox homeostasis of the cell, resulting in oxidative damage to crop plants (AbdElgawad et al. 2016). ROS inside the cell can induce a number of physiological, biochemical, and molecular changes which lead to the formation of different indirect products which can be used for the oxidative biomarkers in crop plants. ROS can counteract the defense capability of the antioxidant system, impairing the essential biomolecules in the cell by oxidizing cell membranes, proteins, nucleic acids, carbohydrates lipids, and many enzymes. Oxidative stress accounts for the formation of cytotoxic compounds as their end products or byproducts.

Peroxidation of polyunsaturated fatty acids (PUFAs) in crop plants is common during the oxidative stress caused by salinity. MDA is the most common aldehyde formed as an end product of the oxidation of lipids in the plants. As the peroxidation of lipids in plants increased due to the oxidative stress, the concentration of these compounds increased inside the cell. MDA in higher concentration is cytotoxic to the cell and promotes programmed cell death. MDA has been most widely used for many years as a most suitable oxidative biomarker for peroxidation of omega-3 and omega-6 fatty acids because of its easy reaction with the thiobarbituric acid (TBA) (Esterbauer and Cheeseman 1990). Apart from MDA content, there are numbers of the products that are formed indirectly and can be used as oxidative biomarkers such as hydrogen peroxide level in cell. Its formation is also triggered by ROS, electrolyte leakage due to disturbance in the permeability of cell membrane due to lipid peroxidation. Enzymatic defense system of plants also gives a rough idea about the oxidative stress under salinity. Most of the plants showed an increased level of enzyme activities, i.e. superoxide dismutase (SOD), catalase (CAT), glutathione reductase (GR), ascorbate peroxidase (APX), and guaiacol peroxidase (GPX) under oxidative stress caused by the salinity to cope with the adverse conditions (AbdElgawad et al. 2016; Sofy et al. 2020). Activity of all these enzymes increases in the salt-tolerant crops with increase in salinity stress.

9.3 Effect of Salt Stress on Crop Plants

Under salinity stress, soil solution contains different dissolved toxic ions, which result both external and internal stress and negatively affect the plants. There are numbers of salts which cause soil salinization, but majority of the salt are constituted by NaCl. Salt stress triggers a multitude of responses in crop plants such as morphological, physiological, biochemical, and molecular changes. High salt concentration in the soil solution causes decreased water potential, resulting osmotic stress in the root zone, and further entry of ions in the cell creates ionic stress. Therefore, salinity stress is a combination of osmotic and ionic stress (Djanaguiraman and Prasad 2014). Salinity affects the crop plants either directly or indirectly. Crop plants are affected indirectly, and many biochemical and molecular changes occur to cope with the salinity stress. Activity of enzymes generally increases to detoxify the cell (AbdElgawad et al. 2016; Sofy et al. 2020). Protein synthesis and their modifications are indirectly affected by salinity (Maheshwari and Dubey 2009). Salinity indirectly changes the gene expression in crop plants to synthesize different bioactive molecules, osmolites, and other phenolic compounds to withstand with the stress (Rossi et al. 2016; Cuong et al. 2020).

In context to direct effect, salinity reduces water potential and causes changes in ion homeostasis and toxicity. This altered water status leads to initial growth reduction and limitation of plant productivity. As high salt concentration involves in both osmotic and ionic stress (Hayashi and Murata 1998), inhibition of growth is directly related to the concentration of soluble salts or osmotic potential of soil solution (Parida and Das 2005). Therefore, an effect of high salt concentration that is observed on whole crop plants can lead to decrease in productivity or death of the plants. Various direct effects of salinity on crop plants are given as follows.

a) **Seed germination**
Osmotic imbalance and oxidative stress caused by salinity affect seed germination (Munns 2002). Salinity can affect the germination of seeds by the toxic effects of Na^+ and Cl^- on embryo viability. The toxic effects include disruption to the structure of enzymes and many other macromolecules, damage to cell organelles and the plasma membrane, the disruption of respiration, photosynthesis, and protein synthesis (Parida and Das 2005; Ibrahim 2016). Venâncio et al. (2020) had reported that seed germination in *Trifolium pratense, Lolium perenne, Festuca arundinacea*, and *Vicia sativa* is inhibited under salinity stress.

b) **Water uptake**
Under high salt concentration, availability of water for growing tissue becomes a limiting factor even in the presence of moisture in the soil resulting in physiological dryness for crop plants. Water uptake by crop plants, hence, attains importance under saline conditions. To overcome the external stress, plant cells tend to readjust their osmotic potential to prevent water losses that can be achieved either by uptake of inorganic ions from the external or by de novo synthesis of compatible solutes. The osmotic readjustment is suddenly induced due to changes in ion fluxes than the synthesis of compatible solutes. Neumann et al. (1988) reported that the salt stress initially inhibits leaf expansion through reduced turgor and may in fact eventually result in increased cell wall extensibility, which counteracts the negative effects of low turgor (Manchanda and Garg 2008). Inhibition in water uptake also alters the process of photosynthesis.

c) **Photosynthesis**
Under saline condition, the rate of photosynthesis is affected by two ways, either directly by decreased intercellular CO_2 availability to crop plant caused by diffusion limitations through the stomata and mesophyll transport of CO_2 or by alterations in photosynthetic metabolism, or secondary such as the oxidative stress arising from the superimposition of multiple stresses (Chaves et al. 2009). Photosynthesis is also affected by water uptake which is poor under saline environment. Photosynthetic pigments in crop plants slightly increased under lower salinity level but tend to reduce at the higher salt concentration. Depletion in chlorophyll content reduces the rate of photosynthesis. It has been reported that abscisic acid (ABA) content increased at higher NaCl concentration which causes reduction of leaves and hence photosynthesis (Kwon et al. 2019).

d) **Transpiration**
The process of water movement through a plant and its evaporation from aerial parts of the plants is known as transpiration which is necessary for leaf cooling. Transpiration from leaves is mainly controlled by the stomata. The rate of transpiration depends on many external and internal factors such as the climate and environment, canopy, stages of growth, and especially, the irrigation regime and quality and availability of water. In

well-watered plants, transpiration represents the most effective means of leaf cooling (Passioura 2010). Under water deficit caused by salinity, the stomata closes and the transpiration rate reduces and only occurs through the cuticle and lenticels. Thus, the temperature tends to increase, which causes increase in stomatal resistance to transpiration and an increase in the rate of tissue respiration. Under such conditions, photosynthetic capacity is limited and the plant uses its own photoassimilates, resulting in decreased growth (Acosta-Motos et al. 2017).

e) **Growth and yield attributes**
Plant growth is directly linked with water uptake, photosynthesis, transpiration, stomatal regulation, nutrient absorption, ionic homeostasis, etc. Under hyper-saline condition, plant growth parameters, such as height of the plant, number of leaves, leaf area, root/shoot ratio, and number of branches, tend to reduce due to poor water availability and reduce nutrient uptake, and ion imbalance (Jimenez et al. 2003). Na^+ and Cl^- at high amounts are toxic to plants, especially if they increase in the cytosol (Kamran et al. 2020). Despite this relevance, not much is known about cytosolic processes that are impaired by excessive concentrations of salt ions. For instance, toxicity effects of chloride in the cytosol remain to be elucidated (Geilfus 2018). Plants that are exposed to excessive concentrations of salt ions are poisoned and eventually die. The ion toxicities have diverse consequences resulting in ionic imbalance, i.e. in terms of uptake competition of Na^+ with K^+, Ca^{2+}, and Mg^{2+} and may accelerate senescence of transpiring leaves, not only because of toxic concentrations of deleterious ions in photosynthetic active tissues but also reduced availability of the beneficial nutrients (Manchanda and Garg 2008). As all the growth attributes and physiological processes of the crop plants are affected by salinity in different manners, they result in decrease in productivity of crop plants (Tomaz et al. 2020).

9.4 Consequences Effect of Oxidative Biomarkers in Crop Plants

Oxidative stress biomarkers can be classified as the molecules which are adapted in the microenvironment by relationships with ROS and antioxidant system molecules which alter in response to the increasing redox stress. Increased ROS levels can disrupt biomolecules including lipids, proteins, carbohydrates, and DNA. Such reactions can modify intrinsic membrane properties such as fluidity, transport of ions, loss of enzyme activity, cross-linking of proteins, inhibition of protein synthesis, harm to DNA, and so on and eventually result in plant cell death (Figure 9.1).

9.4.1 Lipid Peroxidation

Significantly improved lipid peroxidation occurs both in cellular and organelles membranes whenever the ROS level exceeds beyond the maximum, which in turn can influence normal cellular function. Lipid peroxidation provokes oxidative stress by creating radicals derived from lipids which can interfere with it and disrupt proteins and DNA by itself. The rate of lipid peroxidation was commonly used it as a measure of the destruction to cell membranes caused by ROS under stressful conditions. Excessive peroxidation (degradation)

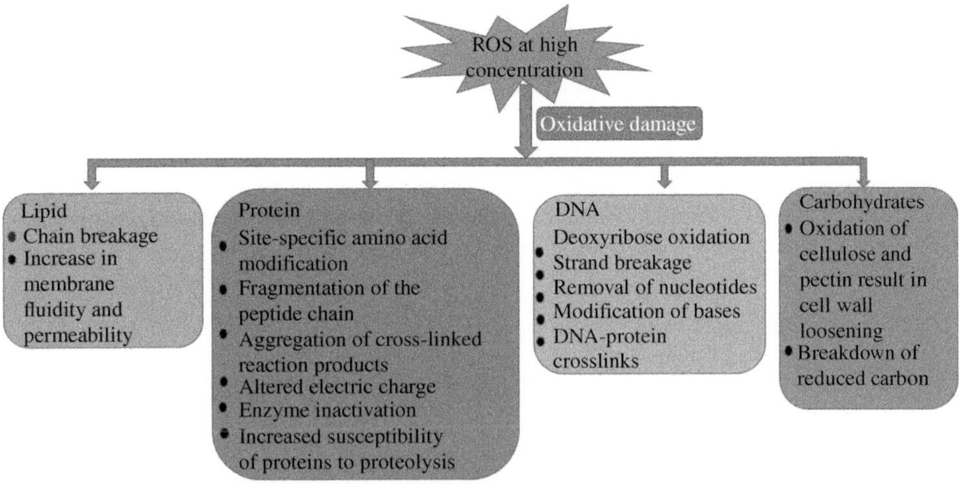

Figure 9.1 Reactive oxygen species (ROS) formed under high salt concentration due to ionic imbalance results in oxidative stress and damage to lipids, proteins, nucleic acids, and carbohydrates. *Source:* Sharma et al. (2012). Licenced under CC BY-3.0.

of lipids in plants growing under environmental stress has been revealed as a production of MDA, end product of lipid peroxidation (Sharma and Dubey 2005; Mishra et al. 2011). The enhanced production of ROS depends on the elevated level of lipid peroxidation under stress. The end products of the peroxidation of unsaturated fatty acids in phospholipids are MDA and causes cell membrane disruption (Halliwell and Gutteridge 1989). The phospholipids molecules basically have two active sites for the attack of ROS, one is unsaturated (double) bond among the two carbon atoms and the other is ester linkage between glycerol and fatty acid. The membrane phospholipids consist of PUFAs which are sensitive to ROS. Production of a single OH is a result of peroxidation of several PUFAs because it is a part of a cyclic chain reaction. In general, the lipid peroxidation cycle includes three distinct stages: stages of initiation, progression, and termination. The initial lipid peroxidation step requires O_2 activation which is rate-limiting. O_2^- and OH must interact with PUFA groups of methylene that produce conjugated dienes, lipid peroxide radicals, and hydroperoxides (Smirnoff 1995):

$$\text{PUFA} - \text{H} + \text{X} \dashrightarrow \text{PUFA} + \text{X} - \text{H}$$

$$\text{PUFA} + O_2 \dashrightarrow \text{PUFA} - \text{OO}$$

The formed peroxy radical is highly reactive and can accelerate the chain reaction:

$$\text{PUFA} - \text{OO} + \text{PUFA} - \text{OOH} \dashrightarrow \text{PUFA} - \text{OOH} + \text{PUFA}$$

Conjugated diene is formed when free radicals target the hydrogen of methylene groups that separate double bonds and thus rearrange the bonds (Recknagal and Glende 1984). According to the following reaction, the lipid hydroperoxides formed (PUFA-OOH) that exhibit reduced metal cleavage, like Fe^{2+}:

$$Fe^{2+}\ \text{complex} + \text{PUFA} - \text{OOH} \dashrightarrow Fe^{3+}\text{complex} + \text{PUFA} - \text{O}$$

The degradation of lipid hydroperoxide can easily produce several reactive species including lipid alcoxyl radicals, aldehydes (malonyldialdehyde, acrolein, and crotonaldehyde), alkanes, lipid epoxides, and alcohols (Davies 2000). The developed lipid alcoxy radical (PUFA-O) can initiate additional chain reactions (Buettner 1993):

$$\text{PUFA} - \text{O} + \text{PUFA} - \text{H} \dashrightarrow \text{PUFA} - \text{OH} + \text{PUFA}$$

Peroxidation of PUFA by ROS attack can respond to chain breakage and thus enhance the fluidity and permeability of the membranes.

9.4.2 Effect on Proteins

ROS assault on proteins can affect protein modification in a number of different ways, a few are direct and others are indirect. Direct modification involves modulating the behavior of a protein via nitrosylation, carbonylation, formation of disulphide bond, and glutathionylation. The breakdown products of the fatty acid peroxidation are involved indirectly in protein modification by amalgamation (Yamauchi et al. 2008). As a result, the enhanced level of ROS formation causes site-specific amino acid modification, fragmentation of the peptide chain, aggregation of cross-linked reaction products, altered electric charge, and increased susceptibility of proteins to proteolysis. Tissues wounded by oxidative stress normally contain increased levels of carbonylated proteins which are widely used protein oxidation markers (Møller and Kristensen 2004). Elevated protein modification has been observed under different kinds of stresses in plants (Romero-Puertas et al. 2002; Sharma and Dubey 2005; Maheshwari and Dubey 2009). In a peptide, the amino acids are different in their proneness to ROS attack. The amino acid-containing thiol groups and sulfur are mostly prone sites for ROS attack. Oxidized oxygen may extract an H atom from cysteine residues to produce a thiyl radical that cross-links to the second thiyl radical to establish a bridge with disulphide. Oxygen may also be combined to methionine in order to build up a derivative of methionine sulphoxide (Brot and Weissbach 1982). Mostly in the presence of ROS, tyrosine is commonly linked to form bityrosine products (Davies 1987). Oxidation of iron-sulfur centers by O_2^- is inconvertible and results in inhibition of the enzyme (Gardner and Fridovich 1991). In such situations, the metal (Fe) on the protein binds to a divalent cation-binding site. The metal (Fe), after that, responds in a Fenton reaction to form a OH that fastly oxidizes a residue of amino acid at or adjacent to the protein's cation-binding site (Stadtman 1986). Oxidizable proteins act as better proteolytic digestive substrates. The oxidation of protein makes it susceptible to ubiquitination, and ultimately it would be a target of proteasomal degradation (Cabiscol et al. 2000). Oxidized proteins were deteriorated more efficiently and metal treatment increased proteolytic activity by 20% (Romero-Puertas et al. 2002). Many reports have shown that even more loss leads to extensively cross-linked and gathered products over a certain degree that are not only weak substrates for destruction but they can also prevent proteases to degrade other oxidized proteins (Grune et al. 1997).

9.4.3 Effects on Carbohydrates

Carbohydrates are the major category of biological substances on the earth and less examined in view of oxidative damage and their role in signaling stresses. They provide mechanical strength and shape of plant cells (cellulose, pectin, etc.), accumulate reduced carbon (starch, sucrose, etc.), monitor the activity of enzymes and osmotic pressure (low molecular weight sugars), confer nonenzymatic antioxidant defense system (flavonoids, mannitol, etc.), and perform other important roles. Carbohydrate oxidation is highly harmful to plants. Cu^{2+} generated OH interacts nonenzymatically toward xyloglucans and pectins, splitting them into the components and causing loosening of cell walls. It may support expansive growth in cells and encourage fruit ripening (Fry et al. 2002). Under the stress condition, this reaction is important, when the catalytic activities of Cu and Fe are kept increasing, it could have pathophysiological effects (Moran et al. 1994). Mono- and disaccharides possibly function as scavengers of ROS (Couée et al. 2006). Their potential to scavenge OH is as: maltose > sucrose > fructose > deoxyribose > sorbitol (Morelli et al. 2003) (electron paramagnetic resonance [EPR] test, i.e. electron paramagnetic resonance, can be used for nondestructive detection of free radicals and high-performance liquid chromatography (HPLC)). However, the metabolism of products synthesized in all these reactions (apart from format) is undefined. Uptake of certain carbohydrates (e.g. mannitol) has been demonstrated to correspond with enhanced oxidative stress resistance in a variety of species such as *Asparagus officinalis, Coffea arabica*, and *Olea europea*. However, the direct connection between oxidative stress triggered carbohydrate alterations and plant physiology is still lacking. Perhaps, they serve as structural, osmotic, nutrient, and signaling agents instead of redox switches or large ROS targets (Couée et al. 2006).

9.4.4 Effect on Polynucleic Acids

DNA oxidative damage is the focus of comprehensive animal physiology studies, since it is a significant cause of cancerogenesis. This problem is not significant to plants of agronomic importance; consequently, this is not completely defined in plant physiology. However, oxidative DNA damage can be leading to seed stock aging and, occasionally, crop plant death (Britt 1996). Oxidative DNA damages mainly consist of three types of lesions that are mismatched bases, double-strand breaks, and chemically modified bases (Yoshiyama et al. 2013). Hydroxyl radicals are a significant damaging agent for polynucleic acids, interacting with them through the insertion of double bonds of nucleotide bases and the removal of H^+ from each of the 2-deoxyribose and thymine methyl group C–H bonds. 8-oxo-7, 8-dihydroguanine (8-oxoG) and 2, 6 diamino-4-hydroxy-5-formamid opyrimidine (FapyG) are frequently identified components for oxidation of OH-induced DNA/RNA (Yoshiyama et al. 2013). In plants, repair systems are available that preclude DNA damage, along with direct repair of damaged molecule portion, in addition to base and substitution of nucleotides (Britt 1996; Tuteja et al. 2001; Yoshiyama et al. 2013). Safety also involves enforcing the protection of antioxidants in both cytosol and organelles. Vanderauwera et al. (2011) have already shown that scavenging enzymes (peroxiredoxin and glutathione) with nuclear ROS are inadequate to protect DNA throughout oxidative stress. CAT and cytosolic APX have been shown to be essential for the safety of nuclear DNA under stress conditions. In addition, DNA damage caused by ROS triggers signaling phenomena and activates specific transcription factors, inducing repair of DNA (Yoshiyama et al. 2013).

9.5 Generation of Self-defense Mitigating Strategy in Crop Plants

Soil salinity poses a significant threat to the production of agricultural crops worldwide. The salt stress plays an important role in the inhibition of plant growth and development. The seed germination, root length, plant height, and fruiting of plants are significantly inhibited under salt treatments (Liang et al. 2014). Plants adopt mainly two strategies to counteract the effects of high salt concentration, namely enzymatic and nonenzymatic strategies (Figure 9.2).

9.5.1 Counteract Mitigating Strategy by Enzymatic Antioxidants System

Along with the nonenzymatic components of the antioxidant system (AOX), the enzymatic components include a dynamic and multifaceted defensive mechanism to maintain ROS homeostasis, both to prevent oxidative damage in plant cells and to promote the growth of plants (Mittler 2017). The enzymatic AOX components include SOD, several peroxidases such as CAT and guaiacol peroxidase (GPX), glutathione S-transferase (GST), and a set of ascorbate–glutathione (AsA–GSH) cycle enzymes including APX, GR, monodehydroascorbate reductase (MDHAR), and dehydroascorbate reductase (DHAR). Taking into consideration their function as ROS scavengers, variability in their activity and/or transcript accumulation are common features as triggered by salinity in plants under biotic and abiotic stress (Ali et al. 2017; Siddiqui et al. 2017).

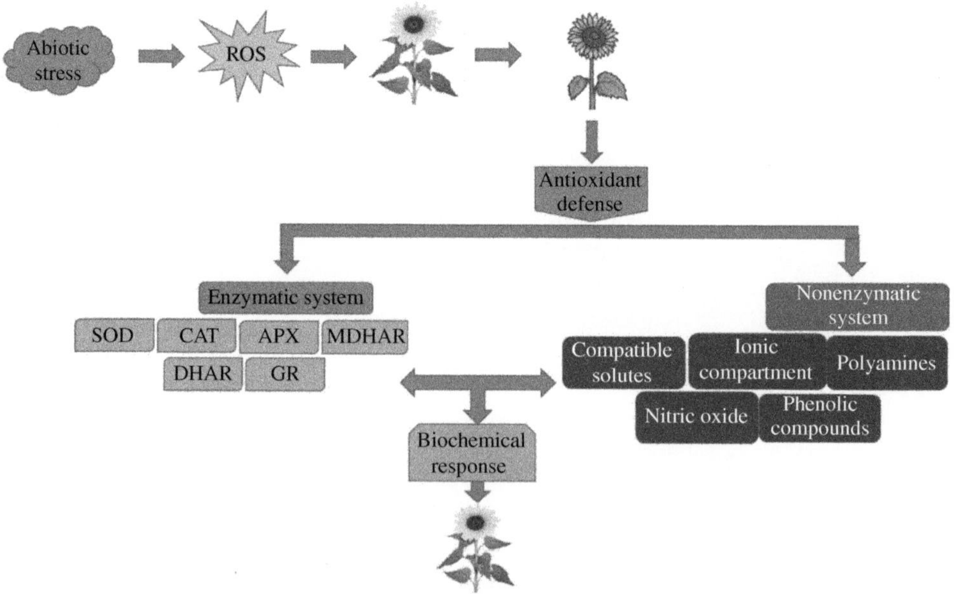

Figure 9.2 Flow diagram showing major antioxidant system (enzymatic and nonenzymatic) for mitigation of high salt concentration in plant system.

9.5.1.1 Superoxide Dismutase (SOD)

SOD, a ubiquitous protein contained in various aerobic species such as microbes, animals, and plants, is known as the first enzymatic line of protection against the oxidative stress (Gratão et al. 2005). SOD catalyzes O_2^- dismutation in H_2O_2 and molecular oxygen; hence, SOD plays a crucial role in ROS detoxification by influencing both O_2^- and H_2O_2 levels and mitigating O_2^--related toxicity (Abouzari and Fakheri 2015). It is classified on the basis of metal ion in their binding site, namely copper and zinc (Cu/ZnSOD), manganese (MnSOD), and iron (FeSOD). Cu/ZnSOD is localized in cytosol and chloroplasts, MnSOD in the matrix of mitochondria and peroxisomes, and FeSOD in certain higher plants' chloroplasts; however, these are also commonly present in prokaryotes (Scandalios 1993). Increased SOD's activity reduces abiotic oxidative stress and plays a significant role in a plant's adaptation to stress environments (Singh et al. 2008; Ben Ahmed et al. 2010).

9.5.1.2 Catalase (CAT)

CATs are heme-containing tetrameric enzymes that are using H_2O_2 as a substrate and transform it to H_2O and O_2, hence mitigating oxidative damage to cells (Sánchez-Casas and Klesseg 1994). CATs occur in peroxisomes, glyoxysomes, and associated organelles with H_2O_2-generating enzymes (Agrawal et al. 2009). CAT does have one of the highest production rates among all the enzymes: one CAT molecule can convert approximately six million H_2O_2 molecules to H_2O and O_2 per minute. CAT is therefore important in the removal of H_2O_2, which is produced in peroxisomes by oxidases involved in β-fatty acid oxidation, photorespiration, and purine catabolism (Gill and Tuteja 2010). This has also been revealed that CAT reacts with certain hydroperoxides besides its reaction with H_2O_2 (Ali and Alqurainy, 2006). CAT activity reveals variable changes such as aging, chlorosis, and senescence under various abiotic stresses (Singh et al. 2008; Hasanuzzaman and Fujita 2011).

9.5.1.3 Ascorbate–Glutathione (AsA–GSH) Cycle Enzymes

The cycle of AsA–GSH is the main defense system for chloroplast, cytosol, mitochondria, peroxisomes, and apoplast against ROS. The AsA–GSH cycle includes four enzymes (APX, MDHAR, DHAR, and GR) as well as AsA, GSH, and nicotinamide adenine dinucleotide phosphate (NADPH) that work along to detoxify H_2O_2 in a sequence of cyclic reactions and regenerate AsA and GSH in addition. In this process, APX catalyzes the reduction of H_2O_2 to H_2O mostly with subsequent production of monodehydroascorbate (MDHA), which is transformed to AsA by the activity of NADPH-dependent MDHAR or disproportionate to AsA and DHA (Asada 1994). DHA performs irreversible hydrolysis to 2,3-diketogulonic acid or is recycled to AsA by DHAR, using GSH as a reducing agent (Chen et al. 2003b). This leads to GSSG generation, which GR regenerates to GSH.

9.5.1.4 Ascorbate Peroxidases (APX)

APX scavenging of H_2O_2 is the first step in the AsA–GSH cycle and it may perform the most critical role in scavenging ROS and helping to protect higher plant cells (Asada 1994). APXs are heme-containing enzymes responsible for water–water scavenging of H_2O_2 and AsA–GSH cycles uses AsA as a substrate material, catalysing the transport of electrons from AsA to H_2O_2, and generating DHA and water (Pang and Wang 2010). The APX family includes at least five distinct isoforms such as the form of mitochondrial (mAPX), thylakoid (tAPX), glyoxisome membrane (gmAPX), stromal soluble chloroplast (sAPX), and cytosolic form (cAPX) (Noctor and Foyer 1998). In response to various abiotic stress conditions, APX activity in plants is enhanced (Singh et al. 2008; Hossain et al. 2010; Hasanuzzaman and Fujita 2011).

9.5.1.5 Monodehydroascorbate Reductase (MDHAR) and Dehydroascorbate Reductase (DHAR)

MDHA is formed by the univalent oxidation of AsA. If MDHAR does not again reduce MDHA to AsA, it will randomly become disproportionate to AsA and DHA. In such a reaction necessitating GSH, DHA is then reduced by DHAR to AsA (Chen et al. 2003a). Gradual regeneration is essential to maintain AsA's antioxidant capacity. In this cycle, AsA regeneration could mainly be regulated by NADPH-dependent MDHAR activity (Mittova et al. 2000) and is therefore essential for AsA regeneration and necessary for maintaining a reduced AsA pool (Martínez and Araya 2010). In spite of the fact that there are also some evidence of MDHAR activity in other oxidative stress-related physiological processes, research on various crops under environmental stress confirmed MDAHR's regulatory role during oxidative stress tolerance and acclimatization (Hossain et al. 2010). The regulations of redox state of AsA under oxidative stress (Eltayeb et al. 2007) are controlled by MDHAR and DHAR equally. DHAR is really a major part of the AsA recycling system (Martínez and Araya 2010) that reproduces AsA from oxidized state (DHA) and controls the AsA redox cellular state. Thus, tolerance to different abiotic stresses resulting to ROS production is essential. Increased DHAR activity in response to various stresses triggering ROS was mentioned (Hossain et al. 2010; Hasanuzzaman and Fujita 2011).

9.5.1.6 Glutathione Reductase (GR)

Glutathione reductase (GR) is a potential AsA–GSH cycle enzyme and plays a key role in the ROS defense system. Enhanced GR activity gives stress tolerance and is capable of altering the redox status of essential components of the electron transport chain (ETC) system. This enzyme catalyses the reduction of GSH, which is involved in many metabolic regulatory and antioxidant processes in plants where GR catalyses GSSG's NADPH-dependent disulphide bond reduction and is therefore necessary to maintain the GSH pool (Rao and Reddy 2008). Therefore, GR also maintains a high ratio of GSH/GSSG in plant cells, which is also essential to promote the H_2O_2 scavenging path, especially under stress conditions (Pang and Wang 2010). GR plays a significant role in characterizing a plant's tolerance under different stresses by maintaining the cell's antioxidant machinery, conferring stress tolerance (Hossain and Fujita 2010; Hasanuzzaman and Fujita 2011).

9.5.1.7 Guaiacol Peroxidase (GPX)

GPX is a widely distributed protein found in animals, microbes, and plants across all living organisms, where it can be found particularly in different plant organs and organelles, particularly in vacuoles, cytosol, and cell walls (Sharma et al. 2012). As with other peroxidase, GPX can control intracellular levels of H_2O_2 using various organic compounds like substrates such as guaiacol or pyrogallol (Gill and Tuteja 2010). In addition, various reports indicate that GPX is the key enzyme in H_2O_2 detoxification due to both its intracellular and extracellular enzymatic activity. This heme-containing enzyme is structurally made up of monomer units of approximately 40–50 kDa and has four conserved disulphide bridges and two structural Ca^{2+} ions (Gill and Tuteja 2010; Das and Roychoudhury 2014). In addition to its role in plant oxidative stress tolerance, GPX also takes part in other important biosynthetic pathways, making a contribution to the lignification of cell walls, indole-3-acetic acid (IAA) catabolism, and ethylene biosynthesis (Sharma et al. 2012). Given its great importance in cellular redox homeostasis, the activation and enhanced activity of GPX have been largely reported in response to various kinds of unfavorable environments (Gill and Tuteja 2010; Sharma et al. 2012).

9.5.2 Counteract Mitigating Strategy by Nonenzymatic Antioxidants System

The nonenzymatic antioxidant system consists of various molecules, particularly low molecular mass metabolites such as glutathione (GSH), ascorbic acid (AsA) and, numbers of phenolic compounds and proline (Pro), which are capable of neutralizing, removing, and/or transforming ROS, enabling the management and sensing of ROS homeostasis to obtain cellular redox equilibrium in stressed plants (Gratão et al. 2005; Mittler 2017). The effectiveness and behavior of nonenzymatic AOX stress machinery depend on a variety of factors, including stress types, exposure time and intensity, plant species and their genotypes, organs, or tissues among others (Borges et al. 2018; Piotto et al. 2018). Hence, different responses and results from nonenzymatic AOX have been recorded for the most part in plants exposed to various stressors such as salinity (Farhangi-Abriz and Torabian 2017). The principle mechanisms of nonenzymatic antioxidant system against salinity stress are as follows:

a) **Compatible solutes accumulation and osmotic protection**
Compatible solutes, also known as compatible osmolytes, are a category of novel chemical organic compounds which are uncharged, polar, and soluble in nature and do not intrude even at high concentrations with the cellular metabolism. There are so many osmolytes but some of them are proline (Hoque et al. 2007), glycine betaine (Khan et al. 2000), sugar (Kerepesi and Galiba 2000), and polyols (Saxena et al. 2013). The synthesis and accumulation of organic osmolyte vary from species to species in plants, e.g. the accumulation of quaternary ammonium compound β-alanine betaine is confined to a few members of Plumbaginaceae (Hanson et al. 1994), whereas the amino acid and proline accumulation occurs in taxonomically diverse sets of plants. The concentration of compatible solutes inside the cell is retained either by irreversible compound synthesis or a combination of both synthesis and breakdown. The biochemical pathways and genes which are involved in these processes have been studied in depth. The accumulation of osmolyte is proportional to the exterior osmolarity, and the main functions are to protect the structure and maintain osmotic equilibrium within the cell through continuous flow of water (Hasegawa et al. 2000). Accumulation of proline in the plant cell is a well-known example and indication for the eradication of the salinity stress. Intracellular-accumulated proline in the plant cell not only plays an important role during salinity stress but also acts as an organic nitrogen reserve during stress recovery (Hoque et al. 2007).

Glycine betaine is an amphoteric quaternary ammonium compound occurring omnipresently in microorganisms and higher plants and animals and is electrically neutral over a wide range of pH. It is highly water-soluble but also contains 3-methyl groups forming nonpolar components. Due to its special structural features, it interacts with both the macromolecules' hydrophobic and hydrophilic domains, such as enzymes and protein complexes. Glycine betaine is an innocuous compound that helps in increasing the osmolarity of the cell under stress condition and also involved in stress reduction. Glycine betaine also protects the cell by osmotic adjustment (Gadallah 1999), stabilizes proteins (Mäkelä et al. 2000), protects the photosynthetic apparatus against stress damage (Chaum and Kirdmanee 2010), and decreases ROS (Ashraf and Foolad 2007).

Polyols are compounds essential for organic reactions, with several hydroxyl functional groups. Sugar alcohol is a category of polyols that act as compatible solutes, having low molecular weight chaperones, and as ROS scavenging compounds (Ashraf and Foolad 2007). Carbohydrates such as sugars (e.g. glucose, fructose, fructans, and trehalose) and starch are accumulated in cell under salt stress (Parida et al. 2004). The major function of these carbohydrates to play a significant role in reducing stress includes the osmoprotection, preservation of energy, and scavenging of ROS.

b) **Ionic compartmentalization**

The plants maintain their own ion homeostasis by ion uptake and compartmentalization, which do not play an important role only in normal growth but also essential for growth during salt stress (Hasegawa 2013). The signaling pathway of SOS (salt overly sensitive) having three major proteins, SOS1, SOS2, and SOS3, is shown in Figure 9.3. Out of these three proteins, SOS1 codifies an antiporter Na^+/H^+ plasma membrane and is essential for cellular-level regulation of Na^+ efflux. This also enables the transport of Na^+ from root to shoot over long distances. High concentrations of this protein in plants provide salt tolerance (Shi et al. 2002). SOS2 gene codes serine/threonine kinase, which is active in the presence of salt stress and also generates Ca^+ signals. This protein consists of two domains, i.e. N-terminal catalytic domain and a C-terminal regulatory domain (Liu et al. 2000;

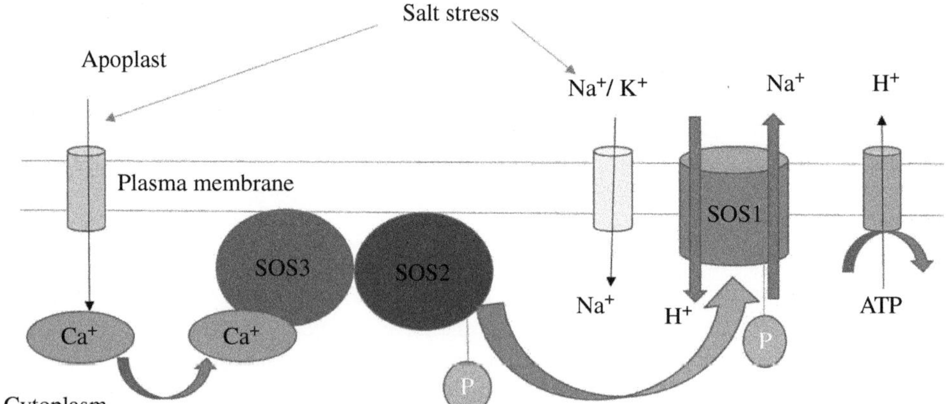

Figure 9.3 Flow diagram showing SOS pathway for mitigation of high salt concentration in plant system. *Source:* Modified from Gupta and Huang (2014).

Martínez-Atienza et al. 2007). The third protein SOS3 is myristoylated Ca^+-binding protein having amyristoylation site at its N-terminus. This site plays an important role in providing salt tolerance in the cells (Ishitani et al. 2000). Thus, there is a sharp increase in intracellular Ca^{2+} level with increase in Na^+ concentration which in turn facilitates its binding with SOS3 protein. Ca^{2+} modulates homeostasis of the Na^+ intracellular along with SOS proteins. The protein SOS3 then interacts with it and activates the protein SOS2 by triggering its self-inhibition. The complex SOS3–SOS2 is then charged to the plasma membrane where it phosphorylates SOS1 (Figure 9.3). The phosphorylated SOS1 results in increased efflux of Na^+, which reduces salt toxicity (Martínez-Atienza et al. 2007).

One way to withstand high salinity is by compartmentalizing inside the cells (Pérez-Tornero et al. 2009). The salt stress results in an imbalance of the nutrients needed for the growth of the plants. Ionic stress is responsible for an increase of Na^+ and Cl^- and decrease of K^+ ions. Even these ion imbalances are harmful to plant growth (Abdel Kader et al. 2011). Increased Na^+ under salinity in cytosol reduces Ca^{+2}, K^+, and Mg^{+2} ion concentrations and disturbs ionic homeostasis (Alemán et al. 2011). Na^+ is highly toxic at high concentrations causing membrane destabilization, loss of functional proteins, slowing cell division and expansion, lower plant metabolism rates, and altering the mineral nutrient homeostasis (Munns and Tester 2008). Na^+ also competes for cellular uptake with potassium when its extracellular concentration is very high (Niu et al. 1995). When the Na^+ ion level in the cytosol increases by more than 100 mM, it becomes toxic (Serrano et al. 1998). Na^+ at high concentration causes a decrease in K^+ concentration accompanied by an imbalance of the K^+/Na^+ ratio and other electrolytes vital for optimal cytosol homeostasis (Shabala and Cuin 2008). Plants preserve ionic homeostasis by normal regulation of Na^+ intake into the cell and its vacuole compartmentalization (Zhu 2003). This strategy is very successful in both halophytes and glycophytes because it helps to lower the concentration of Na^+ and Cl^- within the cell and maintains high concentration of salt growth (Blumwald et al. 2000; Hasegawa et al. 2000).

c) **Roles of polyamines**

Polyamines (PA) are small, omnipresent, polycationic aliphatic molecules and have low molecular weight that are widely distributed throughout the entire plant kingdom. Polyamines play a number of roles in normal growth and development of the plants, including cell proliferation control, somatic embryogenesis, differentiation and morphogenesis, tuber breaking and seed germination, flowers and fruit production, and senescence (Gupta et al. 2013). Polyamines help in abiotic stress tolerance, i.e. in salinity, and concentration level of polyamines corresponds to stress tolerance in plants (Kovács et al. 2010). An increase in the level of endogenous polyamines was recorded when the plant is exposed to salinity stress. The level of the intracellular polyamine is regulated by catabolism of polyamine. Polyamines are catabolized oxidatively by amine oxidases that involve copper-binding diamine oxidases and flavin adenine dinucleotide (FAD)-binding polyamine oxidases. This enzyme also possesses stress tolerance properties (Takahashi and Kakehi 2010).

The changes in the amount of cellular polyamine due to stress have possible impacts on stress but do not give details of their role in combating stress. Thus, for knowing that polyamines really either protect or not protect the cells from stress-induced damages, the application of exogenous polyamines predicted to increase endogenous polyamine has been investigated before or during stress (Navakoudis et al. 2003). The application of exogenous polyamine has been considered to improve the level of endogenous polyamine during stress; the positive effects of polyamines have been associated with maintaining membrane integrity, regulating gene expression for the synthesis of osmotically active solutes, reducing the formation of ROS, and regulating the accumulation of Na^+ and Cl^- ions in various organs (Yiu et al. 2009).

d) **Roles of nitric oxides**

Nitric oxide (NO) is a tiny labile gaseous molecule, having properties of the regulation of various plants growth and developmental processes, such as root growth, respiration, stomata closure, flowering, cell death, seed germination, and stress responses, as well as a stress signaling molecule (Zhao et al. 2009). NO triggers the expression of many redox-regulated genes, directly or indirectly. NO reacts with lipid radicals to preclude lipid oxidation to exert a protective effect by scavenging radical superoxide and to form peroxynitrite that can be neutralized by many other biological processes. This also facilitates in antioxidant enzyme activation (SOD, CAT, GPX, APX, and GR) (Bajguz 2014). The beneficial effects of NO on salinity tolerance or stress reduction were associated with antioxidant activities and ROS detoxification system modulation (Mishra et al. 2011). The exogenous application of NO helps in improving the plant growth under salinity stress and allied with increases in antioxidant enzymes such as SOD, CAT, GPX, APX, and GR (Zhao et al. 2004) and suppression of MDA production or lipid peroxidation (Nalousi et al. 2012). The effects of NO on salinity tolerance are also linked to its regulation of the H^+-ATPase and Na^+/K^+ plasma membrane ratios (Crawford 2006). NO stimulates H^+-ATPase (H^+-PPase), thus generating an H^+ gradient and providing force for the exchange of Na^+/H^+. An increase in the Na^+/H^+ exchange could make a significant contribution to homeostasis in K^+ and Na+ (Zhang et al. 2006). Though, NO behaves under salt stress as a signal molecule and promotes salt tolerance by enhancing the activity of plasma membrane H^+-ATPase.

e) **Phenolic compounds**

Under salinity stress, it has been reported that plants synthesize numbers of phenolic compounds such as gallic acid, chlorogenic acid, rutin, hyperoside, isoquercetine, quercitrine, quercetine, etc. These phenolic compounds help in plant defense system against high salt concentration. ROS generated during oxidative stress is scavenged by these compounds as they donate the electron to the free radicals and neutralize them (Waśkiewicz et al. 2013; Caliskan et al. 2017).

9.6 Conclusion and Future Prospective

Human life is impossible without plants as we depend on plants for number of services such as oxygen, foods, medicines, timbers, and many other things and to make human life easier. Plants in their life face fluctuations in environmental conditions. Salinity is one of the most harmful abiotic stresses to the crop plants as the most of the crop plants are sensitive to salt. Salinity adversely affects the physiological, biochemical, and molecular processes in crop plants and ultimately results in decrease in productivity of the crop plants. Salinity induces ionic imbalance and oxidative stress in crop plants and disturbs the normal physiology of plants. Under salinity stress, ion homeostasis of plants altered and hyper-accumulation of Na^+, Cl^-, and other ions hamper the assimilation, transport, and distribution of other important nutrients. ROS formed by oxidative stress damages the lipids, proteins, nucleic acid, carbohydrates, etc., and account for cell death. Under salt stress conditions, plants defense system starts synthesizing different compatible solutes. Accumulation of different compatible solutes has been reported during salt stress in different plants such as maize, soybean, barley, and alfalfa. These compatible solutes have protective roles during salt stress including contribution to cellular osmotic adjustment, detoxification of ROS, protection of membrane integrity, and stabilization of enzymes and proteins.

Nowadays, we have challenges to overcome the problem of salinity to the crop plants. The anthropogenic activity that leads to salinization of soil should be reduced and salt resistance with high yielding varieties of crop plants should be developed with the help of biotechnology by transgenic technique. The advancement in omics is being used in elucidating important plant processes in response to various abiotic stresses.

References

Abdel Kader, E., Aly, M.S., and Esawy, M.A. (2011). Inoculation of Geobacillus caldoxylosilyticus IRD into maize (*Zea mays* L.) to sustain tolerance against high salt stress. *J. Am. Sci.* 7: 71–79.

AbdElgawad, H., Zinta, G., Hegab, M.M. et al. (2016). High salinity induces different oxidative stress and antioxidant responses in maize seedlings organs. *Front. Plant Sci.* 7: 276.

Abouzari, A. and Fakheri, B.A. (2015). Reactive oxygen species: generation, oxidative damage and signal transduction. *Int. J. Life Sci.* 5: 3–17.

Acosta-Motos, J., Ortuño, M., Bernal-Vicente, A. et al. (2017). Plant responses to salt stress: adaptive mechanisms. *Agronomy* 7: 18.

Agrawal, S.B., Singh, S., and Agrawal, M. (2009). Ultraviolet-B induced changes in gene expression and antioxidants in plants. *Adv. Bot. Res.* 52: 48–86.

Alemán, F., Nieves-Cordones, M., Martínez, V., and Rubio, F. (2011). Root K^+ acquisition in plants: the Arabidopsis thaliana model. *Plant Cell Physiol.* 52 (9): 1603–1612.

Ali, A.A. and Alqurainy, F. (2006). Activities of antioxidants in plants under environmental stress. In: *The Lutein-Prevention and Treatment for Diseases* (ed. N. Motohashi), 187–256. India: Transworld Research Network.

Ali, S., Rizwan, M., Qayyum, M.F. et al. (2017). Biochar soil amendment on alleviation of drought and salt stress in plants: a critical review. *Environ. Sci. Pollut. Res.* 24 (14): 12700–12712.

Anjum, N.A., Sofo, A., Scopa, A. et al. (2015). Lipids and proteins – major targets of oxidative modifications in abiotic stressed plants. *Environ. Sci. Pollut. Res.* 22 (6): 4099–4121.

Asada, K. (1994). Production and action of active oxygen species in photosynthetic tissues. In: *Causes of Photooxidative Stress and Amelioration of Defense Systems in Plants* (eds. C.H. Foyer and P.M. Mullineaux), 77–104. Boca Raton: CRC Press, Taylor and Farncis Group.

Ashraf, M.F.M.R. and Foolad, M.R. (2007). Roles of glycine betaine and proline in improving plant abiotic stress resistance. *Environ. Exp. Bot.* 59 (2): 206–216.

Bajguz, A. (2014). Nitric oxide: role in plants under abiotic stress. In: *Physiological Mechanisms and Adaptation Strategies in Plants Under Changing Environment* (eds. P. Ahmad and M.R. Wani), 137–159. New York: Springer Verlag.

Ben Ahmed, C., Ben Rouina, B., Sensoy, S. et al. (2010). Exogenous proline effects on photosynthetic performance and antioxidant defense system of young olive tree. *J. Agric. Food Chem.* 58 (7): 4216–4222.

Blumwald, E., Aharon, G.S., and Apse, M.P. (2000). Sodium transport in plant cells. *Biochim. Biophys. Acta Biomembr.* 1465 (1-2): 140–151.

Borges, K.L.R., Salvato, F., Alcantara, B.K. et al. (2018). Temporal dynamic responses of roots in contrasting tomato genotypes to cadmium tolerance. *Ecotoxicology* 27: 245–258.

Britt, A.B. (1996). DNA damage and repair in plants. *Annu. Rev. Plant Physiol. Plant Mol. Biol.* 47: 75–100.

Brot, N. and Weissbach, H. (1982). The biochemistry of methionine sulfoxide residues in proteins. *Trends Biochem. Sci.* 7 (4): 137–139.

Buettner, G.R. (1993). The pecking order of free radicals and antioxidants: lipid peroxidation, α-tocopherol, and ascorbate. *Arch. Biochem. Biophys.* 300 (2): 535–543.

Cabiscol, E., Piulats, E., Echave, P. et al. (2000). Oxidative stress promotes specific protein damage in *Saccharomyces cerevisiae*. *J. Biol. Chem.* 275 (35): 27393–27398.

Caliskan, O., Radusiene, J., Temizel, K.E. et al. (2017). The effects of salt and drought stress on phenolic accumulation in greenhouse-grown Hypericum pruinatum. *Italian J. Agro.*

Chaum, S. and Kirdmanee, C. (2010). Effect of glycinebetaine on proline, water use, and photosynthetic efficiencies, and growth of rice seedlings under salt stress. *Turkish J. Agric. Forestry* 34 (6): 517–527.

Chaves, M.M., Flexas, J., and Pinheiro, C. (2009). Photosynthesis under drought and salt stress: regulation mechanisms from whole plant to cell. *Ann. Bot.* 103: 551–560.

Chen, S., Li, J., Wang, S. et al. (2003a). Salt, nutrient uptake and transport, and ABA of *Populus euphratica*; a hybrid in response to increasing soil NaCl. *Trees* 15 (3): 186–194.

Chen, Z., Young, T.E., Ling, J. et al. (2003b). Increasing vitamin C content of plants through enhanced ascorbate recycling. *Proc. Natl. Acad. Sci. U. S. A.* 100: 3525–3530.

Couée, I., Sulmon, C., Gouesbet, G., and El Amrani, A. (2006). Involvement of soluble sugarsin reactive oxygen species balance and responses to oxidative stress in plants. *J. Exp. Bot.* 57: 449–459.

Crawford, N.M. (2006). Mechanisms for nitric oxide synthesis in plants. *J. Exp. Bot.* 57 (3): 471–478.

Cuong, D.M., Kwon, S.J., Nguyen, B.V. et al. (2020). Effect of salinity stress on Phenylpropanoid Genes expression and related gene expression in Wheat Sprout. *Agronomy* 10: 390.

Das, K. and Roychoudhury, A. (2014). Reactive oxygen species (ROS) and response of antioxidants as ROS-scavengers during environmental stress in plants. *Front. Environ. Sci* 2: 53.

Davies, K.J. (1987). Protein damage and degradation by oxygen radicals. I. general aspects. *J. Biol. Chem.* 262 (20): 9895–9901.

Davies, K.J.A. (2000). Oxidative stress, antioxidant defenses, and damage removal, repair, and replacement systems. *IUBMB Life* 50 (4-5): 279–289.

Djanaguiraman, M. and Prasad, P.V.V. (2014). Effects of salinity onion transport, water relations and oxidative damage. In: *Ecophysiology and Responses of Plants Under Salt Stress* (eds. P. Ahmad, M. Azooz and M.N.V. Prasad), 89–114. New York, NY: Springer.

Eltayeb, A.E., Kawano, N., Badawi, G.H. et al. (2007). Overexpression of monodehydroascorbate reductase in transgenic tobacco confers enhanced tolerance to ozone, salt and polyethylene glycol stresses. *Planta* 225: 1255–1264.

Esterbauer, H. and Cheeseman, K.H. (1990). Determination of aldehydic lipid peroxidation products: malonaldehyde and 4-hydroxynonenal. *Methods Enzymol.* 186: 407–421.

Farhangi-Abriz, S. and Torabian, S. (2017). Antioxidant enzyme and osmotic adjustment changes in bean seedlings as affected by biochar under salt stress. *Ecotoxicol. Environ. Saf.* 137: 64–70.

Flowers, T.J., Hajibagheri, M.A., and Clipson, N.J.W. (1986). Halophytes. *Q. Rev. Biol.* 61: 313–337.

Fry, S.C., Miller, J.G., and Dumville, J.C. (2002). A proposed role for copper ions in cell wallloosening. *Plant Soil* 247: 57–67.

Gadallah, M.A.A. (1999). Effects of proline and glycine betaine on *Vicia faba* responses to salt stress. *Biol. Plant.* 42 (2): 249–257.

Gardner, P.R. and Fridovich, I. (1991). Superoxide sensitivity of the *Escherichia coli* 6-phosphogluconate dehydratase. *J. Biol. Chem.* 66 (03): 1478–1483.

Geilfus, C.M. (2018). Chloride$^-$ from nutrient to toxicant. *Plant Cell Physiol.* 59: 877–886.

Gill, S.S. and Tuteja, N. (2010). Cadmium stress tolerance in crop plants: probing the role of sulfur. *Plant Signal. Behav.* 6: 215–222.

Gratao, P.L., Polle, A., Lea, P.J. et al. (2005). Making the life of heavy metal stressed Plants a little easier. *Funct. Plant Biol.* 32: 481–494.

Grune, T., Reinheckel, T., and Davies, K.J.A. (1997). Degradation of oxidized proteins in mammalian cells. *FASEB J.* 11 (7): 526–534.

Gupta, K., Dey, A., and Gupta, B. (2013). Plant polyamines in abiotic stress responses. *Acta Physiol. Plantarum* 35 (7): 2015–2036.

Gupta, B. and Huang, B. (2014). Mechanism of salinity tolerance in plants: physiological, biochemical, and molecular characterization. *Int. J. Genom.*

Halliwell, B. and Gutteridge, J.M.C. (1989). *Free Radicals in Biology and Medicine*. Oxford: Oxford University Press.

Hanson, A.D., Rathinasabapathi, B., Rivoal, J. et al. (1994). Osmoprotective compounds in the Plumbaginaceae: a natural experiment in metabolic engineering of stress tolerance. *Proc. Natl. Acad. Sci.* 91 (1): 306–310.

Hasanuzzaman, M. and Fujita, M. (2011). Selenium pretreatment upregulates the antioxidant defense and methylglyoxal detoxification system and confers enhanced tolerance to drought stress in rapeseed seedlings. *Biol. Trace Elem. Res.* 143 (3): 1758–1776.

Hasegawa, P.M., Bressan, R.A., Zhu, J.K., and Bohnert, H.J. (2000). Plant cellular and molecular responses to high salinity. *Annu. Rev. Plant Biol.* 51 (1): 463–499.

Hasegawa, P.M. (2013). Sodium (Na^+) homeostasis and salt tolerance of plants. *Environ. Exp. Bot.* 92 (19–31): 2013.

Hayashi, H. and Murata, N. (1998). Genetically engineered enhancement of salt tolerance in higher plants. In: *Stress Response of Photosynthetic Organisms: Molecular Mechanisms and Molecular Regulation* (eds. N. Murata and K. Satoh), 133–148. Elsevier Science.

Hoque, M.A., Banu, M.N.A., and Okuma, E. (2007). Exogenous proline and glycine betaine increase NaCl-induced ascorbate glutathione cycle enzyme activities, and proline improves salt tolerance more than glycine betaine in tobacco Bright Yellow-2 suspension-cultured cells. *J. Plant Physiol.* 164 (11): 1457–1468.

Hossain, M.A. and Fujita, M. (2010). Evidence for a role of exogenous glycine betaine and proline in antioxidant defense and methyl glyoxal detoxification systems in mung bean seedlings under salt stress. *Physiol. Mol. Biol. Plants* 16: 19–29.

Hossain, M.A., Hasanuzzaman, M., and Fujita, M. (2010). Coordinate induction of antioxidant defense and glyoxalase system by exogenous proline and glycine betaine is correlated with salt tolerance in mung bean. *Front. Agric. China* 5: 1–14.

Ibrahim, E.A. (2016). Seed priming to alleviate salinity stress in germinating seeds. *J. Plant Physiol.* 192: 38–46.

Ishitani, M., Liu, J., Halfter, U. et al. (2000). SOS3 function in plant salt tolerance requires N-myristoylation and calcium binding. *Plant Cell* 12 (9): 1667–1677.

Jimenez, J.S., Debouck, D.G., and Lynch, J.P. (2003). Growth, gas exchange, water relations, and ion composition of Phaseolus species grown under saline conditions. *Field Crop Res.* 80: 207–222.

Kamran, M., Parveen, A., Ahmar, S. et al. (2020). An overview of Hazardous impacts of soil salinity in crops, tolerance mechanisms, and amelioration through selenium supplementation. *Int. J. Mol. Sci.* 21: 148.

Kerepesi, I. and Galiba, G. (2000). Osmotic and salt stress-induced alteration in soluble carbohydrate content in wheat seedlings. *Crop Sci.* 40 (2): 482–487.

Khan, M.A., Ungar, I.A., and Showalter, A.M. (2000). Effects of sodium chloride treatments on growth and ion accumulation of the halophyte *Haloxylonre curvum*. *Commun. Soil Sci. Plant Anal.* 31 (17-18): 2763–2774.

Kovács, Z., Simon-Sarkadi, L., Szűcs, A., and Kocsy, G. (2010). Differential effects of cold, osmotic stress and abscisic acid on polyamine accumulation in wheat. *Amino Acids* 38 (2): 623–631.

Kumar, I. and Sharma, R.K. (2018). Production of secondary metabolites in plants under abiotic stress: an overview. *Significan. Bioeng. Biosci* 2: 1–5.

Kwon, O.K., Mekapogu, M., and Kim, K.S. (2019). Effect of salinity sress on photosynthesis and related physiological responses in carnation (*Dianthus caryophyllus*). *Hortic. Environ. Biotechnol.* 60: 831–839.

Liang, W., Cui, W., Ma, X. et al. (2014). Function of wheat Ta-UnP gene in enhancing salt tolerance in transgenic Arabidopsis and rice. *Biochem. Biophys. Res. Commun.* 450: 794–801.

Liang, W., Ma, X., Wan, P., and Liu, L. (2018). Plant salt-tolerance mechanism: a review. *Biochem. Biophys. Res. Commun.* 495 (1): 286–291.

Liu, J., Ishitani, M., Halfter, U. et al. (2000). The Arabidopsis thaliana SOS2 gene encodes a protein kinase that is required for salt tolerance. *Proc. Natl. Acad. Sci.* 97 (7): 3730–3734.

Mahajan, S. and Tuteja, N. (2005). Cold, salinity and drought stresses: an overview. *Arch. Biochem. Biophys.* 444: 139–158.

Maheshwari, R. and Dubey, R.S. (2009). Nickel-induced oxidative stress and the role of antioxidant defence in rice seedlings. *Plant Growth Regul.* 59 (1): 37–49.

Mäkelä, P., Kärkkäinen, J., and Somersalo, S. (2000). Effect of glycinebetaine on chloroplast ultrastructure, chlorophyll and protein content, and RuBPCO activities in tomato grown under drought or salinity. *Biol. Plant.* 43 (3): 471–475.

Manchanda, G. and Garg, N. (2008). Salinity and its effects on the functional biology of legumes. *Acta Physiol. Plant.* 30: 595–618.

Martínez, J.P. and Araya, H. (2010). Ascorbate–glutathione cycle: enzymatic and non-enzymatic integrated mechanisms and its biomolecular regulation. In: *Ascorbate-glutathione Pathway and Stress Tolerance in Plants* (eds. N.A. Anjum, M.T. Chan and S. Umar), 303–322. Dordrecht: Springer.

Martínez-Atienza, J., Jiang, X., Garciadeblas, B. et al. (2007). Conservation of the salt overly sensitive pathway in rice. *Plant Physiol.* 143 (2): 1001–1012.

Mishra, S., Jha, A.B., and Dubey, R.S. (2011). Arsenite treatment induces oxidative stress, upregulates antioxidant system, and causes phytochelatin synthesis in rice seedlings. *Protoplasma* 248 (3): 565–577.

Mittler, R. (2017). ROS are good. *Trends Plant Sci.* 22: 11–19.

Mittova, V., Volokita, M., Guy, M., and Tal, M. (2000). Activities of SOD and the ascorbate-glutathione cycle enzymes in subcellular compartments in leaves and roots of the cultivated tomato and its wild salt-tolerant relative *Lycopersicon pennellii. Physiol. Plant.* 110: 42–51.

Møller, I.M. and Kristensen, B.K. (2004). Protein oxidation in plant mitochondria as a stress indicator. *Photochem. Photobiol. Sci.* 3 (8): 730–735.

Moran, J.F., Becana, M., Iturbe-Ormaetxe, I. et al. (1994). Drought induces oxidative stress in pea plants. *Planta* 194: 346–352.

Morelli, R., Russo-Volpe, S., Bruno, N., and Lo Scalzo, R. (2003). Fenton-dependent dam-age to carbohydrates: free radicals scavenging activity of some simple sugars. *J. Agric. Food Chem.* 51: 7418–7425.

Munns, R. and Tester, M. (2008). Mechanisms of salinity tolerance. *Annu. Rev. Plant Biol.* 59: 651–681.

Munns, R. (2002). Comparative physiology of salt and water stress. *Plant Cell Environ.* 25: 239–250.

Nalousi, A.M., Ahmadiyan, S., Hatamzadeh, A., and Ghasemnezhad, M. (2012). Protective role of exogenous nitric oxide against oxidative stress induced by salt stress in bell-pepper (*Capsicum annum* L.). *Am. Eurasian J. Agric. Environ. Sci.* 12: 1085–1090.

Navakoudis, E., Lütz, C., Langebartels, C. et al. (2003). Ozone impact on the photosynthetic apparatus and the protective role of polyamines. *Biochim. Biophys. Acta Gen. Subj.* 1621 (2): 160–169.

Neumann, P.M., Van Volkenburgn, E., and Cleland, R.E. (1988). Salinity stress inhibits bean leaf expansion by reducing turgor, not wall extensibility. *Plant Physiol.* 85: 233–237.

Niu, X., Bressan, R.A., Hasegawa, P.M., and Pardo, J.M. (1995). Ion homeostasis in NaCl stress environments. *Plant Physiol.* 109: 735–742.

Noctor, G. and Foyer, C.H. (1998). Ascorbate and glutathione: keeping active oxygen under control. *Annu. Rev. Plant Physiol. Plant Mol. Biol.* 49: 249–279.

Pang, C.H. and Wang, B.S. (2010). Role of ascorbate peroxidase and glutathione reductase in Ascorbate–Glutathione cycle and stress tolerance in plants. In: *Ascorbate-glutathione Pathway and Stress Tolerance in Plants* (eds. N.A. Anjum, M.T. Chan and S. Umar), 91–113. Dordrecht: Springer.

Parida, A.K., Das, A.B., and Mohanty, P. (2004). Investigations on the antioxidative defence responses to NaCl stress in a mangrove, Bruguiera parviflora: differential regulations of isoforms of some antioxidative enzymes. *Plant Growth Regul.* 42 (3): 213–226.

Parida, A.K. and Das, A.B. (2005). Salt tolerance and salinity effects on plants: a review. *Ecotoxicol. Environ. Saf.* 60: 324–349.

Passioura, J.B. (2010). Scaling up: the essence of effective agricultural research. *Funct. Plant Biol.* 37: 585–591.

Pérez-Tornero, O., Tallón, C.I., Porras, I., and Navarro, J.M. (2009). Physiological and growth changes in micropropagated citrad macrophylla explants due to salinity. *J. Plant Physiol.* 166: 1923–1933.

Piotto, F.A., Carvalho, M.E.A., Souza, L.A. et al. (2018). Estimating tomato tolerance to heavy metal toxicity: cadmium as study case. *Environ. Sci. Pollut. Res.* 25 (27): 27535–27544.

Rao, A.C. and Reddy, A.R. (2008). Glutathione reductase: a putative redox regulatory system in plant cells. In: *Sulfur Assimilation and Abiotic Stress in Plants* (eds. N.A. Khan, S. Singh and S. Umar), 111–147. Berlin, Heidelberg: Springer.

Rasool, S., Ahmad, A., Siddiqi, T.O., and Ahmad, P. (2013). Changes in growth, lipid peroxidation and some key antioxidant enzymes in chickpea genotypes under salt stress. *Acta Physiol. Plant.* 35 (4): 1039–1050.

Recknagal, R.O. and Glende, E.A. (1984). Oxygen radicals in biological systems. *Methods Enzymol.* 105: 331–337.

Romero-Puertas, M.C., Palma, J.M., G′omez, M. et al. (2002). Cadmium causes the oxidative modification of proteins in pea plants. *Plant Cell Environ.* 25 (5): 677–686.

Rossi, L., Borghi, M., Francini, A. et al. (2016). Salt stress induces differential regulation of the phenylpropanoid pathway in *Olea europaea* cultivars Frantoio (salt-tolerant) and Leccino (salt-sensitive). *J. Plant Physiol.* 204: 8–15.

Sánchez-Casas, P. and Klesseg, D.F. (1994). A salicyclic acid-binding activity and a salicyclic acid inhibitable catalase activity are present in a variety of plant species. *Plant Physiol.* 106 (4): 1675–1679.

Saxena, S.C., Kaur, H., Verma, P. et al. (2013). Osmoprotectants: potential for crop improvement under adverse conditions. In: *Plant Acclimation to Environmental Stress* (eds. N. Tuteja and S.S. Gill), 197–232. New York, NY: Springer.

Scandalios, J.G. (1993). Oxygen stress and superoxide dismutases. *Plant Physiol.* 101: 7–12.

Serrano, R., Culiañz-Maciá, F.A., and Moreno, V. (1998). Genetic engineering of salt and drought tolerance with yeast regulatory genes. *Sci. Hortic.* 78 (1-4): 261–269.

Shabala, S. and Cuin, T.A. (2008). Potassium transport and plant salt tolerance. *Physiol. Plant.* 133: 651–669.

Sharma, P. and Dubey, R.S. (2005). Drought induces oxidative stress and enhances the activities of antioxidant enzymes in growing rice seedlings. *Plant Growth Regul.* 46 (3): 209–221.

Sharma, P., Jha, A.B., Dubey, R.S., and Pessarakli, M. (2012). Reactive oxygen species, oxidative damage, and antioxidative defense mechanism in plants under stressful conditions. *J. Bot.* 26: 217037.

Shi, H., Quintero, F.J., Pardo, J.M., and Zhu, J.K. (2002). The putative plasma membrane Na^+/H^+ antiporter SOS1 controls long-distance Na^+ transport in plants. *Plant Cell* 14 (2): 465–477.

Siddiqui, M.N., Mostofa, M.G., Akter, M.M. et al. (2017). Impact of salt-induced toxicity on growth and yield-potential of local wheat cultivars: oxidative stress and ion toxicity are among the major determinants of salt-tolerant capacity. *Chemosphere* 187: 385–394.

Singh, A., Sharma, D.K., Kumar, R. et al. (2018). Soil salinity management in fruit crops: a review of options and challenges. In: *Engineering Practices for Management of Soil Salinity: Agricultural, Physiological, and Adaptive Approaches* (eds. S.K. Gupta, M.R. Goyal and A. Singh), 39–85. Waretown, NJ: CRC Press, Apple Academic Press Inc.

Singh, S., Khan, N.A., Nazar, R., and Anjum, N.A. (2008). Photosynthetic traits and activities of antioxidant enzymes in blackgram (*Vigna mungo* L. Hepper) under cadmium stress. *Am. J. Plant Physiol.* 3: 25–32.

Smirnoff, N. (1995). Antioxidant systems and plant response to the environment. In: *Environment and Plant Metabolism: Flexibility and Acclimation*, 217–243. Bios Scientific Publishers.

Sofy, M.R., Elhawat, N., and Tarek, A. (2020). Glycine betaine counters salinity stress by maintaining high K^+/Na^+ ratio and antioxidant defense via limiting Na^+ uptake in common bean (*Phaseolus vulgaris* L.). *Ecotoxicol. Environ. Saf.* 200: 110732. https://doi.org/10.1016/j.ecoenv.2020.110732.

Srivalli, S. and Khanna-Chopra, R. (2008). Role of glutathione in abiotic stress tolerance. In: *Sulfur Assimilation and Abiotic Stress in Plants* (eds. N.A. Khan, S. Singh and S. Umar), 207–225. Berlin, Heidelberg: Springer.

Stadtman, E.R. (1986). Oxidation of proteins by mixed-function oxidation systems: implication in protein turnover, ageing and neutrophil function. *Trends Biochem. Sci.* 11 (1): 11–12.

Takahashi, T. and Kakehi, J.I. (2010). Polyamines: ubiquitous polycations with unique roles in growth and stress responses. *Ann. Bot.* 105 (1): 1–6.

Tomaz, A., Palma, P., Alvarenga, P., and Gonçalves, M.C. (2020). Soil salinity risk in a climate change scenario and its effect on crop yield. In: *Climate Change and Soil Interactions* (eds. M.N.V. Prasad and M. Pietrzykowski), 351–396. https://doi.org/10.1016/b978-0-12-818032-7.00013-8.

Tuteja, N., Singh, M.B., Misra, M.K. et al. (2001). Molecular mechanisms of DNA damage and repair: progress in plants. *Crit. Rev. Biochem. Mol. Biol.* 36: 337–397.

Vanderauwera, S., Suzuki, N., Miller, G. et al. (2011). Extranuclear protection of chromosomal DNA from oxidativestress. *Proc. Natl. Acad. Sci.* 108: 1711–1716.

Venâncio, C., Pereira, R., and Lopes, I. (2020). The influence of salinization on seed germination and plant growth under mono and polyculture. *Environ. Pollut.* 260: 113993. https://doi.org/10.1016/j.envpol.2020.113993.

Waśkiewicz, A., Muzolf-Panek, M., and Goliński, P. (2013). Phenolic content changes in plants under salt stress. In: *Ecophysiology and Responses of Plants Under Salt Stress* (eds. P. Ahmad, M. Azooz and M.N.V. Prasad), 283–314. New York, NY: Springer.

Yamauchi, Y., Furutera, A., Seki, V. et al. (2008). Malondialdehyde generated from peroxidized linolenic acid causes protein modification in heat-stressed plants. *Plant Physiol. Biochem.* 46 (8-9): 786–793.

Yiu, J.C., Juang, L.D., Fang, D.Y.T. et al. (2009). Exogenous putrescine reduces flooding-induced oxidative damage by increasing the antioxidant properties of Welsh onion. *Sci. Hortic.* 120 (3): 306–314.

Yoshiyama, K.O., Sakaguchi, K., and Kimura, S. (2013). DNA damage response in plants: conserved and variable response compared to animals. *Biology* 2: 1338–1356.

Zhang, Y., Wang, L., Liu, Y. et al. (2006). Nitric oxide enhances salt tolerance in maize seedlings through increasing activities of proton-pump and Na^+/H^+ antiport in the tonoplast. *Planta* 224 (3): 545–555.

Zhao, L., Zhang, F., Guo, J. et al. (2004). Nitric oxide functions as a signal in salt resistance in the calluses from two ecotypes of reed. *Plant Physiol.* 134 (2): 849–857.

Zhao, M.G., Chen, L., Zhang, L.L., and Zhang, W.H. (2009). Nitric reductase-dependent nitric oxide production is involved in cold acclimation and freezing tolerance in Arabidopsis. *Plant Physiol.* 151 (2): 755–767.

Zhu, J.K. (2002). Salt and drought stress signal transduction in plants. *Annu. Rev. Plant Biol.* 53: 247–273.

Zhu, J.K. (2003). Regulation of ion homeostasis under salt stress. *Curr. Opin. Plant Biol.* 6: 441–445.

10

Effects of Salt Stress on Osmolyte Metabolism of Crop Plants and Mitigating Strategy by Osmolyte

Abreeq Fatima[1], Garima Singh[1], Anuradha Patel[1], Sanjesh Tiwari[1], Divya Gupta[1], Dilip Kumar Prajapati[2], Anurag Dubey[3], and Sheo Mohan Prasad[1]

[1] *Ranjan Plant Physiology and Biochemistry Laboratory, Department of Botany, University of Allahabad, Prayagraj, India*
[2] *Ganga River Ecology Research Laboratory, Environmental Sciences Division, Centre of Advanced study in Botany, Institute of Sciences, Banaras Hindu University, Varanasi, India*
[3] *Mycology and Plant pathology Laboratory, School of Biological Sciences, Dr. Hari Singh Gour Central University, Sagar, Madhya Pradesh, India*

10.1 Introduction

Among various abiotic stresses, salinity is a major problem in arid and semiarid regions of the world that negatively reduces the crop productivity. Plants are consistently exposed to harsh environment due to their ecological habitats, and there are a lot of evidences that clearly describe the devastating effect of salinity which is responsible for global agricultural arrest (Wani et al. 2013; Yu et al. 2013; Yadav et al. 2014; Singh and Prasad 2019). To cope up with damaging effects induced by salt stress, plants modify their different physiological and biochemical processes such as signal perceptions and transduction cascades for survival (Faical et al. 2009; Ranganayakulu et al. 2013). Accumulation of salt ions such as Na^+ or Cl^- in the plant may cause drastic changes in various physiological processes such as photosynthesis, transpiration, and respiration which lead toward the decline in the growth and crop productivity (Kaya et al. 2013). Further, it is known that salt significantly reduces the leaf area, pigment content, and photosynthetic efficiency of wheat plants after irrigation with sea water (Aldesuquy et al. 2014). Salt also declines the overall photosynthetic efficiency of plants due to inhibition in the photochemistry of photosystem II (PS II) (Kalaji et al. 2011). Osmotic stress is the major cause of reduction in growth under salt stress and to minimize the osmotic stress, plants either accumulates or enhances the synthesis of osmoprotectants (Shahbaz et al. 2013; Talat et al. 2013; Filippou et al. 2014). Osmoprotectants are small molecules with low molecular weight, electrically neutral, highly soluble, and nontoxic at molar concentrations (Ahn et al. 2011). They assist the plants to survive under extreme osmotic environment (Lang 2007) and stabilize the membrane as well as proteins

Physiology of Salt stress in Plants: Perception, Signalling, Omics and Tolerance Mechanism, First Edition.
Edited by Pratibha Singh, Madhulika Singh, Rajiv Kumar Singh, and Sheo Mohan Prasad.

to maintain the osmotic potential to prevent dehydration inside the cell (Wani et al. 2013). They also play an important role inside the cell in maintaining balance of osmotic difference between the cell's surroundings and the cytosol.

10.2 Groups and Biosynthetic Pathways of Osmolytes in Crop Plant

The osmoprotectants play vital role in improving the adaptation under salt stress by increasing the osmotic pressure in the cytoplasm (Jagesh et al. 2010; Ranganayakulu et al. 2013). The osmoprotectants are broadly categorized into three parts:

i) Osmoprotectants with ammonium compounds (polyamines, glycinebetaine, b-alanine betaine, dimethylsulfoniopropionate, and choline-O-sulfate).
ii) Osmoprotectants containing sugars and sugar alcohols (trehalose, fructan, mannitol, D-ononitol, and sorbitol).
iii) Osmoprotectants containing amino acids (proline and ectoine).

The enhanced production of osmolytes is the manifestation of osmoprotectant responsive genes, which improves stress tolerance in plants (Reguera et al. 2012; Jain 2013). Polyamines (PAs) belong to the biogenic amines group and are widely distributed in plants that regulate various essential cellular functions necessary for the growth and development. PAs are low molecular weight, aliphatic nitrogen-containing compounds, and positively charged and regulate the pH of cellular components (Groppa and Benavides 2008) and have high affinity toward DNA, RNA, and proteins (Kusano et al. 2007). One of the best studied roles of osmolytes is its assistance in adaptation and also in mediating the osmotic adjustment *via* Na^+/K^+ discrimination. Glycine betaine (GB) also acts as antioxidants and stabilizes the membrane integrity. Apart from above-mentioned osmoprotectant, some sugars and their alcohols and amino acids, mainly proline (Pro), also acts as osmolyte. The biosynthesis and detailed function have been discussed briefly in below sections.

10.2.1 Polyamines and Their Biosynthetic Pathways in the Cell Organelles of Crop Plant

Polyamines (PAs) are small aliphatic nitrogenous bases with two or more amino groups having cationic nature at neutral pH, possess potent biological activity (Xu et al. 2009; Vuosku et al. 2018), and is present in both prokaryotes and eukaryotes (Liu et al. 2017; Mustafavi et al. 2018). In plants, PAs exist in three forms: putrescine (Put), a diamine; spermidine (Spd), a triamine; and spermine (Spm), a tetramine. These polyamines have a contributing role in regulating many physiological processes such as flower development, organogenesis, embryogenesis, fruit maturation, development, and senescence (Xu et al. 2014; Xu 2015; Mustafavi et al. 2018). Further, PAs are very effective in increasing the stress tolerance behavior of plants under biotic and abiotic stresses (Reis et al. 2016; Mustafavi et al. 2018). With detailed analysis, it was found that free PAs covalently bound with phenolic compounds with an amide bond called PSCC-PA that also acts as secondary metabolites. These phenolic compounds may be hydroxy cinnamic acid, caffeic acid, ferulic acid, or coumaric acid (Luo et al. 2009; Martin-Tanguy 2010) that constitute largest pools of PAs in plants (Kusano et al. 2008; Bassard et al. 2010). Recent studies confirm that PSCC-PAs play a part not only on external infections in plants causing local allergic reaction (Kumar et al. 1997) but also in plant morphogenesis (De Oliveira et al. 2018; Mustafavi et al. 2018). Biomacromolecules such as nucleic acids, proteins, uronic acids, or lignin are covalently bound with polyamines through ionic or hydrogen bond and form PISCC-PAs. Being ubiquitous, PAs are found even in plant tumors and plant RNA viruses (Liu et al. 2016, 2017), thus showing tissue- and organ-specific scattering patterns in plants. Presence of polyamines in plant ranges from micromolar to milimolar concentrations (Galston and Sawhney 1990). The synthesis of polyamines begins with the decarboxylation of ornithine (Orn) or arginine (Arg) catalyzed by enzyme ornithine/arginine decarboxylases (ODCs/ADCs) and generates diamine Put. Further, Put biosynthesis occurs via ADC pathway which is a process of sequential action of three enzymes: ADC, agmatine iminohydrolase (AIH), and N-carbamoylputrescine amidohydrolase (CPA). To confirm the specific role of ADC and ODC in PAs biosynthesis, inhibitors of these enzymes (difluoromethylarginine (DFMA) and difluoromethylornithine (DFMO) were also used (Grossi et al. 2016; Yamamoto et al. 2016). Spd and Spm are formed from aminopropyl and Put and residues, which are steadily delivered by methionine (Vuosku et al. 2018). Other PAs such as Spd and Spm are also produced by the subsequent addition of aminopropyl group to either Put or Spd mediated by enzyme spermidine synthase/spermine synthase. Alcázar et al. (2006) reported that genes such as SPDS1 and SPDS2 play a significant role in the biosynthesis of Spd which is further required in PAs synthesis. In contrast to the distribution of PAs, Put is abundantly found in leaves, but the content of Spd and Spm is three times higher than Put abundantly present in other organs (Takahashi et al. 2017). Different plants show different distribution of PAs such as Spm was found to accumulate in the cell wall while Put in the cytoplasm of carrot (Cai et al. 2006). In PAs biosynthesis pathway, central product role is played by putrescine having two amino groups, and it is a synthetic precursor of Spm and Spd (Xu et al. 2009). Biosynthesis of Put is carried out by three different routes in plants; two products are formed in the first route when eighth number carbon is detached from Arg with the help of ADC that are agmatine (Agm) and CO_2; Put (NCPA) and NH_3 are formed when second number nitrogen atom is detached from Agm to form N carbamoyl; and finally three different products are formed: Put, CO_2, and NH_3 when NCPA is hydrolyzed with the help of N-carbamoylputreseine amidohydrolase (NCPAH) (Docimo et al. 2012; Pegg 2016). In the second route, Orn is produced from Arg through arginase; and then Put and CO_2 are formed when ODC detaches first carbon atom of the carboxyl group of Orn (Docimo et al. 2012; Pegg 2016). An anomaly of this pathway is the absence of ODC gene from the members of Brassicaceae family and also in *Arabidopsis thaliana* plant (Hanfrey et al. 2010), which shows that this pathway is not essential (Hanfrey et al. 2010). Finally, in third route, Arg is principally converted into citrulline (Cit), which is then decarboxylated through citrulline decarboxylase (CDC) which forms Put (Ouyang et al. 2017; De Oliveira et al. 2018) and the pathway is present in sesame.

10.2.2 Betaine and Their Biosynthetic Pathways in the Cell Organelles of Crop Plant

Glycine betaine (GB) is the quaternary-based ammonium compound and acts as osmoprotectant best known as a compatible solute present in wide range of organism including bacteria, animals, and in some salt-tolerant angiospermic plants (Chen and Murata 2002). In previous studies, it was found that accumulation of GB in plant cells significantly increases the tolerance against osmotic stress (Wyn Jones 1984). Flora such as halophytes, spinach, maize, barley, and sugar beet are the natural accumulators of GB in response to different environmental constrains including salt (Bohnert et al. 1995). Economically important crops such as rice (*Oryza sativa*), tomato (*Solanum lycopersicum*), and potato (*Solanum tuberosum*) are not able to accumulate GB; hence, these species are possible targets for the engineering of *in vivo* GB biosynthesis (McCue and Hanson 1990). Major functions performed by GB are to stabilize the membranes and proteins against the denaturing effects such as abiotic stresses (Yancey 1994). The GB biosynthetic pathway starts with choline and headways with a two-step dehydrogenation of choline and progressive oxygenation of betaine aldehyde (a toxic compound for plants). The biosynthesis of GB in every organism (animals, microorganisms, and plants) occurs by two distinct pathways, and choline or glycine is the major GB precursors (Wani et al. 2013). In GB pathway, GB is produced during two consecutive oxidation processes of choline mediated by enzyme choline monooxygenase (CMO) and betaine-aldehyde dehydrogenase (BADH). These reactions take place in the stroma of chloroplasts and mainly catalyzed by the enzymes CMO and NAD^+-dependent betaine-aldehyde dehydrogenase (BADH) (Wani et al. 2013). The three subsequent methylation of amino residue occurs with the help of enzymes: glycine sarcosine methyltransferase (GSMT) and sarcosine dimethylglycine methyltransferase (SDMT). In this process, S-adenosylmethionine serves as a donor of methyl group (Igamberdiev and Kleczkowski 2018). Choline mono-oxygenase has an active Rieske-categorized [2Fe-2S] in addition to a transit peptide sequence and is frequently confined to the chloroplast or other subcellular sections (Rathinasabapathi et al. 1997). Studies also justifies that BADH which is either dependent on NADC or NADPC is also present in non-GB accumulating plants having higher activity with NADC (Fitzgerald et al. 2009). In cereals, through the use of abscisic acid (ABA), the upregulation of BADH is noticed. But in this process of induction, commanding signal is not recognized by NaCl/ABA nor turgor reduction, albeit an unusual signal coming from roots and other plant parts might be involved (Takabe et al. 1998). In the family Chenopodiaceae, plants like spinach and sugar beet have two main catalyzing enzymes (CMO and BADH) limited to the stroma of chloroplast (Fujiwara et al. 2008). These enzymes are important for the metabolic engineering of economically important plants against osmotic stress due to the biotechnological application of choline and their derivatives in biological fluids (Shimomura et al. 2009; Salvi et al. 2014). GB can also go through a stress-inducible biosynthesis as it is derived from the serine induced by salt tress: (i) the phosphorylated phospho-hydroxypyruvate pathway (Igamberdiev and Kleczkowski 2018), (ii) non-phosphorylated glycerate pathway, and (iii) photorespiratory glycolate pathway (Wingler et al. 2000; Carillo et al. 2008). In the study, the upregulation of gene of CMO and BADH was noticed by two- to threefolds and hence confirms enzyme levels (Rhodes et al. 2002). Xu et al. (2018) acknowledged that CMO and BADH genes are responsive to methyl jasmonate (MeJA) with a CGTCA motif in the promoter region of *Citrullus lanatus,* MeJA activates the cells of *C. lanatus* to synthesize GB even without the experience of osmotic stress, and the new cell is formed from the previously activated ones with a huge GB content deprived of previous stress or MeJA activation. These all findings prove that in GB synthesis, JA signal transduction has played a significant role in osmotic stress, hardening tolerance and osmotic stress tolerance (Xu et al. 2018). Conversely, the collaborating effects of synchronized salinity and other stresses can lessen the countenance of CMO and the specific GB production (Woodrow et al. 2017; Ciarmiello et al. 2018). The step catalyzed by CMO found to be the rate-limiting one in GB biosynthesis which makes it more relevant (Bhuiyan et al. 2007).

10.2.3 The Biosynthetic Pathway of Carbohydrate Sugar, Sugar Alcohol, and Amino Acids in the Cell Organelles of Crop Plant

Salt stress is the accumulation of immoderate form of salt ion contents in the soil which eventually results in the inhibition of plant growth and productivity. Accumulation of osmoprotectants helps to survive the plants in stress condition (Singh et al. 2015). An augmented level of osmoprotectants or osmolyte such as carbohydrate sugar and sugar alcohols has been demonstrated to protect the plant under salinity stress (Koyro et al. 2012). Earlier findings have described the role of carbohydrate sugars like glucose, fructose, sucrose, fructans, and trehalose, etc., as an osmoprotectants under salt stress as well as in drought conditions that regulate the osmotic homeostasis and metabolic mechanism thereby averts membrane damage in crop plants. The oxidized form of trehalose is constituent of microorganisms like bacteria and fungi and provides some desiccation tolerance to plants involved in contending stress condition (Vinocur and Altman 2005). Contrastingly, in crop plants like *Triticum* or *O. sativa*, trehalose acts as an osmoprotectant and protects membrane proteins associated with minimizing accumulation of denatured proteins as well (Koyro et al. 2012; Nounjana et al. 2012). Other compatible sugars such as glucose, sucrose, fructose, and fructans accumulate in tubers and protect plants against dessication (Murakeozy et al. 2003). Previous study also revealed that, during seed germination of wheat (*Triticum*), supplementation of small quantity of sugar such as glucose molecule enhances the germination percentage under salt stress (Hu et al. 2012). Similarly, supplementation of low amount of glucose molecule also promotes the growth of coleoptiles and radical during the seed germination. However, higher concentration of glucose molecule prevents the seed germination process (Hu et al. 2012). On the other hand, under salt stress condition, exogenous application of glucose molecule decreases the destruction of chlorophyll pigments, increases the dry weight, enhances the accumulation of proline, maintains the redox homeostasis condition, inhibits the loss of water, and enhances the antioxidant enzyme (Cuoee et al. 2006; Nemati et al. 2011). The biosynthesis of carbohydrate sugar like trehalose commonly occurs in plant system via TPS-TPP pathways (trehalose-6-phosphate synthase-trehalose-6-phosphate phosphatase). In the first step of this pathway, glucose molecule is catalyzed by TPS and then forms the UDP-glucose-6-phosphate (G6P). Formed U6P then interchanges into an intermediate compound, i.e. trehalose-6-phosphate (T6P) followed by dephosphorylation of TPP converting it into T6P (Elbein et al. 2003).

Similarly, sugar alcohols are organic compounds which are basically derived from the sugar content and are also known as polyols which are mainly characterized on the basis of their chemical structures. It contains two types of structure, i.e. cyclic and linear structure (Tari et al. 2010). Cyclic structure mainly includes pinitol and myoinositol, while linear structure covers sorbitol, mannitol, xylitol, and ribitol. Polyols (sugar alcohols) are generally hydrophilic in nature, and their synthesis occurs due to the reduction of phosphate ester group or aldose group (Tari et al. 2010). In earlier studies, it has been mentioned that various forms of sugar alcohol including mannitol, sorbitol, and inositol activate tolerance mechanism of plants against the saline environment (Williamson et al. 2002). Sorbitol and mannitol are present in *O. sativa* and *Zea mays*, respectively, that have protective role under salt stress (Theerakulpisut and Gunnula 2012; Kaya et al. 2013). Under severe oxidative stress, formation of most toxic hydroxyl (-OH) via the Fenton reaction is quenched by the mannitol which is mainly synthesized by the plant in response to stress factors (Gill and Tuteja 2010). Sugar alcohol (polyols) also regulates various physiological functions during salt stress, maintains the osmotic homeostasis, facilitates water flow in cytoplasm, and protects the cellular structure of membrane proteins and enzymes under the salinity (Kanayama et al. 2006). The biosynthetic pathway of sugar alcohol is an energy-dependent process and it consumes large amount of ATP molecules. G6P acts as a precursor for synthesis of mannitol and sorbitol. In the first step, there is isomerization of G6P into the fructose-6-phosphate (F6P) mediated by enzyme isomerase. The catalysis of mannitol-1-phosphate dehydrogenase (mtlD) leads to the conversion of F6P into mannitol (Stoop et al. 1996). Similarly, in sorbitol biosynthesis, the G6P is converted into the sorbitol-6-phosphate (S6P) in the presence of sorbitol-6-phosphate dehydrogenase (S6PDH) enzyme. The release of $NADP^+$ through dephosphorylation results in the synthesis of sorbitol sugar alcohol (Shen et al. 1999).

Most of the crop plants during abiotic stress conditions such as salinity, drought, and high temperature protect themselves through amino acid metabolism as well. The amino acids such as proline, alanine, Arg, and glycine perform a significant role against exposure to abiotic stresses. Among these amino acids, proline is a protein-genic and most important osmolyte which is found at highest concentration in cytosol in several plants such as tobacco (*Nicotiana*), rice (*O. sativa*), and wheat (*Triticum*) plant (Nounjana et al. 2012; Singh et al. 2015). Accumulation of Pro in the cytosol and chloroplast plays a crucial role in metabolism as well as in defense mechanism under salinity. Proline acts as a molecular chaperone and shows various signaling functions regarding mechanism of regulation of enzyme activities and protection of protein probity. It also acts as an antioxidant enzyme and exhibits reactive oxygen species (ROS)-scavenging properties, protects the membrane proteins, buffers cellular redox potential, and acts as an inducer responsive gene during salinity (Sharma and Dietz 2006; Hoque et al. 2008). In plant, proline is synthesized into the cytosol and plastids and is degraded in the mitochondria (Ashraf and Foolad 2007). It is synthesized by two pathways: (i) Orn pathway and (ii) glutamate pathway, in which Orn and glutamate act as precursor. The glutamate pathway is the major source for proline accumulation, while the Orn pathway is activated into the chloroplasts or cytoplasm, synthesizing proline from glutamic acid through an intermediate compound like pyrroline-5-carboxylate (P5C) under stress condition. The main enzymes which are involved in proline biosynthesis are (i) pyrroline-5-carboxylate synthetase (P5CS) and (ii) pyrroline-5-carboxylate reductase (P5CR) (Nounjana et al. 2012). Enzyme P5CS is bifunctional in nature and it activates the glutamate through the phosphorylation process, while P5CR enzyme reduces an intermediate c-glutamyl phosphate into glutamate semialdehyde (GSA). Likewise, in mitochondria, Orn enzymes are present and are transported to P5C by the mechanism of orn-d-aminotransferase (OAT) (Verbruggen and Hermans 2008). However, proline degradation is mediated by the action of proline dehydrogenase (PDH)

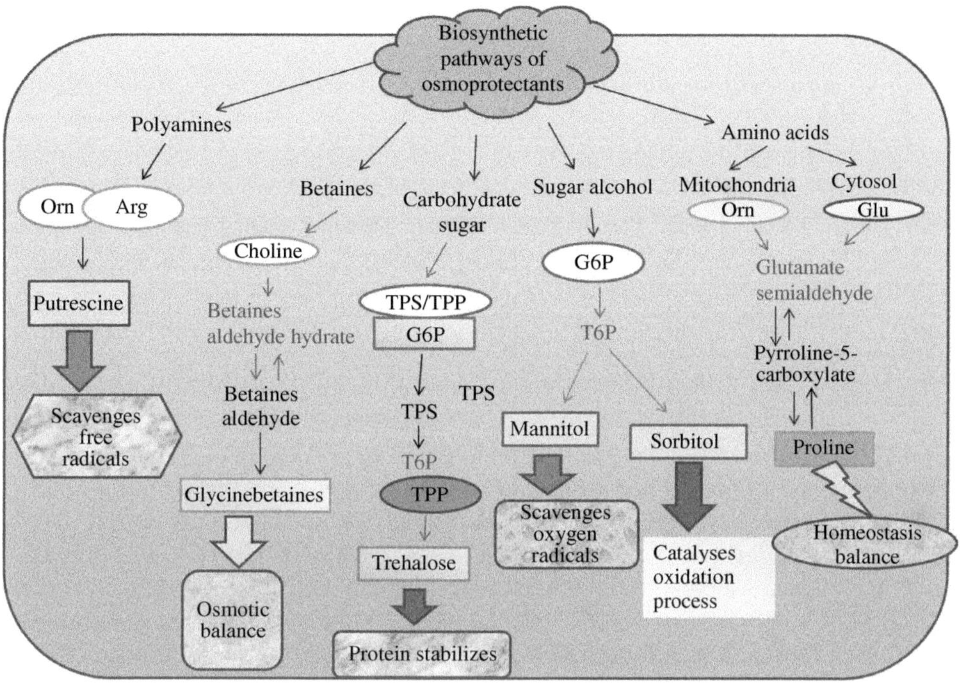

Figure 10.1 Schematic representation of biosynthesis of osmolytes in plants under salt stress.

and P5C dehydrogenase (P5CDH) enzymes. In the glutamate pathway, glutathione act as precursor in the presence of enzyme glutamate-semialdehyde (GSA) and D1-P5C. The enzyme P5CS catalyzes the conversion of glutathione into P5C in the presence of P5CR enzyme, which reduces the P5C to proline (Verbruggen and Hermans 2008). Under stress condition, proline production in the chloroplast performs various types of functions such as maintaining the ratio between NADPH and $NADP^+$ and maintaining the flow of electrons between the photosynthetic excitation centers, i.e. PS II and PS I, and also protect the photosynthetic apparatus from the excess of energy which causes photoinhibition phenomenon (Ashraf and Foolad 2007). Thus, the accumulation of carbohydrates, sugar alcohol, and amino acid regulates the metabolic processes,acts as an osmoprotectant, and works concomitantly to protect the plant against desiccation. The biosynthetic pathway has been shown in Figure 10.1.

10.3 Effect of Salt Stress on Osmolyte Production and Work Action

Increase in the soil salinization (accumulation of ions such as NaCl, KCl, or $MgCl_2$) is a leading environmental problem that significantly reduces the quality and quantity of crops, and on an average around 45 million hectares of irrigated land is under the influence of soil salinity problem (Rengasamy 2010). Salinity stress affects various physiological and metabolic processes (germination of seed, growth, photosynthesis, and nitrogen assimilation), and the severity of stress depends on duration, varies among different plants species, and finally obstructs crop production (Rahnama et al. 2010; James et al. 2011). Major inhibitory effects on plants under soil salinity are due to increase in the osmotic stress associated with ion toxicity (Parida and Das 2005). To combat with the negative effects induced by salt toxicity, plants modify variety of physiological and biochemical mechanisms that include balanced ion uptake and their transport, biosynthesis of osmoprotectants and compatible solutes, upregulation of the antioxidative machinery, increasing the synthesis of secondary signaling molecules (NO, Ca, H_2S), or activating the stress-responsive plant growth regulators (Munns 2002; Munns and Tester 2008). Among various protective mechanisms, the synthesis of osmoprotectants or osmolytes or compatible solutes is a major pathway to minimize the salt stress via the synthesis of these low molecular weight metabolites that do not restrict the cellular metabolism even at high concentration. They mainly include polyamines (Gholami et al. 2013), glycine betaine (Ashraf and Foolad 2007), amino acid proline (Deivanai et al. 2011), and compatible sugars (Kerepesi and Galiba 2000). Polyamines play diverse role in plant growth and development such as regulation of cell proliferation, somatic embryogenesis, differentiation and morphogenesis, breaking dormancy and seed germination, development of flowers and fruit, and senescence (Gupta et al. 2013; Vuosku et al. 2018). Polyamines present in either free forms (F-PAs) or covalently/non-covalently conjugated forms (CC-Pas or NCC-PAs) (Gholami et al. 2013). *A. thaliana* genome express two ADCs, two SPDS, four SAMDCs, and one SPMS encoding genes (*ADC1* and *ADC2*; *SPDS1* and *SPDS2*; *SAMDC1, SAMDC2, SAMDC3*, and *SAMDC4; SPMS1*) (Hanzawa et al. 2002; Janowitz et al. 2003; Urano et al. 2004). The potential role of PAs in plants is identified by the use of PAs biosynthetic inhibitors such as DFMO and DFMA which could restrict the growth and development about 83% as reported by Kumar et al. (2008) in *Coffea canophora*. Plants under the influence of high salt concentrations exhibited loss in membrane integrity correlated with decrease in the structure of photosynthetic apparatus. To overcome these situations, plants accumulated high concentrations of PAs that bind with the phospholipids of membrane, reduce the damaging effects (Li et al. 2008), and increase salt tolerance. Salt (NaCl)-stressed *Bakraii citrus* and *Zoysia japonica* when supplemented with 1 mM Put and 0.3 mM SPD and significantly improved tolerance to salinity, respectively (Li et al. 2016; Khoshbakht et al. 2017). It has been proposed that the level of SPM in plants is an important indicator of salt tolerance (Li and He 2012). Similarly, maize plants under salt stress exhibited upregulation of two genes *Zmodc* and *Zmspds2A* that contribute to PA and SPD synthesis (Rodríguez-Kessler et al. 2006). Further, the positive role of PAs is correlated with upregulating the gene expression for biosynthesis of osmotically active solutes and ROS homeostasis and minimized the overaccumulation of Na^+ and Cl^- ion in various plant parts (Afzal et al. 2009; Yiu et al. 2009; Roychoudhury et al. 2011). Plant *Viola odoarta* showed that deficient genes *ADC1* and *ADC2* are hypersensitive to salt stress (Hussain et al. 2011). Apart from this, exogenous SPD is significantly involved in the regulation of Calvin cycle, protein folding and assembly, and the inhibition of protein proteolysis (Li et al. 2013).

Among osmoprotectants, exogenous supplementation of glycine betaine (GB) is found to be very effective in increasing the salt stress tolerance in plants (Ashraf and Foolad 2007). It is an ammonium-containing amphoteric molecule, is soluble in water, and widely distributed throughout the plant kingdom as well as in microorganisms and animals. Due to their peculiar chemical structure, it can bind either with hydrophobic or hydrophilic molecules with the same affinity and is involved in increasing the osmolarity of plant during stress conditions (Habib et al. 2012; Takabe 2012). Studies demonstrate that plants show accumulation of GB under salt stress conditions and minimize the negative effects (Sithtisarn et al. 2009; Fujiwara et al. 2010). After synthesis in chloroplast, GB is transported to other plant parts facing severe salt stress, and it is believed that both proline and betaine are transported through betaine transporters (Hoque et al. 2008). Some GB transporter proteins such as *AmBet/ProT1–2* present in *Avicennia marina* (Waditee et al. 2000), *BvBet/ProT1* in sugarbeet (Yamada et al. 2009), and *ZmBetProt* in maize are overexpressed under salt stress. Pretreatement of seedlings with GB improves the photosynthetic efficiency and damages of chloroplast and mitochondria (Ashraf and Foolad 2007). Plants species under the severe salt stress exhibited decrease in the accumulation of amino acids such as cysteine, Arg, and methionine, except proline which was found to be increased and considered as major osmolyte (Amirjani 2010; Guo et al. 2015). Besides its major role as osmoprotectant, it is also involved in the direct quenching of ROS in lower plants too, such as fungi (*Aspergillus* and *Saccharomyces*) used proline as source of nitrogen and carbon (Keller and Hohn 1997; Huang and Brandriss 2000).Hence, it also serves as an organic nitrogen reserve during stress recovery. Precursor for Pro biosynthesis is either glutamate or Orn, and in stress conditions glutamate behaves as primary substrate. Conversion of the glutame into Pro is mediated by the combined action of two key enzymes, viz. pyrroline-5-carboxylate synthetase (P5CS) and pyrroline-5-carboxylate reductase (P5CR) (Sairam and Tyagi 2004). Further, conversion of Orn into Pro takes place in the mitochondria mediated by enzyme OAT that causes transamination of Orn into P5C and then transforms into Pro. Accumulation of Pro is considered as salt stress indicator (Sannazzaro et al. 2007), and several workers reported excessive increase in the proline concentration in plants: maize (Kaya et al. 2010), *Plantago* (Al Hassan et al. 2016), and wheat (Darko et al. 2017) exposed to salt stress. Accumulation of Pro under abiotic stress also signifies its antioxidative role as it functions as an O_2 quencher (Matysik et al. 2002). Similarly, exogenous Pro supplementation improved the photosynthetic activity, plant growth, or development and also prevents the plant water status under salinity conditions in olive (*Olea europaea*) (Ben Ahmed et al. 2010). Deivanai et al. (2011) also confirmed that rice seedlings from seeds pretreated with 1 mM proline showed improvement in growth during salt stress. Higher salinity also causes amplified accumulation of amino acid valine (Val) (Nedjimi 2011). Valine has also been reported to accumulate in plants in response to salt stress such as strawberry (Keutgen and Pawelzik 2008) and *Phragmites australis* (Hartzendorf and Rolletschek 2001). Proline and glycine betaine are the major osmolyte responsible for osmoregulation in plants under salt stress. Carbohydrates such as glucose, fructose, sucrose, fructans, and starch are found to be accumulated under salt stress conditions (Parida et al. 2002). They perform diverse functions such as osmoprotection, osmotic adjustment, and carbon storage and also mediate the scavenging of free radiclas. A number of studies demonstrate the excessive formation of reducing sugars (glucose and fructose), sucrose, and fructans in a number of plants under salt stress (Khatkar and Kuhad 2000; Parida and Das 2004; Velarde-Buendia et al. 2012). Accumulation and compartmentalization of these compatible solutes or toxic ions is considered as a primary tolerance mechanism behind salt stress (Türkan and Demiral 2009). A number of plants belonging to different species exhibited increase in the formation of reducing sugars (sucrose and fructans) within the cell (Kerepesi and Galiba 2000). Carbohydrate such as trehalose also found to be accumulated in salt-stressed and protects organisms against several physical and chemical stresses (Ahmad et al. 2013). Similarly, tomato (*S. lycopersicum*) under salt stress showed accumulation of sucrose by activating the enzyme sucrose phosphate synthase (Gao et al. 1998), while leaves of plant *Bruguiera parviflora* showed remarkable increase in reducing or non-reducing sugar despite decrease in the starch content (Parida and Das 2004). In several plants, it is believed that accumulation of sugar content participates in salt stress tolerance; for example, Cha-um et al. (2009) and López et al. (2008) found that high salinity increased the total soluble sugars in rice cultivars and *Medicago* species, respectively. In brief, accumulation of inorganic and organic solutes helps in osmotic adjustment, which is necessary for growth in a saline environment. Inorganic ions are believed to be sequestered in the vacuoles, while organic solutes are assumed to be compartmentalized in the cytoplasm to balance the low osmotic potential in the vacuole (Munns and Tester 2008).

10.4 The Osmotic and Ionic Adjustment Under Salt Stress Tolerance Mechanism

Salinity is one of the most brutal environmental stresses that hamper plant growth and productivity worldwide. It encompasses deviations in various physiological and metabolic processes, depending on severity and duration of the stress. It ultimately inhibits crop production (James et al. 2011). Osmolyte are organic compounds of low molecular weight that encompasses amino acids, tertiary sulphonium and quaternary ammonium compounds, sugars, and polyhydric alcohols (Slama et al. 2015). Osmolytes mount up in the cytoplasm in response to salinity directive to balance the osmotic potential in the vacuole, scavenged the ROS thereby reducing oxidative stress. Salt stress imposes water-deficit symptom, inflection in plasma membranes, damage to photosynthetic machinery, and alteration in nitrogen metabolism and PS II photochemistry (Munns and Tester 2008). The Na^+ gets easily incorporated in enzymatic processes and replaces K^+ that act as a cofactor in several enzymatic processes and inhibits protein synthesis. Contrarily, plants have mechanism for Na^+ and Cl^- sequestration into the vacuole, constructing lower concentrations of these ions and avoiding their inhibitory effects on metabolic processes (Flowers and Colmer 2008; Flowers et al. 2014). Under salt stress, accumulation of Na^+ is potently associated with the increase of Cl^- in shoots as reported in many plants belonging to family Amaranthaceae or Chenopodiaceae. This heaped concentration contributes significantly to the internal osmotic potential and turgor pressure (Flowers et al. 2014). The hypothesis besides compartmentalization is reducing the damaging effects on the cytosol and its organelles. This is the cellular basis of salt tolerance and implies the maintenance of a sufficient uptake of K^+ causing greater K^+/Na^+ ratio, that is the chief determinant of plant stress tolerance.

The accumulated concentration of Na^+ and Cl^- contributes to the osmotic/ionic adjustment in plant species that imparts the tolerance to plants against salt stress. Excessive salt stress leads to the augmented entry of both Na^+ and Cl^- into the cells that cause severe osmotic and ionic imbalance and results in noteworthy physiological alterations:

i) Elevated Na^+ concentration interferes with the transport of K^+ ions thus hampering the growth and development as it is an essential micronutrient.
ii) Production of ROS (James et al. 2011). The accelerated ROS formation ultimately causes oxidative stress and damage to macromolecules such as proteins, lipids, and DNA, interrupting vital cellular functions of plants.

Besides this physiological and biochemical variation, genetic variations also exist. Thus, to combat with the stress situation, plants adopt several strategies that involve ion homeostasis, synthesis of osmoprotectant, antioxidant enzyme, generation of nitric oxide (NO), and hormone modulation. The ion homeostasis is briefly described.

Balancing ion homeostasis via uptake and sectionalization is a necessity of plants growing under salt stress for growth (Hasegawa 2013). The Na^+ ion accumulated in cytoplasm is then transported via Na^+/H^+ exchanger (antiporter pump) to the vacuole. Two forms of H^+ present for transportation, i.e. H^+ATPase and V^+ATPase, the latter primarily exists in plants. Under nonrestrictive condition, V^+ATPase plays pivotal role in ion homeostasis and facilitates the vesicle fusion, thus V^+ATPase is considered for the survivability of plants under salt stress (Dietz et al. 2001). The ion homeostasis is regulated by SOS signaling pathway which involves SOS1, SOS2, and SOS3. SOS1 codes to Na^+/H^+ that regulates the Na^+ efflux and its uptake from root to shoot, SOS2 encodes for serine/threonine kinase, and the SOS3 is the Ca^{2+} protein that possess myristoylated site at N terminal that confers the tolerance response, and upregulation of these genes is responsible for the tolerance response (Liu et al. 2000). Further, it is responsible for the salt tolerance of SOS proteins that regulate pH and facilitate vesicle trafficking, and the overall mechanism is the increased Na^+ concentration that leads to the upsurge in Ca^{2+} level, which activates signaling responses, and SOS protein is phosphorylated which causes efflux of Na^+ thereby reducing Na^+ accumulation and Na^+ toxicity (Martínez-Atienza et al. 2007). The plants have developed this strategy for regulating ion homeostasis and maintaining the low-ion concentration in cytosol under stress condition (Sairam and Tyagi 2004). Different carrier, channel, and antiporter protein are involved in the homeostasis that is key characteristic for the survival of plants under salt stress. Further, the study on *Arabidopsis* points that NADPH regulates the ion homeostasis. Ma et al. (2012) have reported that NADPH regulates the ion homeostasis by activating *AtrbohD* and *AtrbohF* in *Arabidopsis* under salt stress. The membrane-located HKT (histidine kinase transporter) and intracellular NHX (sodium K^+/H^+ transporter) are primarily involved in K^+ uptake as reported in several halophytes, and the HKT imparts tolerance against salinity stress by regulating Na^+ and K^+ uptake. As reported in *A. thaliana*, the HKT transporter prevents excess accumulation of Na^+ and facilitates its excretion out of cell (Schroeder et al. 1987). While NHX are antiporters that have tonoplast, localization involved in K^+ homeostasis and pH regulation and also regulates stomatal and turgor function and establishes homeostasis. Thus, ion homeostasis is responsible for the tolerance response in plants against salinity stress.

10.5 Conclusion

Salt ions such as sodium chloride (NaCl)/ potassium chloride (KCl) play major role in growth and development of plants by maintaining the membrane potential or stabilizing the membrane structure. Due to natural or anthropogenic actions, the concentrations of ions such as Na^+ or Cl^- is found to rise in the plants and thereby adversely affects the plant growth associated with disturbance in physiological and biochemical processes and thus limits the overall crop production and yield (Figure 10.2).

Salt stress also interferes with the uptake of essential nutrients thereby creating nutritional deficiency in plants and also involves in the generation of ROS that causes oxidative stress. Further, the biosynthetic pathways of major osmolyte such as proline, glycine betaine, polyamines, or some synthetic sugar molecules are severely affected under salt stress. Salt-induced toxicity is either alleviated by enhanced antioxidant system or by exogenous supplementation of these major osmolytes. Supplementation of major osmolyte positively regulates the salt-induced toxicity either by enhancing the antioxidant capacity as some osmolyte directly involves in quenching of ROS or enhancing the biosynthesis processes. Integration of mass screening and breeding, marker-assisted selection, exogenous application of osmoprotectants for seeds or growing plants, and cellular-based stress signaling and ion homeostasis mechanisms is functional paradigm for the plant that increases resistance (Figure 10.3). N-carbamoyl Put (NCPA), CGTCA is a motif, UDP is a glucose molecule and SAMDC.

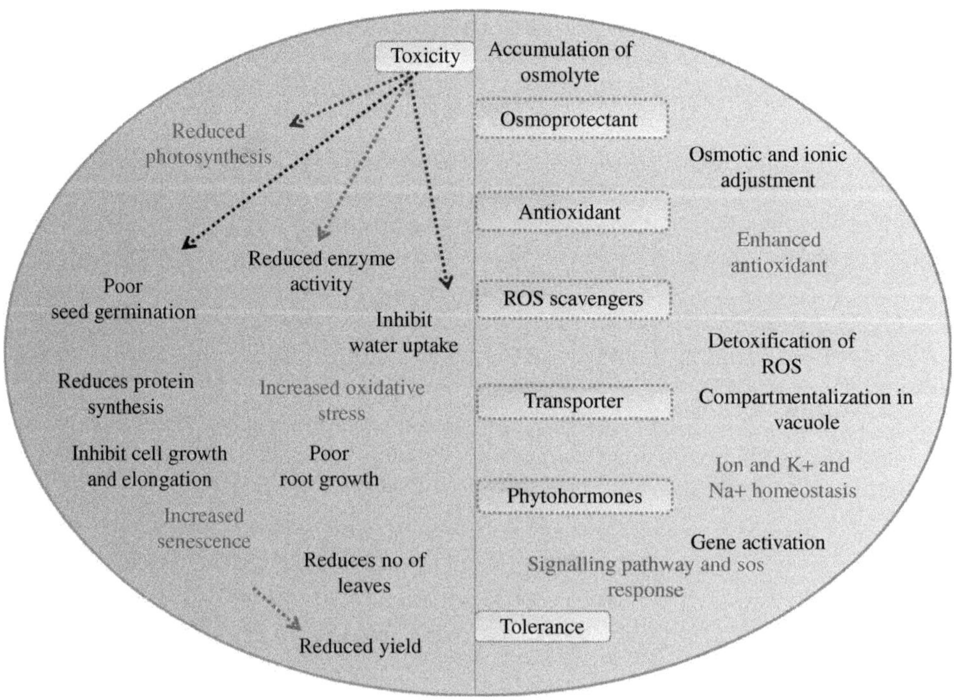

Figure 10.2 Schematic representation of toxicity and tolerance response under salinity stress.

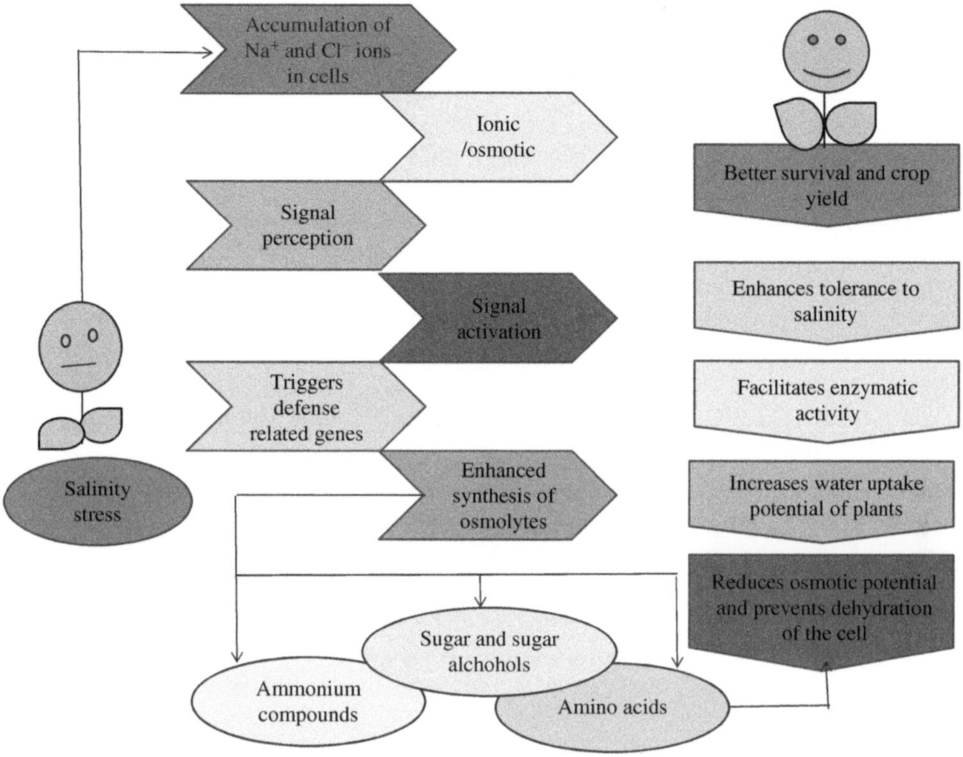

Figure 10.3 A pictorial presentation of osmolyte synthesis and mediated responses under salinity-induced ionic or osmotic toxicity in plants.

References

Afzal, I., Munir, F., Ayub, C.M. et al. (2009). Changes in antioxidant enzymes, germination capacity and vigour of tomato seeds in response of priming with polyamines. *Seed Sci. Technol.* 37: 765–770.

Ahmad, R., Lim, C.J., and Kwon, S.Y. (2013). Glycine betaine: a versatile compound with great potential for gene pyramiding to improve crop plant performance against environmental stresses. *Plant Biotechnol. Rep.* 7: 49–57.

Ahn, C., Park, U., and Park, P.B. (2011). Increased salt and drought tolerance by D-ononitol production in transgenic *Arabidopsis thaliana*. *Biochem. Biophys. Res. Commun.* 415: 669–674.

Al Hassan, M., Pacurar, A., López-Gresa, M.P. et al. (2016). Effects of salt stress on three ecologically distinct plantago species. *PLoS One* 11 (8): e0160236.

Alcázar, R., Marco, F., Cuevas, J.C. et al. (2006). Involvement of polyamines in plant response to abiotic stress. *Biotechnol. Lett.* 28: 1867–1876.

Aldesuquy, H., Baka, Z., and Mickky, B. (2014). Kinetin and spermine mediated induction of salt tolerance in wheat plants: leaf area, photosynthesis and chloroplast ultrastructure of flag leaf at ear emergence. *Egypt. J. Basic Appl. Sci.* 1: 77–87.

Amirjani, M.R. (2010). Effect of salinity stress on growth, mineral composition, proline content, antioxidant enzymes of soybean. *J. Plant Physiol.* 5: 350–360.

Ashraf, M. and Foolad, M.R. (2007). Roles of glycine betaine and proline in improving plant abiotic stress resistance. *Environ. Exp. Bot.* 59: 206–216.

Bassard, J.E., Ullmann, P., Bernier, F., and Werck-Reichhart, D. (2010). Phenolamides: bridging polyamines to the phenolic metabolism. *Phytochemistry* 71: 1808–1824.

Ben Ahmed, C., Ben Rouina, B., Sensoy, S. et al. (2010). Exogenous proline effects on photosynthetic performance and antioxidant defense system of young olive tree. *J. Agric. Food Chem.* 58: 4216–4222.

Bhuiyan, N.H., Hamada, A., Yamada, N. et al. (2007). Regulation of betaine synthesis by precursor supply and choline monooxygenase expression in Amaranthus tricolor. *J. Exp. Bot.* 58: 4203–4212.

Bohnert, H.J., Nelson, D.E., and Jensen, R.G. (1995). Adaptations to environmental stresses. *Plant Cell* 7: 1099–1111.

Cai, Q., Zhang, J., Guo, C., and Al, E. (2006). Reviews of the physiological roles and molecular biology of polyamines in higher plants. *J. Fujian Educ. Coll.* 7: 118–124.

Carillo, P., Mastrolonardo, G., Nacca, F. et al. (2008). Nitrogen metabolism in durum wheat under salinity: accumulation of proline and glycine betaine. *Funct. Plant Biol.* 35: 412–426.

Cha-um, S., Charoenpanich, A., Roytrakul, S., and Kirdmanee, C. (2009). Sugar accumulation, photosynthesis and growth of two indica rice varieties in response to salt stress. *Acta Physiol. Plant.* 31: 477–486.

Chen, T.H.H. and Murata, N. (2002). Enhancement of tolerance to abiotic stress by metabolic engineering of betaines and other compatible solutes. *Curr. Opin. Plant Biol.* 5: 250–257.

Ciarmiello, L.F., Di Maro, A., Woodrow, P. et al. (2018). Unveiling the enigmatic structure of TdCMO transcripts in durum wheat. *Agronomy* 8: 1–11.

Cuoee, I., Sulmon, C., Gouesbet, G., and Amrani, A. (2006). Involvement of soluble sugars in reactive oxygen species balance and responses to oxidative stress in plants. *J. Exp. Bot.* 57: 449–459.

Darko, E., Gierczik, K., Hudak, O. et al. (2017). Differing metabolic responses to salt stress in wheat-barley addition lines containing different 7H chromosomal fragments. *PLoS One* 12 (3): e0174170.

Deivanai, S., Xavier, R., Vinod, V. et al. (2011). Role of exogenous proline in ameliorating salt stress at early stage in two rice cultivars. *J. Stress Physiol. Biochem.* 7 (4): 157–174.

Dietz, K.J., Tavakoli, N., Kluge, C. et al. (2001). Significance of the Vtype ATPase for the adaptation to stressful growth conditions and its regulation on the molecular and biochemical level. *J. Exp. Bot.* 52: 1969–1980.

Docimo, T., Reichelt, M., Schneider, B., and Al, E. (2012). The first step in the biosynthesis of cocaine in Erythroxylum coca: the characterization of arginine and ornithine decarboxylases. *Plant Mol. Biol.* 78: 599–615.

Elbein, A.D., Pan, Y.T., Pastuszak, I., and Carroll, D. (2003). New insights on trehalose: a multifunctional molecule. *Glycobiology* 13 (4): 17R–27R.

Faical, B., Imen, A., Kaouther, F. et al. (2009). Physiological and molecular analyses of seedlings of two Tunisian durum wheat (*Triticum turgidum* L.) varieties showing contrasting tolerance to salt stress. *Acta Physiol. Plant.* 31: 145–154.

Filippou, P., Bouchagier, P., Skotti, E., and Fotopoulos, V. (2014). Proline and reactive oxygen/nitrogen species metabolism is involved in the tolerant response of the invasive plant species *Ailanthus altissima* to drought and salinity. *Environ. Exp. Bot.* 97: 1–10.

Fitzgerald, T.L., Waters, D.L.E., and Henry, R.J. (2009). Betaine aldehyde dehydrogenase in plants. *Plant Biol.* 11: 119–130.

Flowers, T.J. and Colmer, T.D. (2008). Salinity tolerance in halophytes. *New Phytol.* 179: 945–963.

Flowers, T.J., Munns, R., and Colmer, T.D. (2014). Sodium chloride toxicity and the cellular basis of salt tolerance in halophytes. *Ann. Bot.* in press. https://doi.org/10.1093/aob/mcu217.

Fujiwara, T., Hori, K., Ozaki, K. et al. (2008). Enzymatic characterization of peroxisomal and cytosolic betaine aldehyde dehydrogenases in barley. *Physiol. Plant.* 134: 22–30.

Fujiwara, T.S., Mitsuya, H., Miyake, T. et al. (2010). Characterization of a novel glycinebetaine/proline transporter gene expressed in the mestome sheath and lateral root cap cells in barley. *Planta* 232 (1): 133–143.

Galston, A.W. and Sawhney, R.K. (1990). Polyamines in plant physiology. *Plant Physiol.* 94: 406–410.

Gao, Z., Sagi, M., and Lips, S.H. (1998). Carbohydrate metabolism in leaves and assimilate partitioning in fruits of tomato (*Lycopersicon esculentum*L.) as affected by salinity. *Plant Sci.* 135: 149–159.

Gholami, M., Fakhari, A.R., and Ghanati, F. (2013). Selective regulation of nicotine and polyamines biosynthesis in tobacco cells by enantiomers of ornithine. *Chirality* 25: 22–27.

Gill, S.S. and Tuteja, N. (2010). Reactive oxygen species and antioxidant machinery in abiotic stress tolerance in crop plants. *Plant Physiol. Biochem.* 48: 909–930.

Groppa, M.D. and Benavides, M.P. (2008). Polyamines and abiotic stress: recent advances. *Amino Acids* 34: 35–45.

Grossi, M., Phanstiel, O., Rippe, C., and Al, E. (2016). Inhibition of polyamine uptake potentiates the anti-proliferative effect of polyamine synthesis inhibition and preserves the contractile phenotype of vascular smooth muscle cells. *J. Cell. Physiol.* 231: 1334–1342.

Guo, R., Yang, Z., Li, F. et al. (2015). Comparative metabolic responses and adaptive strategies of wheat (Triticum aestivum) to salt and alkali stress. *BMC Plant Biol.* 15: 170.

Gupta, K., Dey, A., and Gupta, B. (2013). Plant polyamines in abiotic stress responses. *Acta Physiol. Plant.* 35: 2015–2036.

Habib, N., Ashraf, M., Ali, Q., and Perveen, R. (2012). Response of salt stressed okra (*Abelmoschus esculentus* Moench) plants to folia-applied glycine betaine and glycine betaine containing sugarbeet extract. *S. Afr. J. Bot.* 83: 151–158.

Hanfrey, C., Sommer, S., Mayer, M.J., and Al, E. (2010). Arabidopsis polyamine biosynthesis: absence of ornithine decarboxylase and the mechanism of arginine decarboxylase activity. *Plant J.* 27: 551–560.

Hanzawa, Y., Imai, A., Michael, A.J. et al. (2002). Characterization of the spermidine synthase-related gene family in *Arabidopsis thaliana*. *FEBS Lett.* 527: 176–180.

Hartzendorf, T. and Rolletschek, H. (2001). Effects of NaCl salinity on amino acid and carbohydrate contents of *Phragmites australis*. *Aquat. Bot.* 69: 195–208.

Hasegawa, P.M. (2013). Sodium (Na+) homeostasis and salt tolerance of plants. *Environ. Exp. Bot.* 92: 19–31.

Hoque, M.A., Banu, M.N.A., Nakamura, Y. et al. (2008). Proline and glycine betaine enhance antioxidant defense and methyl glyoxal detoxification systems and reduce NaCl-induced damage in cultured tobacco cells. *J. Plant Physiol.* 165: 813–824.

Hu, M., Shi, Z., Zhang, Z. et al. (2012). Effects of exogenous glucose on seed germination and antioxidant capacity in wheat seedlings under salt stress. *Plant Growth Regul.* 68: 177–188.

Huang, H.L. and Brandriss, M.C. (2000). The regulator of the yeast proline utilization pathway is differentially phosphorylated in response to the quality of the nitrogen source. *Mol. Cell. Biol.* 20: 892–899.

Hussain, K., Nawaz, K., Majeed, A. et al. (2011). Role of exogenous salicylic acid applications for salt tolerance in violet (*Viola Odorata* L.). *Sarhad J. Agric.* 27: 171–175.

Igamberdiev, A.U. and Kleczkowski, L.A. (2018). The glycerate and phosphorylated pathways of serine synthesis in plants: the branches of plant glycolysis linking carbon and nitrogen metabolism. *Front. Plant Sci.* 9: 318.

Jagesh, K., Tiwari, A.D., Munshi, R.K. et al. (2010). Effect of salt stress on cucumber: Na^+ -K^- ratio, osmolyte concentration, phenols and chlorophyll content. *Acta Physiol. Plant.* 32: 103–114.

Jain, M. (2013). Emerging role of metabolic pathways in abiotic stress tolerance. *J. Plant Biochem. Physiol.* 1: 108.

James, R.A., Blake, C., Byrt, C.S., and Munns, R. (2011). Major genes for Na+ exclusion, Nax1 and Nax2 (wheatHKT1;4 and HKT1;5), decrease Na+ accumulation in bread wheat leaves under saline and waterlogged conditions. *J. Exp. Bot.* 62: 2939–2947.

Janowitz, T., Kneifel, H., and Piotrowski, M. (2003). Identification and characterization of plant agmatine iminohydrolase, the last missing link in polyamine biosynthesis of plants. *FEBS Lett.* 544: 258–261.

Kalaji, H.M., Govindjee, B.K., Koscielniak, J., and Zuk-Gołaszewska, K. (2011). Effects of salt stress on photosystem II efficiency and CO_2 assimilation of two Syrian barley landraces. *Environ. Exp. Bot.* 73: 64–72.

Kanayama, Y., Watanabe, M., Moriguchi, R. et al. (2006). Effects of low temperature and abscisic acid on the expression of the sorbitol-6-phosphate dehydrogenase gene in apple leaves. *J. Jpn. Soc. Hortic. Sci.* 75: 20–25.

Kaya, C., Tuna, A.L., and Okant, A.M. (2010). Effect of foliar applied kinetin and indole acetic acid on maize plants grown under saline conditions. *Turk. J. Agric. For.* 34: 529–538.

Kaya, C., Sonmez, O., Aydemir, S. et al. (2013). Exogenous application of mannitol and thiourea regulates plant growth and oxidative stress responses in salt-stressed maize (*Zea mays* L.). *J. Plant Interact.* 3: 234–241.

Keller, N.P. and Hohn, T.M. (1997). Metabolic pathway gene clusters in filamentous fungi. *Fungal Genet. Biol.* 21: 17–29.

Kerepesi, I. and Galiba, G. (2000). Osmotic and salt stress-induced alteration in soluble carbohydrate content in wheat seedlings. *Crop Sci.* 40: 482–487.

Keutgen, A.J. and Pawelzik, E. (2008). Contribution of amino acids to strawberry fruit quality and their relevance as stress indicators under NaCl salinity. *Food Chem.* 111: 642–647.

Khatkar, D. and Kuhad, M.S. (2000). Short-term salinity induced changesin two wheat cultivars at different growth stages. *Biol. Plant.* 43: 629–632.

Khoshbakht, D., Asghari, M.R., and Haghighi, M. (2017). Influence of foliar application of polyamines on growth, gas exchange characteristics, and chlorophyll fluorescence in Bakraii citrus under saline conditions. *Photosynthetica* 56: 731–742.

Koyro, H.W., Ahmad, P., and Geissler, N. (2012). Abiotic stress responses in plants: an overview. In: *Environmental Adaptations and Stress Tolerance of Plants in the Era of Climate Change* (eds. P. Ahmad and M.N.V. Prasad), 1–28. New York: Springer.

Kumar, A., Taylor, M., Altabella, T., and Al, E. (1997). Recent advances in polyamine research. *Trends Plant Sci.* 2: 124–130.

Kumar, V., Giridhar, P., Chandrashekar, A., and Al, E. (2008). Polyamines influence morphogenesis and caffeine biosynthesis in vitro cultures of *Coffea canephora* P. *Acta Physiol. Plant.* 30: 217–223.

Kusano, T., Yamaguchi, K., Berberich, T., and Takahashi, Y. (2007). Advances in polyamine research in 2007. *J. Plant Res.* 120: 345–350.

Kusano, T., Berberich, T., Tateda, C., and Al, E. (2008). Polyamines: essential factors for growth and survival. *Planta* 228: 367–381.

Lang, F. (2007). Mechanisms and significance of cell volume regulation. *J. Am. Coll. Nutr.* 26: 613S–623S.

Li, Y. and He, J. (2012). Advance in metabolism and response to stress of polyamines in plant. *Acta Agric. Boreali Sinica* 27: 240–245.

Li, Q., Ge, H., and Hu, S. (2008). Effects of exogenous putrescine and calcium on ion uptake of strawberry seedling under NaCI stress. *J. Plant Nutr. Fertil.* 14: 540–545.

Li, B., He, L., Guo, S. et al. (2013). Proteomics reveal cucumber Spd-responses under normal condition and salt stress. *Plant Physiol. Biochem.* 60: 7–14.

Li, S., Jin, H., and Zhang, Q. (2016). The effect of exogenous spermidine concentration on polyamine metabolism and salt tolerance inzoysiagrass (*Zoysia japonica* Steud) subjected to short-term salinity stress. *Front. Plant Sci.* 7: 1221.

Liu, J., Ishitani, M., Halfter, U. et al. (2000). The *Arabidopsis thaliana* SOS2 gene encodes a protein kinase that is required for salt tolerance. *Proc. Natl. Acad. Sci. U. S. A.* 97: 3730–3734.

Liu, Q., Nishibori, N., Imai, I., and Al, E. (2016). Response of polyamine pools in marine phytoplankton to nutrient limitation and variation in temperature and salinity. *Mar. Ecol. Prog. Ser.* 544: 93–105.

Liu, W., Tan, M., Zhang, C., and Al, E. (2017). Functional characterization of murB-potABCD operon for polyamine uptake and peptidoglycan synthesis in *Streptococcus* suis. *Microbiol. Res.* 207: 177–187.

López, M., Tejera, N.A., Iribarne, C. et al. (2008). Trehalose and trehalase in root nodules of *Medicago truncatula* and *Phaseolus vulgaris* in response to salt stress. *Physiol. Plant.* 134: 575–582.

Luo, J., Fuell, C., Parr, A., and Al, E. (2009). A novel polyamine acyltransferase responsible for the accumulation of spermidine conjugates in *Arabidopsis* seed. *Plant Cell* 21: 318–333.

Ma, L., Zhang, H., Sun, L. et al. (2012). NADPH oxidase AtrbohD and AtrbohF function in ROS-dependent regulation of Na^+/K^+ homeostasis in Arabidopsis under salt stress. *J. Exp. Bot.* 63: 305–317.

Martínez-Atienza, J., Jiang, X., Garciadeblas, B. et al. (2007). Conservation of the salt overly sensitive pathway in rice. *Plant Physiol.* 143: 1001–1012.

Martin-Tanguy, J. (2010). Conjugated polyamines and reproductive development: biochemical, molecular and physiological approaches. *Physiol. Plant.* 100: 675–688.

Matysik, J., Alia, A., Bhalu, B., and Mohanty, P. (2002). Molecular mechanisms of quenching of reactive oxygen species by proline under stress in plants. *Curr. Sci.* 82: 525–532.

McCue, K.F. and Hanson, A.D. (1990). Drought and salt tolerance: towards understanding and application. *Trends Biotechnol.* 8: 358–362.

Munns, R. (2002). Comparative physiology of salt and water stress. *Plant Cell Environ.* 25: 239–250.

Munns, R. and Tester, M. (2008). Mechanisms of salinity tolerance. *Annu. Rev. Plant Biol.* 59: 651–681.

Murakeozy, E.P., Nagy, Z., Duhaze, C. et al. (2003). Seasonal changes in the levels of compatible osmolytes in three halophytic species of inland saline vegetation in Hungary. *J. Plant Physiol.* 160: 395–401.

Mustafavi, S.H., Badi, H.N., Sekara, A., and Al, E. (2018). Polyamines and their possible mechanisms involved in plant physiological processes and elicitation of secondary metabolites. *Acta Physiol. Plant.* 40: 102.

Nedjimi, B. (2011). Is salinity tolerance related to osmolytes accumulation in *Lygeum spartum* L. seedlings? *J. Saudi Soc. Agric. Sci.* 10: 81–87.

Nemati, I., Moradi, F., Gholizadeh, S. et al. (2011). The effect of salinity stress on ions and soluble sugara distribution in leaves, leaf sheaths and roots of rice (*Oryza sativa* L.) seedlings. *Plant Soil Environ.* 57: 26–33.

Nounjana, N., Nghiab, P.T., and Theerakulpisuta, P. (2012). Exogenous proline and trehalose promote recovery of rice seedlings from salt-stress and differentially modulate antioxidant enzymes and expression of related genes. *J. Plant Physiol.* 169: 596–604.

de Oliveira, L.F., Navarro, B.V., Cerruti, G., and Al, E. (2018). Polyamines and amino acid related metabolism: the roles of arginine and ornithine are associated with the embryogenic potential. *Plant Cell Physiol.* 59: 1084–1098.

Ouyang, J., Song, C., and Chen, D. (2017). Research progress on heat-tolerance mechanism and transports of polyamfines in plant. *Mol. Plant Breed* 15: 3286–3294.

Parida, A.K. and Das, A.B. (2004). Effects of NaCl stress on nitrogen and phosphorous metabolism in a true mangrove *Bruguiera parviflora* grown under hydroponic culture. *J. Plant Physiol.* 161: 921–928.

Parida, A.K. and Das, A.B. (2005). Salt tolerance and salinity effects on plants: a review. *Ecotoxicol. Environ. Saf.* 60: 324–349.

Parida, A., Das, A.B., and Das, P. (2002). NaCl stress causes changes in photosynthetic pigments, proteins and other metabolic components in the leaves of a true mangrove, *Bruguiera parviflora*, in hydroponic cultures. *J. Plant Biol.* 45: 28–36.

Pegg, A.E. (2016). Functions of polyamines in mammals. *J. Biol. Chem.* 291: 14904–14912.

Rahnama, A., James, R.A., Poustini, K., and Munns, R. (2010). Stomatal conductance as a screen for osmotic stress tolerance in durum wheat growing in saline soil. *Funct. Plant Biol.* 37: 255–263.

Ranganayakulu, G.S., Veeranagamallaiah, G., and Sudhakar, C. (2013). Effect of salt stress on osmolyte accumulation in two groundnut cultivars (*Arachis hypogaea* L.) with contrasting salt tolerance. *Afr. J. Plant Sci.* 12: 586–592.

Rathinasabapathi, B., Burnet, M., Russell, B.L. et al. (1997). Choline monooxygenase, an unusual iron-sulfur enzyme catalyzing the first step of glycine betaine synthesis in plants: prosthetic group characterization and cDNA?cloning. *Proc. Natl. Acad. Sci. U. S. A.* 94: 3454.

Reguera, M., Peleg, Z., and Blumwald, E. (2012). Targeting metabolic pathways for genetic engineering abiotic stress-tolerance in crops. *Biochim. Biophys. Acta* 1819: 186–194.

Reis, R.S., Vale, E.M., Heringer, A.S., and Al, E. (2016). Putrescine induces somatic embryo development and proteomic changes in embryogenic callus of sugarcane. *J. Proteome* 130: 170–179.

Rengasamy, P. (2010). Soil processes affecting crop production in salt-affected soils. *Funct. Plant Biol.* 37: 613–620.

Rhodes, D., Nadolska-Orczyk, A., and Rich, P.J. (2002). Salinity, osmolytes and compatible solutes. In: *Salinity: Environment – Plants – Molecules* (eds. A. Läuchli and U. Lüttge), 181–204. Dordrecht: Springer.

Rodríguez-Kessler, M., Alpuche-Solís, A.G., Ruiz, O.A., and Jiménez-Bremont, J.F. (2006). Effect of salt stress on the regulation of maize (*Zea mays* L.) genes involved in polyamine biosynthesis. *Plant Growth Regul.* 48: 175–185.

Roychoudhury, A., Basu, S., and Sengupta, D.N. (2011). Amelioration of salinity stress by exogenously applied spermidine or spermine in three varieties of indica rice differing in their level of salt tolerance. *J. Plant Physiol.* 168: 317–328.

Sairam, R.K. and Tyagi, A. (2004). Physiology and molecular biology of salinity stress tolerance in plants. *Curr. Sci.* 86: 407–421.

Salvi, F., Wang, Y.F., Weber, I.T., and Gadda, G. (2014). Structure of choline oxidase in complex with the reaction product glycine betaine. *Acta Crystallogr. Sect. D: Biol. Crystallogr.* 70: 405–413.

Sannazzaro, A.I., Echeverria, M., Albertó, E.O. et al. (2007). Modulation of polyamine balance in Lotus glaber by salinity and arbuscular mycorrhiza. *Plant Physiol. Biochem.* 45: 39–46.

Schroeder, I.J., Klaus, R., and Erwin, N. (1987). Voltage dependence of K^+ channels in guard-cell protoplasts. *PNAS* 84 (12): 4108–4112.

Shahbaz, M., Mushtaq, Z., Andaz, F., and Masood, A. (2013). Does proline application ameliorate adverse effects of salt stress on growth, ions and photosynthetic ability of eggplant (*Solanum melongena* L.)? *Sci. Hortic.* 164: 507–511.

Sharma, S.S. and Dietz, K.J. (2006). The significance of amino acids and amino acid-derived molecules in plant responses and adaptation to heavy metal stress. *J. Exp. Bot.* 57: 711–726.

Shen, B., Hohmann, S., Jensen, R.G., and Bohnert, H.J. (1999). Roles of sugar alcohols in osmotic stress adaptation. Replacement of glycerol by mannitol and sorbitol in yeast. *Plant Physiol.* 121 (1): 45–52.

Shimomura, T., Itoh, T., Sumiya, T. et al. (2009). Amperometric determination of choline with enzyme immobilized in a hybrid mesoporous membrane. *Talanta* 78: 217–220.

Singh, S. and Prasad, S.M. (2019). Management of chromium (VI) toxicity by calcium and sulfur in tomato and brinjal: implication of nitric oxide. *J. Hazard. Mater.* 373: 212–223.

Singh, M., Kumar, J., Singh, S. et al. (2015). Roles of osmoprotectants in improving salinity and drought tolerance in plants: a review. *Rev. Environ. Sci. Biotechnol.* https://doi.org/10.1007/s11157-015-9372-8.

Sithtisarn, S., Harinasut, P., Pornbunlualap, S., and Cha-Um, S. (2009). Accumulation of glycinebetaine and betaine aldehyde dehydrogenase activity in Eucalyptus camaldulensis clone T5 under in vitro salt stress. *Kasetsart J. Nat. Sci.* 43 (5): 146–152.

Slama, I., Abdelly, C., Bouchereau, A. et al. (2015). Diversity, distribution and roles of osmoprotective compounds accumulated in halophytes under abiotic stress. *Ann. Bot.* 115 (3): 433–447.

Stoop, J.M.H., Williamson, J.D., and Pharr, D.M. (1996). Mannitol metabolism in plants: a method for coping with stress. *Trends Plant Sci.* 1 (5): 139–144.

Takabe, T. (2012). Engineering of betaine biosynthesis and transport for abiotic stress tolerance in plants. *J. Plant Biochem. Biotechnol.* 21 (1): 58–62.

Takabe, T., Nakamura, T., Nomura, M. et al. (1998). 8 - Glycinebetaine and the genetic engineering of salinity tolerance in plants. In: *Stress Responses of Photosynthetic Organisms* (eds. K. Satoh and N. Murata), 115–131. Amsterdam: Elsevier.

Takahashi, Y., Tahara, M., Yamada, Y., and Al, E. (2017). Characterization of the polyamine biosynthetic pathways and salt stress response in *Brachypodium distachyon*. *J. Plant Growth Regul.* 37: 625–634.

Talat, A., Nawaz, K., Hussian, K. et al. (2013). Foliar application of proline for salt tolerance of two wheat (Triticum aestivum L.) cultivars. *World Appl. Sci. J.* 4: 547–554.

Tari, I., Kiss, G., Deér, A.K. et al. (2010). Salicylic acid increased aldose reductase activity and sorbitol accumulation in tomato plants under salt stress. *Biol. Plant.* 54: 677–683.

Theerakulpisut, P. and Gunnula, W. (2012). Exogenous sorbitol and trehalose mitigated salt stress damage in salt-sensitive but not salt tolerance rice seedlings. *Asian J. Crop Sci.* 4: 165–170.

Türkan, I. and Demiral, T. (2009). Recent developments in understanding salinity tolerance. *Environ. Exp. Bot.* 67: 2–9.

Urano, K., Yoshiba, Y., Nanjo, T. et al. (2004). *Arabidopsis* stress-inducible gene for arginine decarboxylase AtADC2 is required for accumulation of putrescine in salt tolerance. *Biochem. Biophys. Res. Commun.* 313: 369–375.

Velarde-Buendía, A.M., Shabala, S., Cvikrova, M. et al. (2012). Salt-sensitive and salt-tolerant barley varieties differ in the extent of potentiation of the ROS-induced K+ efflux by polyamines. *Plant Physiol. Biochem.* 61: 18–23

Verbruggen, N. and Hermans, C. (2008). Proline accumulation in plants: a review. *Amino Acids* 35: 753–759.

Vinocur, B. and Altman, A. (2005). Cellular basis of salinity tolerance in plants. *Environ. Exp. Bot.* 52: 113–122.

Vuosku, J., Karppinen, K., Muilu-Mäkelä, R. et al. (2018). Scots pine aminopropyltransferases shed new light on evolution of the polyamine biosynthesis pathway in seed plants. *Ann. Bot.* 121: 1243–1256.

Waditee, R., Hibino, T., Tanaka, Y. et al. (2000). Functional characterization of betaine/proline transporters in betaine accumulating mangrove. *J. Biol. Chem.* 277 (21): 18373–18382.

Wani, S.H., Singh, N.B., Haribhushan, A., and Mir, J.I. (2013). Compatible solute engineering in plants for abiotic stress tolerance-role of glycine betaine. *Curr. Genomics* 14: 157–165.

Williamson, J.D., Jennings, D.B., Guo, W.W. et al. (2002). Sugar alcohols, salt stress and fungal resistance: polyols: multifunctional plant protection? *J. Am. Soc. Hortic. Sci.* 127: 467–473.

Wingler, A., Lea, P.J., Quick, W.P., and Leegood, R.C. (2000). Photorespiration: metabolic pathways and their role in stress protection. *Philos. Trans. R. Soc. Lond. Ser. B Biol. Sci.* 355: 1517–1529.

Woodrow, P., Ciarmiello, L.F., Annunziata, M.G. et al. (2017). Durum wheat seedling responses to simultaneous high light and salinity involve a fine reconfiguration of amino acids and carbohydrate metabolism. *Physiol. Plant.* 159: 290–312.

Wyn Jones, R.G. (1984). Phytochemical aspects of osmotic adaptation. *Recent Adv. Phytochem.* 18: 55–78.

Xu, L. (2015). The effect of polyamineon flower bud differentiation and bud germination of chrysanthemum. *Shandong Agric. Univ.*: 31–36. https://doi.org/10.3389/fpls.2018.01469.

Xu, C., Wu, X., and Zhang, H. (2009). Impact of D-Arg on drought resistance and endogenous polyamines in mycorrhizal *Pinus massoniana*. *J. Nanjing For. Univ.* 33: 019–023.

Xu, L., Xing, S.T., and Sun, X. (2014). Effects of polyamines on hormones contents and the relationship with the flower bud differentiation in chrysanthemum. *Plant Physiol. J.* 50: 1195–1202.

Xu, Z., Sun, M., Jiang, X. et al. (2018). Glycine betaine biosynthesis in response to osmotic stress depends on jasmonate signaling in watermelon suspension cells. *Front. Plant Sci.* 9: 1469.

Yadav, G., Srivastava, P.K., Singh, V.P., and Prasad, S.M. (2014). Light intensity alters the extent of arsenic toxicity in Helianthus annuus L. seedlings. *Biol. Trace Elem. Res.* 158: 410–421.

Yamada, N., PromdenW, Y.K., Tamagake, H. et al. (2009). Preferential accumulation of betaine uncoupled to choline monooxygenase in young leaves of sugar beet importance of long distance translocation of betaine under normal and salt stressed conditions. *J. Plant Physiol.* 166 (18): 2058–2070.

Yamamoto, A., Shim, I.S., and Fujihara, S. (2016). Inhibition of putrescine biosynthesis enhanced salt stress sensitivity and decreased spermidine content in rice seedlings. *Biol. Plant.* 61: 385–388.

Yancey, P.H. (1994). Compatible and counteracting solutes. In: *Cellular and Molecular Physiology of Cell Volume Regulation* (ed. S.K. Strange), 81–109. Boca Raton: CRC Press.

Yiu, J.C., Juang, L.D., Fang, D.Y.T. et al. (2009). Exogenous putrescine reduces flooding-induced oxidative damage by increasing the antioxidant properties of Welsh onion. *Sci. Hortic.* 120: 306–314.

Yu, G.H., Li, W., Yuan, Z.Y. et al. (2013). The effects of enhanced UV-B radiation on photosynthetic and biochemical activities in super-high-yield hybrid rice *Liangyoupeijiu* at the reproductive stage. *Photosynthetica* 51: 33–44.

11

Salt Stress Toxicity Amelioration by Phytohormones, Synthetic Product, and Nutrient Amendment Practices

Divya Gupta[1], Garima Singh[1], Sanjesh Tiwari[1], Anuradha Patel[1], Abreeq Fatima[1], Anurag Dubey[2], Neha Naaz[3], Jitendra Pandey[3], and Sheo Mohan Prasad[1]

[1] *Ranjan Plant Physiology and Biochemistry Laboratory, Department of Botany, University of Allahabad, Prayagraj, Uttar Pradesh, India*
[2] *Mycology and Plant pathology Laboratory, School of Biological Sciences, Dr. Hari Singh Gour Central University, Sagar, Madhya Pradesh, India*
[3] *Ganga River Ecology Research Laboratory, Environmental Sciences Division, Centre of Advanced study in Botany, Institute of Sciences, Banaras Hindu University, Varanasi, India*

11.1 Introduction

Across the world, the adverse environmental factors equally affect the terrestrial as well as aquatic ecosystem, causing lesser agricultural production and declining crop yield (Ashraf et al. 2008; Ali et al. 2017). Among various stress factors, salt stress has swamped 20% of the world's arable land and, in this continuous manner, it is supposed to reach 50% by 2050 (Lakhdar et al. 2009). In one of the surveys, it was stated that around 45 million hectares of irrigated land across worldwide has been greatly affected by salt stress and 1.5 million hectares land became barren for the production due to excessive salinity levels in the soil (Munns and Tester 2008). High solubility of salt in water is the major factor to reduce crop yield as it gets accumulated in arable soils of arid and semi-arid regions and acts as a key factor for crop production. The physicochemical properties of agricultural soil was greatly influenced by excessive salt ions like decreased soil aeration, water movement, and soil porosity that imposed serious threats to the plants growth and development. Excessive salt ions disturb intra- and inter-cellular water level and prevent the cell expansion which is tracked by failure in stomatal activity (Izadi et al. 2014; Gong et al. 2018). Further, nutrients uptake is greatly influenced by salt stress that leads either deficiencies of many essential nutrients or surplus amount of Na^+ displays major osmotic and ion-specific effects (Kaya et al. 2002; Eraslan et al. 2007). One of major riots is the disturbance in calcium levels, resulting in the disparity in the passive ion fluxes (Gao and Zhang 2019). Increase in the accumulation of salt ions in plants causes reduction in the contents of photosynthetic

Physiology of Salt stress in Plants: Perception, Signalling, Omics and Tolerance Mechanism, First Edition.
Edited by Pratibha Singh, Madhulika Singh, Rajiv Kumar Singh, and Sheo Mohan Prasad.

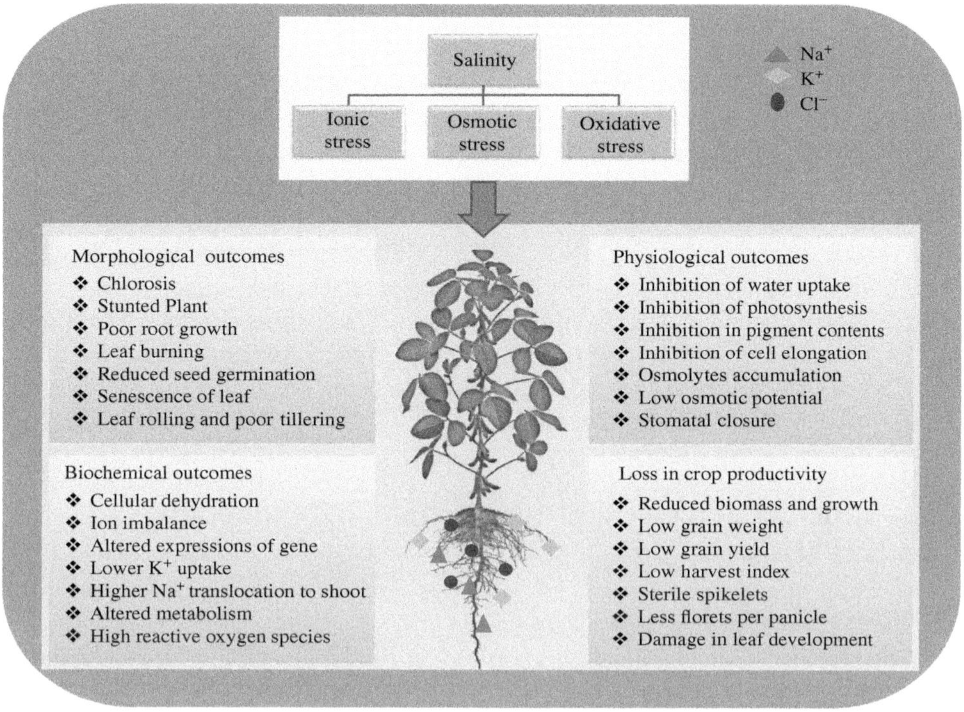

Figure 11.1 Pictorial representation of the types of salinity stress and its outcomes in crop plants. *Source:* Adapted from Srivastava and Kumar (2015) and Yang et al. (2019b).

pigments (mainly chlorophyll and carotenoid), and to the greater extent, it obstructs the most important ribulose-1, 5-bisphophate enzyme (RuBiSco) which ultimately impairs photosynthetic process. Consequently, higher uptake of salt in the plant tissue system leads to the release of reactive oxygen species (ROS) which primarily damages the macromolecules such as nucleic acid, proteins, and lipids that obstruct certain biochemical and metabolic functions of plants (Shrivastava and Kumar 2015; Yang et al. 2019a) Figure 11.1.

Efficiency of such soils can be re-established by way of practical management that not only get involved in the improvement of crop health but also curtails the global pressure for cultivatable land resources food demand and supply. So to overcome the adverse effects of salinity, different novel and vital strategies have been employed in these directions and, among them, one ameliorating approach is either exogenous supplementation of phytohormones such as gibberellic acid (GA), cytokinin (CK), jasmonic acid (JA), brassinosteroids (BRs), or supplementation of synthetic products.

Plant hormones are dynamic members of the signal compounds involved in the induction of plant stress responses (Pedranzani et al. 2003) and also known to regulate and upgrade many of the cellular and biochemical processes in plant cell. They behave as signaling molecule hence, transfer signals in a way altering cellular activities (Vob et al. 2014). Phytohormone gives a wide platform for biotechnologist to develop crop improvement techniques either economically or nutritionally. Beside this, supplementation of synthetic products also remediates saline soils' salinity stress (Wichern et al. 2020). Several organic substances could modify the soil properties by speeding up the process of nutrients leaching specially plummeting exchangeable sodium percentage (ESP) (Wichern et al. 2020). Adding organic manure upturns plant accessible nutrients while drops the absorptions of sodium, bicarbonates, and carbonates leading to improved plant growth, making it an ameliorating approach against salinity stress (Niamat et al. 2019). The nutrients are generally categorized into two parts that are macronutrients (e.g., K, Mg, Ca, S, N, and P) and micronutrients (i.e., Cu, Mn, Fe, Ni, Mo, Co, B, Cl, and Zn) along with beneficial elements (e.g., Si, Se, Na, and V) that play a major and vital role, beginning from the emergence, development, metabolism, productivity, and getting to the progression and protection of plants. So, exogenous supplementation of these nutrients could be a new approach which will be economically cost-effective for farmers to get rid of crop manufacturing problems triggered by salinity. So, this chapter mainly discusses the various ameliorating approaches to cope up the loss of crop yield under saline environment.

11.2 Structure and Mechanism of Action of Phytohormones Under Salt Stress

Plants have developed various physiological and biochemical mechanisms in order to survive under adverse environmental factors (Javid et al. 2011; Prakash et al. 2019). Principle mechanisms involved in salt stress regulation are (i) ion homeostasis and compartmentalization, (ii) ion transport and uptake, (iii) biosynthesis of osmoprotectants and compatible solutes, (iv) activation of antioxidant enzyme and synthesis of antioxidant compounds, (v) synthesis of polyamines, (vi) generation of nitric oxide (NO), and (vii) hormone modulation (Gupta and Huang 2014; Xiao et al. 2017). Here, the action mechanism of phytohormones is briefly dealt with reference to salinity.

Phytohormones are essential growth regulators and are best known for their prominent impact on metabolism and stimulation of defense responses. Plenty of research reports are available advocating the alleviative role of phytohormones under abiotic stress. Since, exogenous application of phytohormones has long been adopted to improve the quality and quantity of major cash crops exposed to harsh climatic condition such as salinity (Kazan 2013; Egamberdieva et al. 2017). The salinity stress alters the endogenous levels of hormones in plants. It includes alteration in endogenous level of phytohormone such as auxin, abscisic acid, gibberellins, jasmonic acid, salicylic acid, brassinosteroids etc. causing disturbance in the growth and development of the plant.

Phytohormone supplementation improves germination rate, increases water uptake, reduces stomatal resistance, facilitates ion partitioning followed by sequestration, and improves antioxidant machinery under salt stress. Phytohormones regulate ion homeostasis and growth of the plant exposed to salinity. They are also accountable for modulating several physiological processes required for stress tolerance *via* stress-mediated signal pathways and feedback mechanisms (Iqbal and Ashraf 2013; Khan et al. 2014). The application of phytohormone under salt stress significantly increased the stress tolerance of plants (Peres et al. 2019). They also play a critical role in ROS detoxification by enhancing the efficiency of antioxidant machinery, such as the enzyme superoxide dismutase (SOD), catalase (CAT), glutathione peroxidase (GPX), ascorbate peroxidase (APX), and glutathione reductase (GR) work concomitantly to reduce oxidative stress (Gupta et al. 2005; Singh et al. 2018; Singh and Prasad 2019). For instance, Iqbal and Ashraf (2013) reported the alleviation of salinity induced toxicity in wheat through seed priming with indole acetic acid (IAA) and also induces synthesis of salicylic acid (SA) and ion homeostasis in the plant. To cope up with the stress situation, plant raises the level of stress responsive hormone ABA that maintains the root or shoot growth by modulating the ethylene synthesis as reported in *Zea mays*. It has also been reported that phytohormone upregulates the antioxidative gene and promotes the synthesis of osmolyte which results into maintenance of turgor potential, relative water content (RWC), and enhanced tolerance against salinity stress (Ahmad 2010; Egamberdieva et al. 2017). On the other hand, phytohormones also exhibit crosstalk with other hormones and elicit a signaling cascade and mediate tolerance mechanism under salinity stress. For example, the synthesis of GAs can also be promoted *via* application of auxin (Wolbang et al. 2004) or synthesis of GAs enhances ABA catabolism. Respectively, Khan et al. (2014) also reported a decline in the endogenous ethylene level due to SA application which signifies towards a crosstalk between the two phytohormones in *Vigna radiata* exposed to salt stress.

11.3 Structural, Physiological, and Biochemical Nature of Phytohormones Under Salt Stress

Phytohormones play a potential role in plant metabolism and regulate key metabolic pathways by maintaining growth under adverse environmental conditions. The crosstalk between different phytohormones induces a complex response in plants (Fahad et al. 2015). Phytohormones regulate an array of physiological or biochemical processes such as cell division, germination, photomorphogenesis, gene expression, flowering, fruit development, senescence, and stress responses (Ahanger et al. 2018). The interactions between the different phytohormones trigger effective defense against the biotic and abiotic stresses in a very precise way (Kumar et al. 2014; Nolan et al. 2020). The physiological and biochemical nature has been discussed below in brief.

11.3.1 Abscisic Acid (ABA)

It is a naturally occurring sesquiterpenoid group of phytohormones and also known as stress hormone that regulates growth under stress environments. There is evidence on the positive implication of exogenously applied ABA in reversing the salinity-induced ill effects (Gomez et al. 2002; Egamberdieva et al. 2017). Enhanced exposure of plants to salinity induced proportional increase in ABA concentration. Kang et al. (2005) reported that salt under different concentrations (20 and 40 mm NaCl) leads to a sharp increase in the concentrations of ABA in rice (Hossain and Dietz 2016). Salt-induced synthesis of ABA is reported to mediate the inhibition of leaf expansion and accumulation of Na^+ and Cl^- in leaves and delays the effect of NaCl in sorghum (Amzallag et al. 1990; Cabot et al. 2009; Nawaz et al. 2017). ABA level is also upregulated under salinity stress and induces genes involved in salt and osmotic stress alleviation (Wang et al. 2001; Prakash et al. 2019). Furthermore, exogenously applied ABA was reported to increase the nutrient uptake, nitrogen fixation, and growth of common bean exposed to salinity (Khadri et al. 2006; Egamberdieva et al. 2017). A positive relationship between ABA accumulation and salinity tolerance has been attributed to the accumulation of K^+, Ca^{2+}, or other compatible solutes, such as proline and sugars in vacuoles of roots of various plants (Gurmani et al. 2011; Prakash et al. 2019). Also, Shi and Zhu (2002) have reported the tissue distribution and regulation of AtNHX1 expression by ABA and salt stress in *Arabidopsis* which plays a significant role in salinity tolerance.

11.3.2 Cytokinins (CKs)

The hormone cytokinin (CK) is known to mediate several aspects of growth and developmental processes of plants such as cell division, apical dominance, chloroplast biogenesis, nutrient mobilization, leaf senescence, vascular differentiation, photomorphogenic development, shoot differentiation, and anthocyanin production (Tognetti et al. 2017; Singh et al. 2018). Naturally occurring cytokinins are N6-substituted purine derivatives. Isopentenyladenine (iP), zeatin (Z), and dihydrozeatin (DZ) are the predominant cytokinins found in higher plants. The biologically active compounds of cytokinins are thought to be its free bases and their ribosides (iPR, ZR, DZR) (Werner et al. 2001). It is produced in the root tips and developing seeds of plants (Zahir et al. 2001; Ahanger et al. 2018) and translocating to the shoot by xylem from roots. Under salinity stress, reduction in the level CK changes the gene expression and elicits stress response (Hare et al. 1997). Cytokinin enhances resistance in cucumber under salinity stress (Barciszewski et al. 2000; Fan et al. 2013). Seed priming with CKs also helps in increasing salt tolerance (Iqbal et al. 2006). Exogenously applied kinetin helps in protecting plants from salt-induced oxidative stress through upregulation of antioxidant system, ascorbate glutathione cycle, and glyoxalase system (Ahanger et al. 2018). However, reduced CKs lead to ABA-induced stomatal closure which substantially decreases the carbon uptake and assimilation under stress condition such as salinity. The upregulation of cytokinin oxidase may also limit carbon metabolism. The receptor genes of cytokinin are regulated by osmotic changes and function in the osmotic stress response in plants (Merchan et al. 2007) and act as a limiting factor during stress.

11.3.3 Gibberellins (GAs)

Gibberellins (GAs) are naturally occurring tetracyclic diterpenoid carboxylic acids, synthesized by the terpenoid pathway in plastids (Hedden and Sponsel 2015), involved in growth and development by controlling seed germination, shoot elongation, and leaf expansion under salinity stress (Magome et al. 2004; Ahmad 2010; Davière and Achard 2015). It is reported to alleviate the negative effects induced by salinity (Yamaguchi and Kamiya 2000; Javid et al. 2011). Exogenously applied GAs in tomato were reported to decrease stomatal resistance at lower salinity level and also enhance growth and yield of wheat and rice under salinity (Prakash and Prathapasenan 1990). GA3-priming induces increase in wheat grain yield due to its priming-induced modulation of ion uptake and partitioning in plants and hormone homeostasis under saline conditions (Iqbal and Ashraf 2010; Parihar et al. 2015). It also interacts with various other hormones to regulate abiotic stress. Further, GA supplementation was found to enhance the osmotic components and maintains tissue water content up to several fold in wheat and maize exposed to salinity (Tuna et al. 2008; Ahmad 2010; Manjili et al. 2012; Egamberdieva et al. 2017).

11.3.4 Auxins (AUXs)

Auxins are essential hormones and are synthesized from tryptophan. Among various auxins, indole acetic acid (IAA) is known to play a major role in regulating the plant growth. It is responsible for cell elongation, vascular tissue differentiation, and apical dominance (Wang et al. 2001; Bashri and Prasad 2016). It has an imperative role under salinity, and salinity is reported to reduce around 75% endogenous IAA level in tomato (Liu et al. 2015). The study on pretreatment of wheat seeds with IAA is known to alleviate the growth inhibiting effect of salt stress (Afzal et al. 2005). The enhanced level of IAA has been largely correlated with reduced growth (Ribaut and Pilet 1994; Singh et al. 2018). Therefore, it is speculated that reduced growth could be an outcome of altered hormonal balance, and the application of auxin increased the biomass of plants under salinity (Akbari et al. 2007; Lavy and Estelle 2016). Further, Jung and Park (2011) established a link between auxin signaling and salt stress which was developed by the involvement of auxin in modulating membrane bound transcription factor NTM2.The study was further consolidated by the overexpression studies in IAA 30 gene of NTM2.

11.3.5 Brassinosteroids (BRs)

BRs are classified as the sixth class of essential phytohormones, and it is steroidal in nature, first isolated from pollens of *Brassica,* and is ubiquitously present in the plant kingdom (Sharma et al. 2019). BRs also serve properties of plant growth regulators (Bishop and Yokota 2001; Tian et al. 2018) and have pleiotropic effects. BRs influence diverse physiological processes such as leaf abscission, senescence, and rhizogenesis (Peres et al. 2019) under regular as well as environmental stress (Nolan et al. 2020). BRs can regulate abiotic stress by increasing the synthesis of osmoprotectants and BRL3 (vascular-enriched member of the BR receptor family) that regulate the biosynthesis of all osmoprotectants (Sharma et al. 2019). The accumulation of proline might be due to enhanced gene expression of biosynthetic genes. Exogenously applied BRs under salinity are found to increase growth by alleviating salt-induced effect on seed germination in *Oryza sativa*. It can also restore disturbed chlorophylls and nitrate reductase (NR) activity under salinity stress. Subsequently, it was found that BR-induced growth under salt stress conditions was associated with enhanced levels of nucleic acids and soluble proteins (Anuradha and Rao 2001). Similarly, two cultivars of *Brassica juncea* (Varuna and RH-30) when treated with proline and 24-epibrassinolide were found to resist the negative effects of salinity with better photosynthetic efficiency and enhanced yield. The study clearly suggests the role of BRs in the synthesis of osmolyte which protects the plant under salinity stress.

11.3.6 Salicylic Acid (SA)

Salicylic acid is (SA) a phenolic compound which participates in the regulation of physiological processes and increases the stress tolerance through modulation of antioxidative enzymes (da Silva et al. 2017). The role of SA in the alleviation of salt stress was deeply studied by Afzal et al. (2005) and Caarls et al. (2015). In many crops such as bean (Azooz et al. 2011), tomato (Tari et al. 2002), and maize (Gunes et al. 2007) when supplemented with SA reduces ion toxicity by promoting the sequestration and partitioning of deleterious ions in plants under salinity. The enhanced synthesis and accumulation of proline and ABA (0.05 mM) were documented in salinity-stressed wheat seedlings, contributing to better growth and yield (Shakirova et al. 2003). Khan et al. (2014) observed that treatment of *V. radiata*, L. seeds with SA helped to mitigate salinity-induced changes in terms of promotion of cell division, higher photosynthetic rate, and increased antioxidant enzyme activity and biomass accumulation (da Silva et al. 2017). Salinity stress-induced alterations in the rate of photosynthesis and membrane stability of barley were reversed by the exogenous application of SA (Janda et al. 2012). The results obtained in the past few years indicate that SA is a promising compound for the tolerance response against salinity stress.

11.3.7 Jasmonic Acid (JA)

Hormone methyl jasmonate (MeJA) and jasmonic acid (JA) are collectively referred as jasmonates (Ávila et al. 2019) derived from linolenic acid in lipoxygenase-dependent process. According to Creelman and Mullet (1995), the chemical structure of JA is somewhat similar to mammalian eicosanoids (derived from lipids through the action of lipoxygenase). Further, biosynthesis of JA occurs *via* enzyme allene oxide synthase (AOS). They are known to play significant roles such as seed germination, fertility, fruit ripening, senescence, and others (Wasternack and Hause 2002; Yang et al. 2019b). In a study, in salt-stressed tomato, there is change in the endogenous JA levels (Ahmad et al. 2018). In an earlier study, a subsequent increase in the concentration of JA was observed in salt tolerant cultivar *Hellfrucht fruhstamm*, whereas in salt-sensitive cultivar Pera, JA level was decreased after 24 hours of salt treatment (Pedranzani et al. 2003). A rapid increase in endogenous JA content was also noticed in leaves of barley subjected to osmotic stress with sorbitol or mannitol, although endogenous jasmonate remains unchanged under NaCl (Kramell et al. 1995; El-Esawi 2016). However, it is speculated that JA acts additively as well as synergistically with other phytohormones to induce tolerance. The structure of some phytohormone and their respective function is presented in Table 11.1.

11.4 Salt Stress Toxicity Amelioration by Exogenous/Endogenous Phytohormones

Plant hormones are essential compounds that play a key role in growth and also maintain the basal physiological activities. Under stress conditions such as salt, heat, drought, and high temperature, these phytohormones are positive regulators and behave as alleviator agents and required in low concentration but essential (Fahad et al. 2014). Under salinity stress, endogenous plant growth hormones regulate metabolic reactions such as seed germination, growth metabolism, photosynthesis, and respiration (Fahad et al. 2014). In earlier studies, these have also been demonstrated that, under salt stress condition, *Lycopersicon esculentum* reduces endogenous IAA concentration. Moreover, auxin intensifies the mechanism toward retardation of senescence process (Ghanem et al. 2008).

Table 11.1 Structure and function of different phytohormones.

S. No.	Name	Structure	Response	References
1.	Gibberellic acid	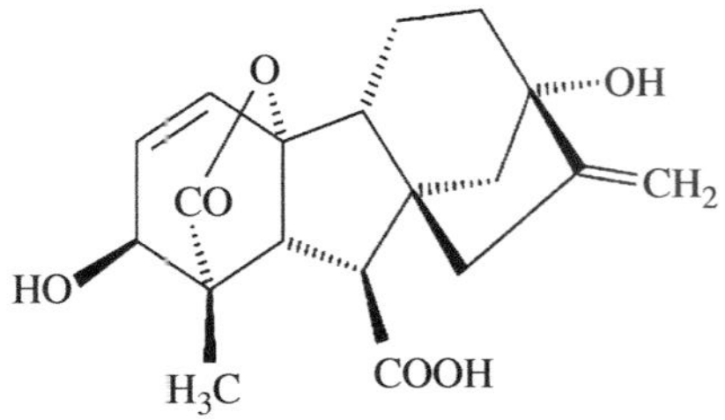	• GA_3 increases germination percentage and shoot and root length, total fresh and dry weight, tissue water content and seedling vigor index • GA_3 improves accumulation of proline which maintained membrane permeability and increased macro and micronutrient levels	Chauhan et al. 2019 Tuna et al. 2008
2.	Abscisic Acid (ABA)	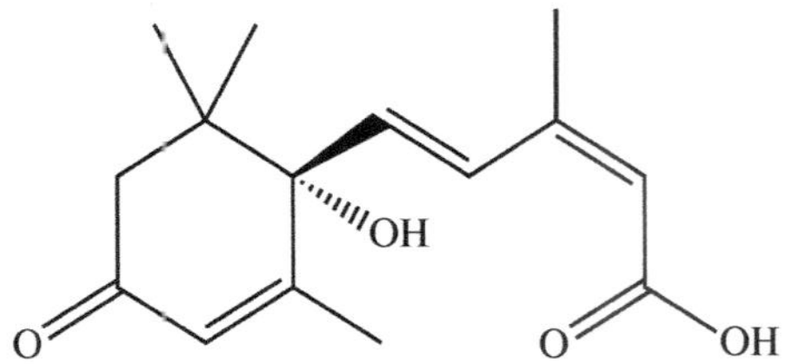	• ABA showed relationship with strigolactone during the induction of H_2O_2-mediated tolerance • Signals for stomatal closure by depolymerization of cortical actin filaments	Ren et al. 2018 Jae-Ung and Youngsook (2001)
3.	Indole acetic Acid (IAA)	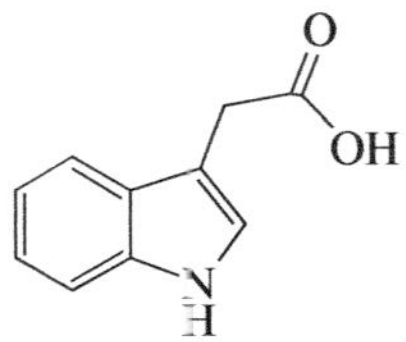	• Provides nutrients, maintains high potassium and sodium ratio, increases accumulation of osmolytes, enhanced photosynthesis and activity of antioxidant enzymes	Bhise and Dandge 2019

(Continued)

Table 11.1 (Continued)

S. No.	Name	Structure	Response	References
4.	Cytokinins (CKs)	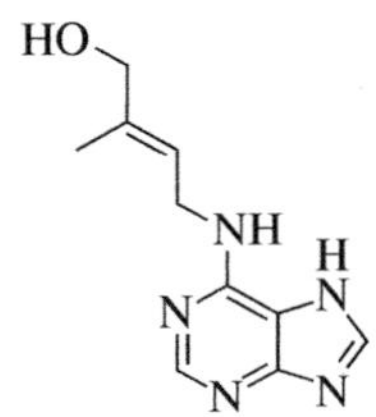	• Breaks stress-induced dormancy during germination • Acts as free radical scavenger • Modulates physio-hormonal attributes and isoflavone contents	Chakrabarti and Mukherji 2003 Hamayun et al. 2015
5.	Jasmonates (JA)	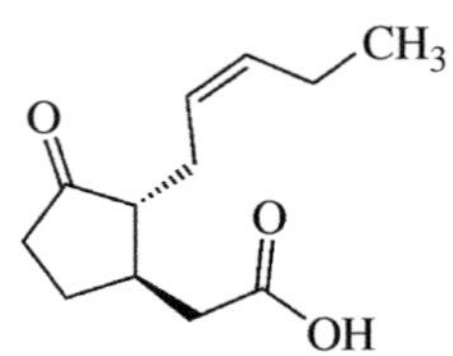	• Maintains homeostasis among ROS • Improves Na^+ exclusion in the root by decreasing Na^+ uptake • Enhances the concentrations of antioxidative compounds	Abouelsaad et al. 2018 Hamayun et al. 2015 Wang et al. 2020
6.	Triazoles	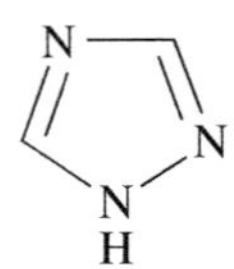	• Maintaining favorable Na^+:K^+ ratio, water use efficiency and osmoprotection • Enhanced photosynthetic efficiency, preservation of cell ultrastructure	Kapoor et al. 2019 Fletcher et al. 2010
7.	Ethylene (ET)	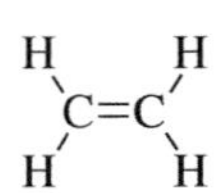	• Ethylene acts as a signaling pathway in improving salt tolerance • Maintaining negative feedback mechanism of coordinating plant growth and salinity response	Tao et al. 2015

8.	Brassinosteroids (BRs)	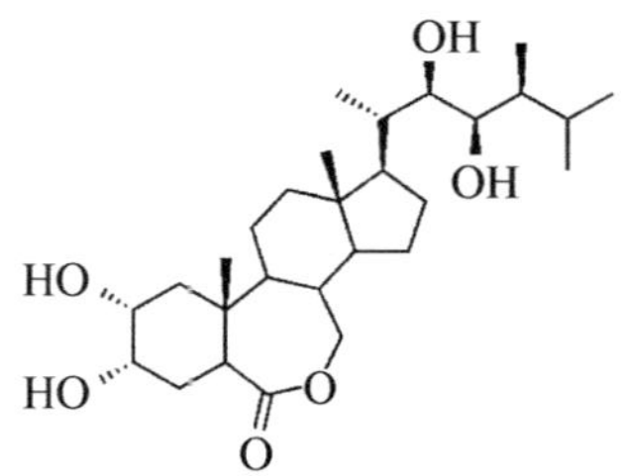	• Protected nucleus and chloroplast ultrastructure • Enhanced the ability of resistance to osmotic stress by increasing free proline, soluble protein content, chlorophyll content, root activity, H^+-ATPase activity, and antioxidant enzyme activities	Vázquez et al. 2019 Dong et al. 2017
9.	Salicyclic acid (SA)	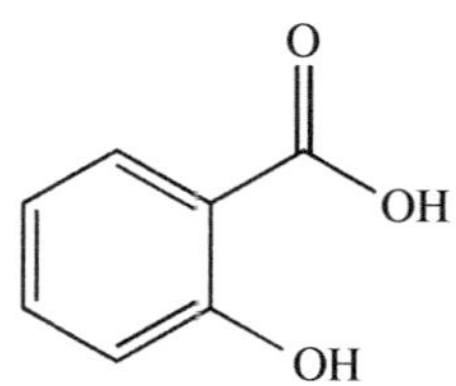	• Modulating the cell redox balance and protecting the plants from oxidative damage • Enhanced nitrate reductase activity, chlorophyll fluorescence, stomatal conductance, net photosynthetic, and transpiration rates and plant water status • Reduced accumulation of Na^+ and Cl^- ions	Husen et al. 2018 Jini and Joseph 2017
10.	Strigolactones (SL)	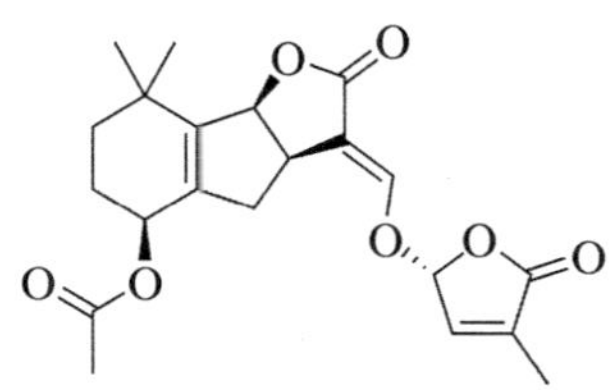	• Enhanced transpiration rate, stomatal conductance, and intercellular CO_2 concentrations, net photosynthetic rate • MAD, POD, and SOD activities were significantly increased by GR24 (synthetic strigolactone)	Ling et al. 2020

However, salt stress also reduces the plant productivity. In present scenario, the exogenous supplementation of plant growth hormones alleviates the salt toxicity and showing vital role in minimizing the stress by improving several metabolic processes like improving seed metabolism, maintaining the homeostasis redox reaction, and other physiological mechanisms (Gul et al. 2000; Afzal et al. 2005).

11.4.1 Auxin

Auxin is the first identified plant growth hormone having morphogen properties, and it is involved in several metabolic functions (Fahad et al. 2014). The auxin is classified into two forms, i.e. natural as well as synthetic auxin. Natural auxin includes indole acetic acid (IAA), indole butyric acid (IBA) etc., while synthetic auxin covers indole-3-propionic acid, indole-3-pyruvic acid, 2,4 Dichloro phenoxy acetic acid (2, 4-D) etc. Earlier findings have also been mentioned that during adverse situation such as salinity, auxin shows similar response like abscisic acid (ABA) that mitigate the oxidative stress condition (Ribaut and Pilet 1991). However, higher amount of auxin hormone reduces the growth of plant as well (Ribaut and Pilet 1994). Therefore, enhanced level of IAA causes the hormonal disparity during salinity condition in plant. Further, the exogenous application of IAA minimizes the oxidative factors along with salt toxicity. While on the similar note, it has also been mentioned that the salt stress condition causes significant reduction of endogenous IAA in *O. sativa* leaves (Nilsen and Orcutt 1996). Auxin shows signaling mechanism regarding to salinity condition, and it has performed a vital role to reduce the stress. Further, it participates in metabolic reaction such as seed germination and is associated through the transcription factors which are bound in membrane such as NTM2 factor by overexpression of gene *IAA30* in Arabidopsis (Park et al. 2011). It has also been demonstrated that auxin also with gibberellins hormone during the seed priming process in salt stress condition mainly by the GA3 (active) form of gibberellin which shows the salt tolerance. Auxin (IAA, IBA) also in association with endogenous ABA hormone in seed priming process alleviates the toxicity of salt in *Triticum aestivum* crop plant (Iqbal and Ashraf 2013).

11.4.2 Gibberellins

Gibberellic acids (GAs) are a plant growth regulator as well as stimulating agent that cover tetracyclic di-terpenoid nature (Olszewski et al. 2002). It regulate plants development and metabolism under adverse condition (Yamaguchi and Kamiya 2000; Olszewski et al. 2002). The earlier studies have demonstrated that, in *O. Sativa,* internode elongation occurred due to the inactivation mechanism of GA1 and GA4 that is mediated by cytochrome p450 mono oxygenase enzyme (Zhu et al. 2006). Under salt stress condition, GA rapidly accumulates and shows tolerance mechanism against salinity stress, for example, in *T. aestivum,* GA enhanced the growth and physiology of plant (Parasher and Varma 1988; Prakash and Prathapasenan 1990). In abiotic stress condition, GA triggers the signaling response that primarily regulates the mechanism of translocation, i.e. source to sink pathway (Iqbal et al. 2011). Due to greater salinity stress, large amount of sucrose content accumulates into the leaves (source) which directly affect the photosynthesis process through the feedback mechanism that reduces the rate of photosynthetic, while on similar note, GA also decreases the activity of enzymes which is mainly found in sink (Iqbal et al. 2011). However, during salt stress condition, DELLA protein of GA getting destabilized in salinity condition along with the other plant hormones has also been involved such as auxin and ethylene and shows interactive or cross-talk behavior at the molecular level (Achard et al. 2006). Under low salt concentration, tomato plants decrease resistance capacity of stomatal cells and enhanced the activity of water uptake mechanism in the presence of GA3 (Maggio et al. 2010). The GA3 also achieves the priming technique in wheat plant to increase the crop yield and enhanced the activity of ion uptake mechanism (Iqbal and Ashraf 2010). Gibberellic acid (GA) regulates the development as well as metabolic reaction in Glycine max (soya bean) (Hamayun et al. 2010). GA along with the nitrogen molecule also reduces salinity in *Brassica* sp. (mustard) crop plant (Siddiqui et al. 2008). Research findings also reported that, at the molecular level, protein content and level of RNA molecules were maintained by the use of exogenous application of gibberellic acid (GA) and it shows the positive impacts on its related mechanism under the salt stress condition (Kumar and Singh 1996).

11.4.3 Cytokinin

CKs are called as master regulator of plant growth as it maintains the plant wellbeing, although their biosynthesis rate decreases under stress condition (Chernyad'ev 2005). Numerous studies have been reported that showcase CKs induced senescence-accompanying changes. They are generally considered to be antagonists of ABA, with the two hormones having opposing effects. Under salinity stress, the CK levels drop and it cuts the supply and alters the gene expression and thus elicits the toxicity response, whereas applying exogenous CK delays leaf senescence (Chernyad'ev 2005) and capable of breaking dormancy during germination of seed as reported in tomato, barley, and cotton seeds (Bozcuk 1981). The possible reason behind the reduction in growth and developmental processes is that CKs could be a limiting factor beneath stress conditions and thus illuminate the fact that an exogenous application of kinetin causes enhancement in growth of plants. The declining CK content is an early response to salt stress; however, in tolerant species, the endogenous level remains unchanged. Exogenous application of CKs overwhelmed the negative effects of salinity stress on the growth and development of wheat seedlings (Zahir et al. 2001) and potato (Abdullah and Ahmad 1990). CKs act as a direct free radical scavenger and involved in antioxidative mechanism and also in transgenic production of tolerant plant species (Chakrabarti and Mukherji 2003).

11.4.4 Brassinosteroids

BRs stimulate various physiological processes such as growth, germination of seed, root development, senescence, and abscission of leaf. The fundamental role of BRs is to impart tolerance to plant under abiotic stress condition. The signal transduction pathway displays that the cell surface receptor brassinosteroid insensitive 1 (BRI1) triggers brassinazole resistant 1 (BZR1) and bri1-ems suppressor 1 (BES1) transcription factors that is responsible for the tolerance behavior. Exogenous BRs bind to receptor BRI1 that leads to association with the receptor (BAK1), and the sequential trans-phosphorylation between BRI1 and

BAK1 leads to activation of BRI1 and enhances BRI1 suppressor 1 (BSU1) activity. Thus, BSU1 *via* dephosphorylation activates transcription factors BZR1 and BZR2. These transcription factors thus enter the nucleus and enhance plant stress tolerance by increasing the capacity of antioxidant enzymes, regulating accumulation of endogenous hormones and upregulation of gene involved (Vardhini and Anjum 2015; Wu et al. 2013). BR regulates the level of ROS accumulation in plants under salt stress; it also activates CBFs (C-repeat binding factors) and key transcriptional factors, which are responsible for controlling the regulation of genes that respond to abiotic stress condition (Anwar et al. 2018).

11.4.5 Salicylic Acid

SA has a fundamental role in the defense mechanisms against stress factors; it is also responsible for the hypersensitive response against the pathogenic attack as well as an important signaling molecule that play a key role in plant growth and development (Cameron 2000). It initiates the signal transduction pathway and regulates proline metabolism, and upregulation of compounds involves in defense such as osmolyte. Exogenous SA stimulates the growth as it facilitates the ion movement and maintains the homeostasis, making it survive under saline environment; maintenance of ion homeostasis is dependent on the transmembrane protein such as H^+ translocating ATPases, pyrophosphatases, and Ca^{2+} ATPases (Sze et al. 1999). It also activates the SOS pathway, which involve: SOS-3; Ca^{2+} sensor, SOS-2, a serine/threonine protein kinase; and SOS-1, a plasma membrane Na^+/H^+ antiporter (Liu et al. 2000; Shi et al. 2000). SA initiates compartmentalization of accumulated Na^+ into the vacuoles or its efflux across the plasma membrane *via* Na^+/H^+ antiporter as well as of V-type H^+ ATPase and H^+ PPase (Plett and Moller 2010). Several physiological and morphological studies have illuminated that SOS pathway maintains ion equilibrium, ensuing sequestration into vacuole (Karasavina 2007). Recently, the fact came into light that SA influences anion and cation uptake through changes in the transmembrane electrical potential of plants. Thus, salicylic acid improves acclimation to salinity stress by ion exclusion and/or compartmentalization, ionic adjustments, decrease in lipid peroxidation, stress-induced synthesis of protein kinases (SIPK), and regulation of oxidative system to combat the lethal oxidative events.

11.5 Salt Toxicity Amelioration by Exogenous Synthetic Products

Numbers of workers have done tremendous work on minimizing the salt-induced negative effects by exogenous supplementation of osmoprotectants, plant hormones, antioxidants, signaling molecules, synthetic products, polyamines, and trace elements in different plant species. These protectants increase tolerance by augmenting plant germination, growth, development, photosynthesis, antioxidative capacities, and yield. In this section, we have discussed the use of synthetic products in counteracting the salt stress.

The exogenous supplementation or endogenous accumulation of proline which is an amino acid and a well-known product enhances behavior of salt-stressed plants. It is believed that the concentration of accumulated Pro in plants is considered as selection criteria for the tolerance of most species to stressed conditions (Ahmad et al. 2010). Under stressful conditions, supplementation of Pro stabilizes the structure of the membrane and minimized the oxidative stress by quenching reactive oxygen species (ROS) (Ashraf and Foolad 2007). In addition to its role as osmolyte, some lower plants such as fungi used Pro as a rich source of nitrogen and carbon. Further, Pro supplementation also enhances the uptake or accumulation of essential micro- or macro-nutrients (Ali et al. 2008) and increases the anti-oxidative defense system correlated with dropping cellular damage (Banu et al. 2009). In the study of Alam et al. (2016), exogenous proline supplementation at 25 mM concentration significantly improved the growth and yield of maize plants exposed with 50 mM NaCl stress. Similarly, improved growth and development have been demonstrated by various authors in different salt-stressed plants subsequently supplemented with proline (Banu et al. 2009; Sobahan et al. 2012; Talat et al. 2013). Similarly, 100 mM NaCl-stressed *Cucumis melo* L. plants when supplemented with 0.2 mM Pro showed remarkable increase in fresh/dry mass, rate of photosynthesis as well as increase the activity of enzymatic antioxidants such as SOD, POD, APX, CAT, and DHAR (Yan et al. 2011). Proline biosynthetic enzymes and their transcript levels of P5CS and P5CR are found to be upregulated under exogenous Pro supplementation (10 nM Pro) as reported in salt-stressed *O. sativa* (Nounjan et al. 2012). The effectiveness of Pro is depended on its concentration such as pre-treatment of Pro with a concentration of 1 mM was found to be effective and stimulated cellular activities, whereas 10 mM Pro was ineffective in improving plant growth under high level of salt (300 and 400 mM NaCl) (Deivanai et al. 2011).

To minimize the salt-induced toxic effects, exogenous use of osmo-protectants plays an important role; among them, glycine betaine (GB) plays diverse role which is a quaternary amine synthesized from choline or glycine (Giri 2011). It is a water soluble compound and non-toxic even at higher concentrations inside the cell having protective role (Ahmad et al. 2013). Salt adversely affect the growth, yield, and development of plant such as lettuce (Ekinci et al. 2012), and supplementation of GB plays a significant role in enhancing the plant's tolerance against salinity (Chen and Murata 2011). A number of metabolic process such as osmotic adjustment, protein stability, protein synthesis, enzyme activity, photosynthesis, oxidative stress response, and plant growth and development are directly influenced by GB application in stressed plants (Giri 2011; Chen and Murata 2011; Wani et al. 2013; Kotb and Elhamahmy 2014). Application of GB in plants follows different routes; it is accumulated in cytosol when sprayed on roots and then translocated toward the meristematic regions (flower buds and shoot apices) (Park et al. 2006), and the protective role of GB primarily acts on actively growing tissues (Annunziata et al. 2019). In a study of Hossain and Fujita (2010) and Yildirim et al. (2015), exogenous supplementation of GB to salt stress mung bean and lettuce (*Lactuca sativa* L.) caused a significant increase in the cellular metabolite content, GSH associated with the higher activities of correlative enzymes involved in the ROS detoxification and simultaneous decrease in the content of GSSG, H_2O_2, and lipid peroxidation that suggests a protective role of GB. Further, GB supplementation also enhances the shoot/root growth, relative water content, efficient photosynthetic apparatus, and better antioxidants as reported in NaCl-stressed ryegrass (*Lolium perenne*) (Hu et al. 2012). Similar results were also obtained in salt-stressed canola plants, where foliar-applied GB improved the water-use efficiency, CO_2 fixation, stomatal conductance, and also protects the oxygen evolving complex (OEC) of PS II and maintained its activity (Athar et al. 2015). Apart from Pro and GB, new osmo-protectant compound is identified named trehalose (Tre) present in bacteria, fungi, arthropods, and plants (Elbein et al. 2003). It is a disaccharide molecule where two monomers (glucose) joined by glycosidic bond (α, β-1,1-; β, β-1,1-; and α,α-1,1) and first reported in *Selaginella lepidophylla*. In plants system, there are five trehalose biosynthetic pathways but the most common is the TPS-TPP pathway; in this system, glucose is transferred from UDP-glucose to glucose-6-phosphate (G6P) to form intermediate trehalose-6-phosphate (T6P) catalyzed by enzyme trehalose-6-phosphate synthase (TPS) (Avonce et al. 2006). Adverse environmental factors including salinity negatively affect the growth and development of plants, and application of trehalose significantly increases the stress tolerance (Nounjan et al. 2012; Ma et al. 2013). A number of works have been performed to show that the exogenous Tre supplementation significantly alleviates the salt-induced toxicity in different plants such as maize (Zeid 2009), *Arabidopsis* (Yang et al. 2014), *Catharanthus roseus* (Chang et al. 2014), *Raphanus sativus* (Akram et al. 2016), rice varieties (Abdallah et al. 2016), and strawberry plants (Samadi et al. 2019). Trehalose majorly inhibited the active uptake of sodium ions associated with lowering Na/K ratio and their transportation toward chloroplast and also stabilized the interaction between protein and lipid bilayer, thereby decreased the membrane damage (Elbein et al. 2003; Chang et al. 2014). To overcome the toxic effects of ROS, studies describe its protective role in quenching of hydroxyl radicals (Couee et al. 2006) and Tre pre-treated plants express better enzymatic antioxidant capacity (Shahbaz et al. 2017). During salt stress, Tre application also participates in enhancing the bio-synthesis of proline and other compatible solutes (Abdallah et al. 2016). Under salt stress, there is an increase in carotenoid content when plants are subjected to Tre supplementation and this will protect the photochemical inhibition of photosynthetic apparatus as well as reduced the PS II protein degradation (Habibi 2017). Although varied studies have shown that exogenously applied Tre shields plants from stress, the specific mechanisms of Tre-mediated adaptation to salt stress have been studied only in a number of model plants. Among various bio-synthetic products, a phenolic compound due to its inherent anti-oxidant capacity is majorly used to increase the salt tolerance. A number of studies demonstrate the peculiar role of phenolic compounds as quencher of ROS or minimize the oxidative stress (Bhardwaj et al. 2015). Mathew et al. (2015) tested the role of some phenolic compounds and found that these compounds have inherent 85% scavenging activity which is further intensified by position and concentration of ROS. Phenolics are secondary metabolites of plants and categorized into two, i.e. hydroxycinnamic acids (*viz.*, ferulic, p-coumaric, sinapic, and caffeic acid) and hydroxybenzoic acids; both showed strong anti-oxidant activity because they have greater hydrogen donating and radical stabilization ability of CH=CH–COOH group (Michalak 2006). Previous works signify the exogenous supplementation of phenolic acids, especially salicylic acid and its derivatives, to ameliorate the oxidative stress generated under stressed conditions (Kang et al. 2012; Wang et al. 2016). Jiang et al. (2016) observed that exogenous melatonin (N-acetyl-5-methoxytryptamine), which is amphibolic, low-molecular-weight biological hormone reduced electrolyte leakage in salt-stressed maize seedlings and concluded that different exogenous phenolic acids play an important role in salinity tolerance. Besides phenolics, polyphenols are another group of secondary metabolite that plays an important role in defense system against adverse environmental factors (Agati et al. 2012; Brunetti et al. 2013). Most abundant polyphenol is the epigallocatechin-3-gallate (EGCG) that belongs to class flavonoid and especially present in tea plant (Hong et al. 2015), synthesized through flavonoid biosynthetic pathway (Petrussa et al. 2013) and also have strong anti-oxidant capacity (Hernandez et al. 2004). In *Arabidopsis thaliana,* even very low concentration of polyphenols such as catechins showed better growth performance, leaf thickness, net photosynthetic rate, stomatal conductance, and also involve in IAA metabolism (Rani et al. 2011). Exogenous supplementation of flavonoids actively participates in ROS detoxification due to the presence of anti-oxidant nature, hence increase in the stress tolerance to plants (Mouradov and Spangenberg 2014). Exogenous application of epigallocatechin-3-gallate (EGCG) improves the overall growth of wheat plants in terms of seed germination, root/shoot growth, by decreasing the generation of ROS (SOR and H_2O_2) and its indices (lipid peroxidation) by upregulating the important antioxidant such as SOD, CAT, APX, and POD. Among different bio-synthetic products and their role in salt stress tolerance, polyamines (PAs) are a kind of plant growth regulator having low molecular weight, aliphatic in nature, and performed a variety of physiological processes associated with growth and development and in some cases a well-known stress alleviator stresses (Pal et al. 2015). In plant system, there are mainly three types of PAs *viz* putrescine (Put), spermidine (Spd), and spermine (Spm) (Terui et al. 2005). Several studies have revealed the participation of PAs in salt stress tolerance in plants (Tang and Newton 2005; Li et al. 2016). Under salinity stress, exogenous PA supplementation significantly upregulates the PA biosynthesis-related genes and subsequently increases the tolerance behavior of plants (Bor and Özdemir 2018; Khare et al. 2018). In brief, repercussion of PAs in the defense against salinity is due to their inherent ability to stabilize the different macromolecules such as DNA, RNA, proteins, and bilayer components (phospholipids and glycolipids) and also involve in ROS detoxification (Hussain et al. 2011). Further, exogenous PAs also involved in enhancing the synthesis of other compatible solutes such as proline, trehalose, and glycine betaine.

11.6 Salt Toxicity Amelioration by Exogenous Nutrient Amendment Practices

Salt toxicity induced alterations in growth and development by inducing the ion toxicity, osmotic stress, essential nutrients (N, Ca, K, P, Fe, Zn) deficiency, inhibition of electron transport chain associated with oxidative stress that limits the activation of anti-oxidative machinery, and water uptake from soil (Javid et al. 2011; Horie et al. 2012). In agricultural fields, the salt toxicity is minimized by exogenous supplementation of plant hormones, osmo-protectants or directly by nutrient management practices. According to Munns (2002), salinity significantly hampered the uptake of essential nutrients, for example, salinity reduces P uptake (Bano and Fatima 2009). Hence, nutrient management significantly improved the growth, physiological, and biochemical modifications as well as quality characteristics of the salt-stressed plants. Salt primary inhibits the seed germination which is considered as a grave challenge for moneymaking legume production, and the intensity of salt-induced toxicity is intensified with plant ontogeny (Farooq et al. 2015). Hence, in early stage, there is need of seed priming to increase the percentage of seed germination under adverse environmental factors. According to Kaur et al. (2003), seeds of chickpea pre-treated with 4% mannitol solution for an hour showed better seed germinating capacity as compared to non-promimg seeds. Seed priming technology not only improves the seed germination capacity but also increases the tolerance behavior of plants by improving their antioxidant potential, osmotic adjustment, photosynthesis, and nutrient uptake (Entesari et al. 2012; Jisha and Puthur 2014; Jisha and Puthur 2015; Ghezal et al. 2016). Salt stress significantly reduces the uptake of nutrients and generates nutrient deficiency in plants, and among them, nitrogen (N) availability is one of the major limiting factors that hampers the growth and development specially the N metabolic process (Duan et al. 2015). Further, it is not very well clear that exogenous N supplementation alleviates the salt toxicity, but some of previous works signifies role of N in ameliorating salt stress of many annual herbaceous species (Villa-Castorena et al. 2003; Chen et al. 2010). In black gram, combination of 25 kg/ha of N along with 2% urea and 0.1 ppm brassinolide significantly enhanced the growth at vegetative stage, flowering stage, pod filling stage, and harvest stage (Mishra 2016). Based on these findings, Duan and Chang (2017) also evaluate the effect of exogenous N supplementation to salt-stressed lodge pole pine and white spruce and concluded that N fertilization was effective in mitigating the detrimental effects of salinity on seedling growth based on the physiological and growth responses. Plants that grow in the saline soils, sufficient nutrient supply plays an important role in increasing the salt tolerance mechanism. Similar to N, exogenous supplementation of potassium (K^+) ions improved growth, water uptake, decrease the osmotic stress, and increased the productivity of crops or plants facing salt stress conditions (Kabir et al. 2004). Supplementation of K mainly increases the plant water status under salt stress due to their higher binding capacity on Na^+/K^+ transport proteins as compared to Na^+ ions (Capula-Rodriguez et al. 2016). In early growth stage, the root system is ill developed for uptake of nutrients and during this time, salt accumulation found to be higher in germinating seeds. Hence, at this stage, foliar spray of K^+ or phosphorus (P^+) fertilizers is a good agent to mitigate the salt stress during early stage of growth (Mallarino et al. 2001). Foliar application of K was found to be very effective in mitigating salt stress in a number of plants such as eggplant (Elwan 2010), tomato (Amjad et al. 2014), sugar beat (Zaki et al. 2014), and wheat (Bybordi 2015) by increasing the K^+/Na^+ ratios. Further, the application Ca^{2+} lessened the damaging effects of salt stress on lentil growth and thus improved dry weight (Kafi et al. 2012). In another study of Burman et al. (2007), three different levels of P (0, 20, and 40 kg/ha) in salt stress clusterbean significantly improved the net photosynthetic rates and contents of photosynthetic pigments, starch, soluble protein, and total free amino acids as well as nitrate reductase activity as compared to control plants at both vegetative and flowering stages. Further, combination of nutrient medium N, P_2O_5, and K_2O in ratio of 20 : 50 : 50 and ratio of $ZnSO_4$ to $FeSO_4$ is in 10 : 10 improved the tolerance mechanism of salt-stressed plant, recorded higher seed yield (Fathima and Sudha 2016). Under salinity conditions, plant growth is severely affected by silicon (Si) deficiency as Si is the second most abundant element in the earth's crust and plants take Si in the form of (Si $(OH)_4$) (Luyckx et al. 2017). Supplementation of Si has been proved to be beneficial to plants under various adverse environmental conditions such as salinity (Manivannan et al. 2016). Supplementation with Si to salt stressed plant; there is a significant decrease in the movement of toxic ions such as Na or Cl *via* apo-plastic pathway (Yeo et al. 1999; Shi et al. 2013). Involvement is the formation of cuticle layer; Si also decreases the excessive transpiration thereby improved water content of cell correlated with increased rates of photosynthesis and growth (Verma et al. 2019). Further, degradation of proteins related to photosynthetic apparatus such as photosystem I (PS I), photosystem II (PS II), CO_2 fixing enzyme; RuBisCO and other chloroplast-related proteins are prevented by Si supplementation during salt stress (Muneer et al. 2014). Several studies reported that Si supplementation also enhances the antioxidative machinery by activating some specific enzymes as superoxide dismutase (SOD), catalase (CAT), guaiacol peroxidase (GPX), and ascorbate peroxidase (APX) to overcome the redox imbalance (Zhu et al. 2004; Manivannan et al. 2015; Soundararajan et al. 2015). Apart from exogenous nutrient supplementation, humus are rich in variety of growth increasing substances including important micro- or macro-nutrients and humic substances that can also be used as a substitute of direct nutrient management practices to ameliorate the deleterious effects of salt stress (Ouni et al. 2014). The addition of humic acids along with some nutrients such as P, K, Ca, or Si significantly increased the salt tolerance mechanisms is different vegetable crops such as bean (Aydin et al. 2012; Bargaz et al. 2016), okra (Paksoy et al. 2010), and pepper (Bacilio et al. 2016). Reason behind tolerance after humic acid application is that humic substances enhanced K^+/Na^+ and Ca^{2+}/Na^+ ratios. Similarly, use of bio-fertilizers instead of chemical fertilizers (urea) is also involved in mitigating the negative effects induced abiotic or biotic factors. Bio-fertilizers lead to improvement in growth and also solubilize the important nutrients from the soil that are easily available to plants (Vessey 2003) and also mitigate salinity effects on vegetables and reduce soil salinization. Best-known bio-fertilizers are cyanobacteria (blue green algae) that are capable to fix the atmospheric nitrogen and increased the soil fertility (Zehr 2011). Apart of this, plant growth promoting rhizobacteria (PGPRs), endo- and ectomycorrhizal fungi, and many other useful microscopic organisms are important bio-fertilizers that actively improved the nutrient uptake, plant growth, and plant tolerance to salt stress. During the seed germination of vegetable crops (tomato, pepper, bean, and lettuce), the inoculation of strains of PGPRs can upshot the root and shoot growth, dry weight, fruit, seed yield, and enhanced tolerance of plants to salt stress (Egamberdieva and Lugtenberg 2014). Some phosphate solubilizing bacterial strains such as *Pantoea agglomerans* and *Pseudomonas putida* have unique property to dissolve the insoluble organic or inorganic phosphates into soluble forms their uptake *via* phosphate transporters. Further, some fungal mycelium of arbuscular mycorrhiza (AM) has also capable in alleviating the salt stress as reported in tomato, onion, and lettuce (Cantrell and Linderman 2001; Latef and Chaoxing 2011; Aroca et al. 2013). Verma et al. (2014) studied the effect of vermin-compost and sulfur on growth, yield, and nutrient uptake of salt-stressed fenugreek and noticed higher plant height, uptake of N, K, and S. In brief, it is concluded that salinity significantly impaired the uptake and transport of essential plant nutrients such as nitrogen, calcium, potassium, magnesium, iron, and zinc leading to diminish plant growth and productivity. Application of these elements has the potential to counteract to the toxicity of salinity stress.

11.7 Future Prospective and Conclusion

Agricultural soil is deteriorating day by day due to excessive accumulation of ions such as Na^+, Cl^-, and K^+ that leads a significant decrease in the overall crop productivity especially in the arid or semi-arid regions where irrigation is mainly carried out with saline waters. Grain legumes are potential source of protein and other valuable compounds that fulfill the food demand of overgrowing population. Due to their ecological habitats, plants experience adverse environmental factors in their life and among them salt is very important factor that regulates the growth and development. Salinity alters the growth in plants by inducing osmotic stress, reactive oxygen species, specific ion effects, impaired hormone regulation, nutritional imbalances, and reduced carbon fixation due to stomatal and non-stomatal limitations in plants; all of which lead to impaired crop yield and quality. To cope up with damaging effects induced by salt, plants modified its physiological or biochemical activities either by increasing the antioxidant defense system or increasing the endogenous hormone (that elicit signaling response) or exogenous supplementation of bio-synthetic products (osmoprotectants) or exogenous nutrient management which is a cost-effective approach. Plant hormones such as gibberellins, cytokinins, salicylic acid, and brassinosteroids, these phytohormones are known for promoting plant growth and productivity under saline environment. Similarly biosynthetic products such as proline, glycine betaine, polyamines, and nutrient supplementation also increase the salinity tolerance. Further, analysis of upstream and downstream elements and detailed analysis of gene expression at different developmental stages with newer technologies will help in understanding and elucidating this complex trait.

References

Abdallah, M.S., Abdelgawad, Z.A., and El-Bassiouny, H.M. (2016). Alleviation of the adverse effects of salinity stress using trehalose in two rice varieties. *S. Afr. J. Bot.* 103: 275–282.

Abdullah, Z. and Ahmad, R. (1990). Effect of pre- and post-kinetin treatments on salt tolerance of different potato cultivars growing on saline soils. *J. Agron. Crop Sci.* 165: 94–102.

Abouelsaad, I., Renault, S. (2018). Enhanced oxidative stress in the jasmonic acid-deficient tomato mutant def-1 exposed to NaCl stress. *J. Plant Physiol.* 226: 136–144.

Achard, P., Cheng, H., De Grauwe, L. et al. (2006). Integration of plant responses to environmentally activated phytohormonal signals. *Science* 311: 91–94.

Afzal, I., Basra, S., and Iqbal, A. (2005). The effect of seed soaking with plant growth regulators on seedling vigor of wheat under salinity stress. *J. Stress Physiol. Biochem.* 1: 6–14.

Agati, G., Azzarello, E., Pollastri, S., and Tattini, M. (2012). Flavonoids as antioxidants in plants: location and functional significance. *Plant Sci.* 196: 67–76.

Ahanger, M.A., Alyemeni, M.N., Wijaya, L. et al. (2018). Potential of exogenously sourced kinetin in protecting *Solanum lycopersicum* from NaCl-induced oxidative stress through up-regulation of the antioxidant system, ascorbate glutathione cycle and glyoxalase system. *PLoS One* 13 (9): 0202175.

Ahmad, P. (2010). Growth and antioxidant responses in mustard (*Brassica juncea*L.) plants subjected to combined effect of gibberellic acid and salinity. *Arch. Agron. Soil Sci.* 56: 575–588.

Ahmad, P., Jaleel, C.A., and Sharma, S. (2010). Antioxidative defence system, lipid peroxidation, proline metabolizing enzymes and biochemical activity in two genotypes of *Morus alba* L. subjected to NaCl stress. *Russ. J. Plant Physiol.* 57: 509–517.

Ahmad, R., Lim, C.J., and Kwon, S.Y. (2013). Glycine betaine: a versatile compound with great potential for gene pyramiding to improve crop plant performance against environmental stresses. *Plant Biotechnol. Rep.* 7: 49–57.

Ahmad, P., Ahanger, M.A., Alyemeni, M.N. et al. (2018). Mitigation of sodium chloride toxicity in *Solanum lycopersicum* L. by supplementation of jasmonic acid and nitric oxide. *J. Plant Interact.* 13: 64–72. https://doi.org/10.1080/17429145.2017.1420830.

Akbari, G., Sanavy, S.A., and Yousefzadeh, S. (2007). Effect of auxin and salt stress (NaCl) on seed germination of wheat cultivars (*Triticum aestivum* L.). *Pak. J. Biol. Sci.* 10: 2557–2561.

Akram, N.A., Waseem, M., Ameen, R., and Ashraf, M. (2016). Trehalose pretreatment induces drought tolerance in radish (*Raphanus sativus* L.) plants: some key physio-biochemical traits. *Acta Physiol. Plant.* 38: 3.

Alam, R., Das, D.K., Islam, M.R. et al. (2016). Exogenous proline enhances nutrient uptake and confers tolerance to salt stress in maize (*Zea mays* L.). *Progress. Agric.* 27: 409–417.

Ali, Q., Ashraf, M., and Shahbaz Humera, H. (2008). Ameliorative effect of foliar applied proline on nutrient uptake in water stressed maize (*Zea mays* L.) plants. *Pak. J. Bot.* 40: 211–219.

Ali, M.A., Naveed, M., Mustafa, A., and Abbas, A. (2017). The good, the bad, and the ugly of rhizosphere microbiome. In: *Probiotics and Plant Health* (eds. V. Kumar et al.), 253–290. Singapore: Springer.

Amjad, M., Akhtar, J., Haq, M.A.U. et al. (2014). Soil and foliar application of potassium enhances fruit yield and quality of tomato under salinity. *Turk. J. Biol.* 38: 208–218.

Amzallag, G.N., Lerner, H.R., and Poljakoff-Mayber, A. (1990). Exogenous ABA as a modulator of response of sorghum to high salinity. *J. Exp. Bot.* 41: 1389–1394.

Annunziata, M.G., Ciarmiello, L.F., Woodrow, P. et al. (2019). Spatial and temporal profile of glycinebetaine accumulation in plants under abiotic stresses. *Front. Plant Sci.* 10: 230.

Anuradha, S. and Rao, S.S.R. (2001). Effect of brassinosteroids on salinity stress induced inhibition of seed germination and seedling growth of rice (*Oryza sativa* L.). *Plant Growth Regul.* 33: 151–153.

Anwar, A., Liu, Y., and Dong, R. (2018). The physiological and molecular mechanism of brassinosteroid in response to stress: a review. *Biol. Res.* 51: 46. https://doi.org/10.1186/s40659-018-0195-2.

Aroca, R., Ruiz-Lozano, J.M., Zamarreño, Á.M. et al. (2013). Arbuscular mycorrhizal symbiosis influences strigolactone production under salinity and alleviates salt stress in lettuce plants. *J. Plant Physiol.* 170: 47–55.

Ashraf, M. and Foolad, M.R. (2007). Roles of glycinebetaine and proline in improving plant abiotic stress resistance. *Environ. Exp. Bot.* 59: 206–216.

Ashraf, M., Athar, H.R., Harris, P.J.C., and Kwon, T.R. (2008). Some prospective strategies for improving crop salt tolerance. *Adv. Agron.* 97: 45–110.

Athar, H.U.R., Zafar, Z.U., and Ashraf, M. (2015). Glycinebetaine improved photosynthesis in canola under salt stress: evaluation of chlorophyll fluorescence parameters as potential indicators. *J. Agron. Crop Sci.* 201: 428–442.

Ávila, A.C., Ochoa, J., Proaño, K., and Martínez, M.C. (2019). Jasmonic acid and nitric oxide protects naranjilla (*Solanum quitoense*) against infection by *Fusarium oxysporum* f. sp. *quitoense* by eliciting plant defense responses. *Physiol. Mol. Plant Pathol.* 106: 129–136.

Avonce, N., Mendoza-Vargas, A., Morett, E., and Iturriaga, G. (2006). Insights on the evolution of trehalose biosynthesis. *BMC Evol. Biol.* 6: 109.

Aydin, A., Canan, K., and Metin, T. (2012). Humic acid application alleviate salinity stress of bean (*Phaseolus vulgaris* L.) plants decreasing membrane leakage. *Afr. J. Agric. Res.* 7: 1073–1086.

Azooz, M.M., Youssef, A.M., and Ahmad, P. (2011). Evaluation of salicylic acid (SA) application on growth, osmotic solutes and antioxidant enzyme activities on broad bean seedlings grown under diluted seawater. *Int. J. Plant Physiol. Biochem.* 3: 253–264.

Bacilio, M., Moreno, M., and Bashan, Y. (2016). Mitigation of negative effects of progressive soil salinity gradients by application of humic acids and inoculation with Pseudomonas stutzeri in a salt-tolerant and a salt-susceptible pepper. *Appl. Soil Ecol.* 107: 394–404.

Bano, A. and Fatima, M. (2009). Salt tolerance in *Zea mays* (L.) following inoculation with Rhizobium and Pseudomonas. *Biol. Fertil. Soils* 45: 405–413.

Banu, M.N.A., Hoque, M.A., Sugimoto, M.W. et al. (2009). Proline and glycinebetaine induce antioxidant defense gene expression and suppress cell death in cultured tobacco cells under salt stress. *J. Plant Physiol.* 166: 146–156.

Barciszewski, J., Siboska, G., Rattan, S.I.S., and Clark, B.F.C. (2000). Occurrence, biosynthesis and properties of kinetin (N6-furfuryladenine). *Plant Growth Regul.* 32: 257–265.

Bargaz, A., Nassar, R.M.A., Rady, M.M. et al. (2016). Improved salinity tolerance by phosphorus fertilizer in two phaseolus vulgaris recombinant inbred lines contrasting in their P-efficiency. *J. Agron. Crop Sci.* 202: 497–507.

Bashri, G. and Prasad, S.M. (2016). Exogenous IAA, differentially affects growth, oxidative stress and antioxidants system in Cd stressed *Trigonella foenum-graecum* L. seedlings: toxicity alleviation by up-regulation of ascorbate-glutathione cycle. *Ecotoxicol. Environ. Saf.* 132: 329–338.

Bhardwaj, R., Sharma, A., Sharma, H., and Srivastva, P. (2015). Role of gallic acid pre-treatment in inducing the antioxidant response of two wheat cultivars differing in drought tolerance. *Ind. J. Agric. Biochem.* 28: 155–165.

Bishop, G.J. and Yokota, T. (2001). Plants steroid hormones, brassinosteroids: current highlights of molecular aspects on their synthesis/metabolism, transport, perception and response. *Plant Cell Physiol.* 42: 114–120.

Bor, M. and Özdemir, F. (2018). Manipulating metabolic pathways for development of salt-tolerant crops. In: *Salinity Responses and Tolerance in Plants*, vol. 1 (eds. V. Kumar, S. Wani, P. Suprasanna and L.S. Tran), 235–256. Cham: Springer International Publishing.

Bozcuk, S. (1981). Effect of kinetin and salinity on germination of tomato, barley and cotton seeds. *Ann. Bot.* 48: 81–84.

Brunetti, C., Di Ferdinando, M., Fini, A. et al. (2013). Flavonoids as antioxidants and developmental regulators: relative significance in plants and humans. *Int. J. Mol. Sci.* 14: 3540–3555.

Burman, U., Garg, B.K., and Kathju, S. (2007). Interactive effects of phosphorus, nitrogen, and thiourea on clusterbean (*Cyamopsis tetragonoloba* L.) under rainfed conditions of the Indian arid zone. *J. Plant Nutr. Soil Sci.* 170: 803–810.

Bybordi, A. (2015). Influence of exogenous application of silicon and potassium on physiological responses, yield, and yield components of salt-stressed wheat. *Commun. Soil Sci. Plant Anal.* 46: 109–122.

Caarls, L., Pieterse, C.M.J., and Van Wees, S.C.M. (2015). How salicylic acid takes transcriptional control over jasmonic acid signaling. *Front. Plant Sci.* 6: 170. https://doi.org/10.3389/fpls.2015.00170.

Cabot, C., Sibole, J.V., Barcelo, J., and Poschenrieder, C. (2009). Abscisic acid decreases leaf Na^+ exclusion in salt-treated *Phaseolus vulgaris* L. *J. Plant Growth Regul.* 28: 187–192.

Cameron, R.K. (2000). Salicylic acid and its role in plant defense responses: what do we really know? *Physiol. Mol. Plant Pathol.* 56: 91–99.

Cantrell, I.C. and Linderman, R.G. (2001). Preinoculation of lettuce and onion with VA mycorrhizal fungi reduces deleterious effects of soil salinity. *Plant Soil* 233: 269–281.

Capula-Rodriguez, R., Valdez-Aguilar, L.A., Cartmill, D.I. et al. (2016). Supplementary calcium and potassium improves the response of tomato (*Solanum lycopersicum* L.) to simultaneous alkalinity, salinity and boron stress. *Commun. Soil Sci. Plant Anal.* 47: 505–511.

Chakrabarti, N. and Mukherji, S. (2003). Effect of phytohormone pretreatment on nitrogen metabolism in *Vigna radiata* under salt stress. *Biol. Plant.* 46: 63–66.

Chang, B., Yang, L., Cong, W. et al. (2014). The improved resistance to high salinity induced by trehalose is associated with ionic regulation and osmotic adjustment in *Catharanthus roseus*. *Plant Physiol. Biochem.* 77: 140–148.

Chen, T.H.H. and Murata, N. (2011). Glycinebetaine protects plants against abiotic stress: mechanisms and biotechnological applications. *Plant Cell Environ.* 34: 1–20.

Chen, W., Hou, Z., Wu, L. et al. (2010). Effects of salinity and nitrogen on cotton growth in arid environment. *Plant Soil* 326: 61–73.

Chernyad, ev, I.I. (2005). Effect of water stress on the photosynthetic apparatus of plants and the protective role of cytokinins: a review. *Appl. Biochem. Microbiol.* 41: 115–128. https://doi.org/10.1007/s10438-005-0021-9.

Couee, I., Sulmon, C., Gouesbet, G., and El-Amrani, A. (2006). Involvement of soluble sugars in reactive oxygen species balance and responses to oxidative stress in plants. *J. Exp. Bot.* 57: 449–459.

Creelman, R.A. and Mullet, J.E. (1995). Jasmonic acid distribution and action in plants: Regulation during development and response to biotic and abiotic stress. *Proc. Natl. Acad. Sci.* 92: 4114–4119.

Davière, J. and Achard, P. (2015). A pivotal role of DELLAs in regulating multiple hormone signals: Review Article. *Mol. Plant* 9: 10–20. https://doi.org/10.1016/j.molp.2015.09.011.

Deivanai, S., Xavier, R., Vinod, V. et al. (2011). Role of exogenous proline in ameliorating salt stress at early stage in two rice cultivars. *J. Stress Physiol. Biochem.* 7: 157–174.

Dong, Y., Wang, W., Hu, G., Chen, W., et al.(2017). Role of exogenous 24-epibrassinolide in enhancing the salt tolerance of wheat seedlings. *J Soil Sc Plant Nutr.* 17:554–569. doi:10.4067/s0718-95162017000300001.

Duan, M. and Chang, X. (2017). Nitrogen fertilization improves the growth of lodgepole pine and white spruce seedlings under low salt stress through enhancing photosynthesis and plant nutrition. *Forest Ecol. Manag.* 404: 197–204.

Duan, M., House, J., and Chang, S.X. (2015). Limiting factors for lodgepole pine (*Pinus contorta*) and white spruce (*Picea glauca*) growth differ in some reconstructed sites in the Athabasca oil sands region. *Ecol. Eng.* 75: 323–331.

Egamberdieva, D. and Lugtenberg, B. (2014). Use of plant growth-promoting rhizobacteria to alleviate salinity stress in plants. In: *Use of Microbes for the Alleviation of Soil Stresses*, vol. 1 (ed. M. Miransari), 73–96. New York, NY: Springer.

Egamberdieva, D., Wirth, S.J., Alqarawi, A.A. et al. (2017). Phytohormones and beneficial microbes: essential components for plants to balance stress and fitness. *Front. Microbiol.* 8: 2104. https://doi.org/10.3389/fmicb.2017.02104.

Ekinci, M., Yildirim, E., Dursum, A., and Turan, M. (2012). Mitigation of salt stress in lettuce (Lettuce sativa L. var. Crispa) by seed and foliar 24-epibrassinolide treatments. *Sci. Hortic.* 114: 5–10.

Elbein, A.D., Pan, Y.T., Pastuszak, I., and Carroll, D. (2003). New insights on trehalose: a multifunctional molecule. *Glycobiology* 13: 17–27.

El-Esawi, M.A. (2016). Jasmonic acid: genetic pathway, signal transduction and action in plant development and defence. *Biochem. Physiol.* 5: 2. https://doi.org/10.4172/2168-9652.1000e14.

Elwan, M.W. (2010). Ameliorative effects of di-potassium hydrogen orthophosphate on salt-stressed eggplant. *J. Plant Nutr.* 33: 1593–1604.

Entesari, M.F., Sharif-Zadeh, S., Zare, S., and Farhangfar, M. (2012). Effect of seed priming on mung bean (*Vigna radiata* L.) cultivars with salicylic acid and potassium nitrate under salinity stress. *Int. J. Agric. Res. Rev.* 2: 926–932.

Eraslan, F., Inal, A., Savasturk, O., and Gunes, A. (2007). Changes in antioxidative system and membrane damage of lettuce in response to salinity and boron toxicity. *Sci. Hortic.* 114: 5–10.

Fahad, S., Hussain, S., Matloob, A. et al. (2014). Phytohormones and plant responses to salinity stress: a review. *Plant Growth Regul.* 75: 391–404. https://doi.org/10.1007/s10725-014-0013-y.

Fahad, S., Hussain, S., Bano, A. et al. (2015). Potential role of phytohormones and plant growth-promoting rhizobacteria in abiotic stresses: consequences for changing environment. *Environ. Sci. Pollut. Res. Int.* 22: 4907–4921.

Fan, H.F., Du, C.X., and Guo, S.R. (2013). Nitric oxide enhances salt tolerance in cucumber seedlings by regulating free polyamine content. *Environ. Exp. Bot.* 86: 52–59.

Farooq, M., Hussain, M., Wakeel, A., and Siddique, K.H.M. (2015). Salt stress in maize effects resistance mechanisms and management: a review. *Agron. Sustain. Dev.* 35: 461–481.

Fathima, K.J. and Sudha, T. (2016). Effect of major and micro nutrients on seed yield, quality and economics of clusterbean (*Cyamopsis tetragonoloba* L. Taub). *J. Farm Sci.* 29: 273–275.

Fletcher, R.A., Gilley, A., Sankhla, N., and Davis, TD. (2010). Triazoles as Plant Growth Regulators and Stress Protectants. *Horticul. Reviews*, 55: 138. doi:10.1002/9780470650776.

Gao, Y. and Zhang, G.A. (2019). Calcium sensor calcineurin B-like 9 negatively regulates cold tolerance via calcium signaling in *Arabidopsis thaliana*. *Plant Signal. Behav.* 14: e1573099.

Ghanem, M.E., Albacete, A., Martínez-Andújar, C. et al. (2008). Hormonal changes during salinity-induced leaf senescence in tomato (*Solanum lycopersicum* L.). *J. Exp. Bot.* 59: 3039–3050.

Ghezal, N., Rinez, I., Sbai, H. et al. (2016). Improvement of *Pisum sativum* salt stress tolerance by bio-priming their seeds using *Typha angustifolia* leaves aqueous extract. *S. Afr. J. Bot.* 105: 240–250.

Giri, J. (2011). Glycinebetaine and abiotic stress tolerance in plants. *Plant Signal. Behav.* 6: 1746–1751.

Gomez, C.A., Arbona, V., Jacas, J. et al. (2002). Abscisic acid reduces leaf abscission and increases salt tolerance in citrus plants. *J. Plant Growth Regul.* 21: 234–240.

Gong, D.H., Wang, G.Z., Si, W.T. et al. (2018). Effects of salt stress on photosynthetic pigments and activity of ribulose-1,5-bisphosphate carboxylase/oxygenase in *Kalidium foliatum*. *Russ. J. Plant Physiol.* 65: 98–103.

Gul, B., Khan, M.A., and Weber, D.J. (2000). Alleviation salinity and dark enforced dormancy in *Allenrolfea occidentalis* seeds under various thermoperiods. *Aust. J. Bot.* 48: 745–752.

Gunes, A., Inal, A., Alpaslam, M. et al. (2007). Salicylic acid induced changes on some physiological parameters symptomaticfor oxidative stress and mineral nutrition in maize (*Zea mays* L.) grown under salinity. *J. Plant Physiol.* 164: 728–736.

Gupta, B. and Huang, B. (2014). Mechanism of salinity tolerance in plants: physiological, biochemical, and molecular characterization. *Int. J. Genom.* 2014: 1–18. https://doi.org/10.1155/2014/701596.

Gupta, K.J., Stoimenova, M., and Kaiser, W.M. (2005). In higher plants, only root mitochondria, but not leaf mitochondria reduce nitrite to NO, in vitro and in situ. *J. Exp. Bot.* 56: 2601–2609.

Gurmani, A.R., Bano, A., Khan, S.U. et al. (2011). Alleviation of salt stress by seed treatment with abscisic acid (ABA), 6-benzylaminopurine (BA) and chlormequat chloride (CCC) optimizes ion and organic matter accumulation and increases yield of rice (*Oryza sativa* L.). *Aus. J. Crop Sci.* 5: 1278–1285.

Habibi, G. (2017). Physiological, photochemical and ionic responses of sunflower seedlings to exogenous selenium supply under salt stress. *Acta Physiol. Plant.* 39: 213.

Hamayun, M., Khan, S.A., Khan, A.L. et al. (2010). Exogenous gibberellic acid reprograms soybean to higher growth and salt stress tolerance. *J. Agric. Food Chem.* 58: 7226–7232.

Hamayun, M., Hussain, A., Khan, S.A., et al. (2015). Kinetin modulates physio-hormonal attributes and isoflavone contents of Soybean grown under salinity stress. *Front. Plant Sci.* 6:377. doi: 10.3389/fpls.2015.00377.

Hare, P.D., Cress, W.A., and van Staden, J. (1997). The involvement of cytokinins in plant responses to environmental stress. *Plant Growth Regul.* 23: 79–103.

Hedden, P. and Sponsel, V. (2015). A century of gibberellin research. *Plant Growth Regul.* 34: 740–760.

Hernandez, I., Alegre, L., and Munne-Bosch, S. (2004). Drought-induced changes in flavonoids and other low molecular weight antioxidants in *Cistus clusii* grown under Mediterranean field conditions. *Tree Physiol.* 24: 1303–1311.

Hong, G., Wang, J., Hochstetter, D. et al. (2015). Epigallocatechin- 3-gallate functions as a physiological regulator by modulating the jasmonic acid pathway. *Physiol. Plant.* 153: 432–439.

Horie, T., Karahara, I., and Katsuhara, M. (2012). Salinity tolerance mechanisms in glycophytes: an overview with the central focus on rice plants. *Rice* 5: 1–18.

Hossain, M.S. and Dietz, K.J. (2016). Tuning of redox regulatory mechanisms, reactive oxygen species and redox homeostasis under salinity stress. *Front. Plant Sci.* 10: 00548.

Hossain, M.A. and Fujita, M. (2010). Evidence for a role of exogenous glycinebetaine and proline in antioxidant defense and methylglyoxal detoxification systems in mung bean seedlings under salt stress. *Physiol. Mol. Biol. Plants* 16: 19–29.

Hu, L., Hu, T., Zhang, X. et al. (2012). Exogenous glycinebetaine ameliorates the adverse effect of salt stress on perennial ryegrass. *J. Am. Soc. Hortic. Sci.* 137: 38–46.

Husen, A., Iqbal, M.,Sohrab, SS., Ansari, MKA. (2018). Salicylic acid alleviates salinity-causeddamage to foliar functions, plant growth and antioxidant system in Ethiopian mustard (Brassica carinata A. Br.) *Agric. Food Secur.* 7: 44. https://doi.org/10.1186/s40066-018-0194-0.

Hussain, S.S., Ali, M., Ahmad, M., and Siddique, K.H.M. (2011). Polyamines: natural and engineered abiotic and biotic stress tolerance in plants. *Biotechnol. Adv.* 29: 300–311.

Iqbal, M. and Ashraf, M. (2010). Gibberellic acid mediated induction of salt tolerance in wheat plants: growth, ionic partitioning, photosynthesis, yield and hormonal homeostasis. *Environ. Exp. Bot.* 10: 002.

Iqbal, M. and Ashraf, M. (2013). Gibberellic acid mediated induction of salt tolerance in wheat plants: growth, ionic partitioning, photosynthesis, yield and hormonal homeostasis. *Environ. Exp. Bot.* 86: 76–85.

Iqbal, M., Ashraf, M., and Jamil, A. (2006). Seed enhancement with cytokinins: changes in growth and grain yield in salt stressed wheat plants. *Plant Growth Regul.* 50: 29–39.

Iqbal, N., Nazar, R., Iqbal, M.R.K. et al. (2011). Role of gibberellins in regulation of source sink relations under optimal and limiting environmental conditions. *Curr. Sci.* 100: 998–1007.

Izadi, M.H., Rabbani, J., Emam, Y. et al. (2014). Effects of salinity stress on physiological performance of various wheat and barley cultivars. *J. Plant Nutr.* 37: 520–531.

Jae-Ung, H. and Youngsook, L. (2001). Abscisic acid-induced actin reorganization in guard cells of day flower is mediated by cytosolic calcium levels and by protein kinase and protein phosphatase activities. *Plant Physiol.* 125: 2120–2128.

Janda, K., Éva, S.G.H., Kovács, L., and Janda, T. (2012). Salicylic acid may indirectly influence the photosynthetic electron transport. *J. Plant Physiol.* 169: 971–978.

Javid, M.G., Sorooshzadeh, A., Moradi, F. et al. (2011). The role of phytohormones in alleviating salt stress in crop plants. *AJCS* 5: 726–734.

Jiang, X., Li, H., and Song, X. (2016). Seed priming with melatonin effects on seed germination and seedling growth in maize under salinity stress. *Pak. J. Bot.* 48: 1345–1352.

Jini, D., and Joseph, B. (2017). Physiological mechanism of salicylic acid for alleviation of salt stress in Rice. *Rice Science*, 24: 97–108. doi:10.1016/j.rsci.2016.07.007.

Jisha, K.C. and Puthur, J.T. (2014). Halopriming of seeds imparts tolerance to NaCl and PEG induced stress in *Vigna radiata* (L.) Wilczek varieties. *Physiol. Mol. Biol. Plants* 20: 303–312.

Jisha, K.C. and Puthur, J.T. (2015). Seed priming with BABA (b-amino butyric acid): a costeffectiven method of abiotic stress tolerance in *Vigna radiata* (L.) Wilczek. *Protoplasma* 253: 277–289.

Jung, J. and Park, C. (2011). Auxin modulation of salt stress signaling in *Arabidopsis* seed germination. *Plant Signal. Behav.* 6: 1198–1200.

Kabir, M.E., Karim, M.A., and Azad, M.A.K. (2004). Effect of potassium on salinity tolerance of mungbean (*Vigna radiata* L. Wilczek). *J. Biol. Sci.* 4: 103–110.

Kafi, M., Goldani, M., and Jafari, M.H.S. (2012). Effectiveness of nutrient management in managing saline agro-ecosystems: a case study of *Lens culinaris* Medik. *Pak. J. Bot.* 44: 269–274.

Kang, D.J., Seo, Y.J., Lee, J.D. et al. (2005). Jasmonic acid differentially affects growth, ion uptake and abscisic acid concentration in salt-tolerant and salt-sensitive rice cultivars. *J. Agron. Crop Sci.* 191: 273–282.

Kang, G., Li, G., Xu, W. et al. (2012). Proteomics reveals the effects of salicylic acid on growth and tolerance to subsequent drought stress in wheat. *J. Proteome Res.* 11: 6066–6079.

Karasavina, M.S. (2007). Effect of salicylic acid on solute transport in plants. In: *Salicylic Acid a Plant Hormone* (eds. S. Hayat and A. Ahmad), 25–68. Springer.

Kapoor, R., Evelin, H., Devi, TS., and Gupta, S. (2019). Mitigation of salinity stress in plants by arbuscular mycorrhizal symbiosis: current understanding and new challenges. *Front. Plant Sci.* 10: 470. doi:10.3389/fpls.2019.00470.

Kaur, S., Gupta, A.K., and Kaur, N. (2003). Priming of chickpea seeds with water and mannitol overcomes the effect of salt stress on seedling growth. *Int. Chickpea Pigeon Pea Newsl.* 10: 18–19.

Kaya, C., Kirnak, H., Higgs, D., and Saltali, K. (2002). Supplementary calcium enhances plant growth and fruit yield in strawberry cultivars grown at high (NaCl) salinity. *Sci. Hortic.* 93: 65–74.

Kazan, K. (2013). Auxin and the integration of environmental signals into plant root development. *Ann. Bot.* 112: 1655–1665.

Khadri, M., Tejera, N.A., and Lluch, C. (2006). Alleviation of salt stress in common bean (*Phaseolus vulgaris*) by exogenous abscisic acid supply. *J. Plant Growth Regul.* 25: 110–119.

Khan, M.I.R., Asgher, M., and Khan, N.A. (2014). Alleviation of salt-induced photosynthesis and growth inhibition by salicylic acid involves glycine betaine and ethylene in mungbean (*Vigna radiata* L.). *Plant Physiol. Biochem.* 80: 67–74.

Khare, T., Srivastav, A., Shaikh, S., and Kumar, V. (2018). Polyamines and their metabolic engineering for plant salinity stress tolerance. *Springer Int. Pub. Cham* 10: 413.

Kotb, M.A. and Elhamahmy, M.A. (2014). Improvement of wheat productivity and their salt tolerance by exogenous glycinebetaine application under saline soil condition for long-term. *Zagazig J. Agric. Res.* 41: 1127–1143.

Kramell, R., Atzorn, R., Schneider, G. et al. (1995). Occurrence and identification of jasmonic acid and its amino acid conjugates induced by osmotic stress in barley leaf tissue. *J. Plant Growth Regul.* 14: 29–36.

Kumar, B. and Singh, B. (1996). Effect of plant hormones on growth and yield of wheat irrigated with saline water. *Ann. Agric. Res.* 17: 209–212.

Kumar, R., Khurana, A., and Sharma, A.K. (2014). Role of plant hormones and their interplay in development and ripening of fleshy fruits. *J. Exp. Bot.* 65: 4561–4575.

Lakhdar, A., Rabhi, M., Ghnaya, T. et al. (2009). Effectiveness of compost use in salt-affected soils. *J. Hazard. Mater.* 171: 29–37.

Latef, A.A.H.A. and Chaoxing, H. (2011). Effect of arbuscular mycorrhizal fungi on growth, mineral nutrition, antioxidant enzymes activity and fruit yield of tomato grown under salinity stress. *Sci. Hortic.* 127: 228–233.

Lavy, M. and Estelle, M. (2016). Mechanism of auxin signaling. *Development* 143: 3223–3229.

Li, S., Jin, H., and Zhang, Q. (2016). The effect of exogenous spermidine concentration on polyamine metabolism and salt tolerance in zoysiagrass (*Zoysia japonica steud*) subjected to short-term salinity stress. *Front. Plant Sci.* 7: 1221.

Liu, J., Ishitani, M., Halfter, U. et al. (2000). The Arabidopsis thaliana SOS2 geneencodes a protein kinase that is required for the salt tolerance. *Proc. Natl. Acad. Sci. U. S. A.* 97: 3730–3734.

Liu, W., Li, R.J., Han, T.T. et al. (2015). Salt stress reduces root meristem size by nitric oxide-mediated modulation of auxin accumulation and signaling in *Arabidopsis*. *Plant Physiol.* 168: 343–356.

Luyckx, M., Hausman, J.F., Lutts, S., and Guerriero, G. (2017). Silicon and plants: current knowledge and technological perspectives. *Front. Plant Sci.* 8: 411.

Ma, C., Wang, Z., Kong, B., and Lin, T. (2013). Exogenous trehalose differentially modulate antioxidant defense system in wheat callus during water deficit and subsequent recovery. *Plant Growth Regul.* 70: 275–285.

Maggio, A., Barbieri, G., Raimondi, G., and De Pascale, S. (2010). Contrasting effects of GA3 treatments on tomato plants exposed to increasing salinity. *J. Plant Growth Regul.* 29: 63–72.

Magome, H., Yamaguchi, S., Hanada, A. et al. (2004). Dwarf and delayed-flowering 1, a novel Arabidopsis mutant deficient in gibberellins biosynthesis because of over expression of a putative AP2 transcription factor. *Plant J.* 37: 720–729.

Mallarino, A.P., Haq, M.U., Wittry, D., and Bermudez, M. (2001). Variation in soybean response to early season foliar fertilization among and within fields. *Agron. J.* 93: 1220–1226.

Manivannan, A., Soundararajan, P., Arum, L.S. et al. (2015). Silicon-mediated enhancement of physiological and biochemical characteristics of *Zinnia elegans* 'Dreamland Yellow' grown under salinity stress. *Hortic. Environ. Biotechnol.* 56: 721–731.

Manivannan, A., Soundararajan, P., Muneer, S. et al. (2016). Silicon mitigates salinity stress by regulating the physiology, antioxidant enzyme activities, and protein expression in *Capsicum annuum* 'Bugwang'. *BioMed Res. Int.* 2016: 3076357.

Manjili, F.A., Sedghi, M., and Pessarakli, M. (2012). Effects of phytohormones on proline content and antioxidant enzymes of various wheat cultivars under salinity stress. *J. Plant Nutr.* 35: 1098–1111.

Mathew, S., Abraham, T.E., and Zakaria, Z. (2015). Reactivity of phenolic compounds towards free radicals under in vitro conditions. *J. Food Sci. Technol.* 52: 5790–5798.

Merchan, F., de Lorenzo, L., Rizzo, S.G. et al. (2007). Identification of regulatory pathways involved in the reacquisition of root growth after salt stress in *Medicago truncatula*. *Plant J.* 51: 1–17.

Michalak, A. (2006). Phenolic compounds and their antioxidant activity in plants growing under heavy metal stress. *Pol. J. Environ. Stud.* 15: 523–530.

Mishra, B.P. (2016). Effects of nitrogen and growth regulators on yield *Phaseolus mungo* L. *Int. J. Adv. Res. Dev.* 1: 39–42.

Mouradov, A. and Spangenberg, G. (2014). Flavonoids: a metabolic network mediating plants adaptation to their real estate. *Front. Plant Sci.* 5: 620.

Muneer, S., Park, Y.G., Manivannan, A. et al. (2014). Physiological and proteomic analysis in chloroplasts of Solanum lycopersicum L. under silicon efficiency and salinity stress. *Int. J. Mol. Sci.* 15: 21803–21824.

Munns, R. (2002). Comparative physiology of salt and water stress. *Plant Cell Environ.* 25: 239–250.

Munns, R. and Tester, M. (2008). Mechanisms of salinity tolerance. *Annu. Rev. Plant Biol.* 59: 651–681.

Nawaz, F., Shabbir, R.N., Shahbaz, M. et al. (2017). Cross talk between nitric oxide and phytohormones regulate plant development during abiotic stresses. In: *Phytohormones-Signaling Mechanisms and Crosstalk in Plant Development and Stress Responses* (ed. M. El Esawi). In Tech https://doi.org/10.5772/intechopen.69812.

Niamat, B., Naveed, M., Ahmad, Z. et al. (2019). Calcium-enriched animal manure alleviates the adverse effects of salt stress on growth, physiology and nutrients homeostasis of *Zea mays* L. *Plants* 8: 480.

Nilsen, E.T. and Orcutt, D.M. (1996). *The Physiology of Plants Under Stressabiotic Factors*, 118–130. New York: Wiley.

Nolan, T.M., Vukasinović, N., Liu, D. et al. (2020). Brassinosteroids: multidimensional regulators of plant growth, development, and stress responses. *Plant Cell* 32: 295–318.

Nounjan, N., Nghia, P.T., and Theerakulpisut, P. (2012). Exogenous proline and trehalose promote recovery of rice seedlings from salt-stress and differentially modulate antioxidant enzymes and expression of related genes. *J. Plant Physiol.* 169: 596–604.

Olszewski, N., Sun, T.P., and Gubler, F. (2002). Gibberellin signaling, biosynthesis, catabolism, and response pathways. *Plant Cell* 14: 561–580.

Ouni, Y., Ghnaya, T., Montemurro, F. et al. (2014). The role of humic substances in mitigating the harmful effects of soil salinity and improve plant productivity. *Int. J. Agron. Plant Prod.* 8: 353–374.

Paksoy, M., Türkmen, Ö., and Dursun, A. (2010). Effects of potassium and humic acid on emergence, growth and nutrient contents of okra (*Abelmoschus esculentus* L.) seedling under saline soil conditions. *Afr. J. Biotechnol.* 9: 5343–5346.

Pal, M., Szalai, G., and Janda, T. (2015). Speculation: polyamines are important in abiotic stress signaling. *Plant Sci.* 237: 16–23.

Parasher, A. and Varma, S.K. (1988). Effect of pre-sowing seed soaking in gibberellic acid on growth of wheat (*Triticum aestivum* L.) under different saline conditions. *Indian J. Biol. Sci.* 26: 473–475.

Parihar, P., Singh, S., Singh, R. et al. (2015). Effect of salinity stress on plants and its tolerance strategies: a review. *Environ. Sci. Pollut. Res.* 22: 4056–4075.

Park, E.J., Jeknic, Z., and Chen, T.H.H. (2006). Exogenous application of glycinebetaine increases chilling tolerance in tomato plants. *Plant Cell Physiol.* 47: 706–714.

Park, J., Kim, Y.S., Kim, S.G. et al. (2011). Integration of auxin and salt signals by the NAC transcription factor NTM2 during seed germination in *Arabidopsis*. *Plant Physiol.* 156: 537–549.

Pedranzani, H., Racagni, G., Alemano, S. et al. (2003). Salt tolerant tomato plants show increased levels of jasmonic acid. *Plant Growth Regul.* 41: 149–158.

Peres, A.L.G.L., Soares, J.S., Tavares, R.G. et al. (2019). Brassinosteroids, the sixth class of phytohormones: A molecular view from the discovery to hormonal interactions in plant development and stress adaptation. *Int. J. Mol. Sci.* 20: 331. https://doi.org/10.3390/ijms20020331.

Petrussa, E., Braidot, E., Zancani, M. et al. (2013). Plant flavonoids–biosynthesis, transport and involvement in stress responses. *Int. J. Mol. Sci.* 14: 14950–14973.

Plett, D.C. and Moller, I.S. (2010). Na^+ transport in glycophytic plants: what we know and would like to know. *Plant Cell Environ.* 33: 612–626.

Prakash, L. and Prathapasenan, G. (1990). NaCl and gibberellic acid induced changes in the content of auxin, the activity of cellulose and pectin lyase during leaf growth in rice (*Oryza sativa*). *Ann. Bot.* 365: 251–257.

Prakash, V., Singh, V.P., Tripathi, D.K. et al. (2019). Crosstalk between nitric oxide (NO) and abscisic acid (ABA) signalling molecules in higher plants. *Environ. Exp. Bot.* 161: 41–49.

Rani, A., Kumar Vats, S., Sharma, M., and Kumar, S. (2011). Catechin promotes growth of *Arabidopsis thaliana* with concomitant changes in vascular system, photosynthesis and hormone content. *Biol. Plant.* 55: 779.

Ribaut, J.M. and Pilet, P.E. (1991). Effect of water stress on growth, osmotic potential and abscisic acid content of maize roots. *Physiol. Plant.* 81: 156–162.

Ribaut, J.M. and Pilet, P.E. (1994). Water stress and indole-3-acetic acid content of maize roots. *Planta* 193: 502–507.

Samadi, S., Habibi, G., and Vaziri, A. (2019). Exogenous trehalose alleviates the inhibitory effects of salt stress in strawberry plants. *Acta Physiol. Plant.* 41: 112.

Shahbaz, M., Abid, A., Masood, A., and Waraich, E.A. (2017). Foliar-applied trehalose modulates growth, mineral nutrition, photosynthetic ability, and oxidative defense system of rice (*Oryza sativa* L.) under saline stress. *J. Plant Nutr.* 40: 584–599.

Shakirova, F.M., Sakhabutdinova, A.R., Bezrukova, M.V. et al. (2003). Changes in the hormonal status of wheat seedlings induced by salicylic acid and salinity. *Plant Sci.* 164: 317–322.

Sharma, A., Shahzad, B., Kumar, V. et al. (2019). Phytohormones regulate accumulation of osmolytes under abiotic stress. *Biomolecules* 9: 285. https://doi.org/10.3390/biom9070285.

Shi, H.Z. and Zhu, J.K. (2002). Regulation of expression of the vacuolar Na^+/H^+ antiporter gene *AtNHX1* by salt stress and abscisic acid. *Plant Mol. Biol.* 50: 543–550.

Shi, H., Ishiani, M., Kim, C., and Zhu, J.K. (2000). The Arabidopsis thaliana salt tolerance gene SOS 1 encodes a putative Na^+/H^+ antiporter. *Proc. Natl. Acad. Sci. U. S. A.* 97: 6896–6901.

Shi, Y., Wang, Y., Flowers, T.J., and Gong, H. (2013). Silicon decreases chloride transport in rice (*Oryza sativa* L.) in saline conditions. *J. Plant Physiol.* 170: 847–853.

Shrivastava, P. and Kumar, R. (2015). Soil salinity: a serious environmental issue and plant growth promoting bacteria as one of the tools for its alleviation. *Saudi J. Biol. Sci.* 22: 123–131.

Siddiqui, M.H., Khan, M.N., Mohammad, F., and Khan, M.M.A. (2008). Role of nitrogen and gibberellic acid (GA3) in the regulation of enzyme activities and in osmoprotectant accumulation in *Brassica juncea* L. under salt stress. *J. Agron. Crop Sci.* 194: 214–224.

da Silva, A.S., Suassuna, J.F., de Melo, A.S. et al. (2017). Salicylic acid as attenuator of drought stress on germination and initial development of sesame. *Rev. Bras. Eng. Agríc. Ambient.* 21: 156–162.

Singh, S. and Prasad, S.M. (2019). Management of chromium (VI) toxicity by calcium and sulfur in tomato and brinjal: implication of nitric oxide. *J. Hazard. Mater.* 373: 212–223.

Singh, S., Singh, A., Srivastava, P.K., and Prasad, S.M. (2018). Cadmium toxicity and its amelioration by kinetin in tomato seedlings vis-àvis ascorbate-glutathione cycle. *J. Photochem. Photobiol.* 178: 76–84.

Sobahan, M.A., Akter, N., Ohno, M. et al. (2012). Effects of exogenous proline and glycinebetaine on the salt tolerance of rice cultivars. *Biosci. Biotechnol. Biochem.* 76: 1568–1570.

Soundararajan, P., Manivannan, A., Park, Y.G. et al. (2015). Silicon alleviates salt stress by modulating antioxidant enzyme activities in *Dianthus caryophyllus* 'Tula'. *Hortic. Environ. Biotechnol.* 56: 233–239.

Sze, H., Li, X., and Palmgren, M.G. (1999). Energrization of plant cell membrane by H^+ pumping ATPases; regulation and biosynthesis. *Plant Cell* 11: 677–689.

Talat, A., Nawaz, K., Hussian, K. et al. (2013). Foliar application of proline for salt tolerance of two wheat (*Triticum aestivum* L.) cultivars. *World Appl. Sci. J.* 22: 547–554.

Tang, W. and Newton, R.J. (2005). Polyamines reduce salt-induced oxidative damage by increasing the activities of antioxidant enzymes and decreasing lipid peroxidation in Virginia pine. *Plant Growth Regul.* 46: 31–43.

Tao, JJ., Chen, HW., Ma, B., Zhang, WK., et al. (2015). The Role of ethylene in plants under salinity stress. *Front. Plant Sci.* 6: 1059. doi: 10.3389/fpls.2015.01059.

Tari, I., Csisaa1, J., Szalai, G. et al. (2002). Acclimation of tomato plants to salinity stress after a salicylic acid pre-treatmen. *Acta Biol. Szegediensis* 46: 55–56.

Terui, Y., Ohnuma, M., Hiraga, K. et al. (2005). Stabilization of nucleic acids by unusual polyamines produced by an extreme thermophile, *Thermus thermophilus*. *Biochem. J.* 388: 427–433.

Tian, H., Lv, B., Ding, T. et al. (2018). Auxin-BR interaction regulates plant growth and development. *Front. Plant Sci.* 8: 2256.

Tognetti, V.B., Bielach, A., and Hrtyan, M. (2017). Redox regulation at the site of primary growth: auxin, cytokinin and ROS crosstalk. *Plant Cell Environ.* 40: 2586–2605.

Tuna, A.L., Kaya, C., Dikilitas, M., and Higgs, D. (2008). The combined effects of gibberellic acid and salinity on some antioxidant enzyme activities, plant growth parameters and nutritional status in maize plants. *Environ. Exp. Bot.* 63: 1–9.

Vardhini, B.V. and Anjum, N.A. (2015). Brassinosteroids make plant life easier under abiotic stresses mainly by modulating major components of antioxidant defense system. *Front. Environ. Sci.* 2: 67.

Vázquez, M.N., Guerrero, Y.R., Noval, W.T.D.L. et al. (2019). Advances on exogenous applications of brassinosteroids and their analogs to enhance plant tolerance to salinity: a review. *AJCS* 13: 115–121.

Verma, S.R., Shivran, A.C., Bhanwaria, R., and Singh, M. (2014). Effect of vermicompost and sulphur on growth, yield and nutrient uptake of fenugreek (*Trigonella foenum-graecum* L.). *The Bioscan* 9: 667–670.

Verma, K.K., Liu, X.H., Wu, K.C. et al. (2019). The impact of silicon on photosynthetic and biochemical responses of sugarcane under diffrent soil moisture levels. *Silicon* 12: 1355–1367.

Vessey, J.K. (2003). Plant growth promoting rhizobacteria as biofertilizers. *Plant Soil* 255: 571.

Villa-Castorena, M., Ulery, A.L., Catalán-Valencia, E.A., and Remmenga, M.D. (2003). Salinity and nitrogen rate effects on the growth and yield of Chile pepper plants. *Soil Sci. Soc. Am. J.* 67: 1781–1789.

Vob, U., Bishopp, A., Farcot, E., and Bennett, M.J. (2014). Modeling hormonal response and development. *Trends Plant Sci.* 19: 311–319.

Wang, Jia., Song, Li., Gong, Xue., Jinfan, Xu., Minhui, Li. (2020). Functions of Jasmonic Acid in Plant Regulation and Response to Abiotic Stress. *Int. J. Mol. Sci.* 21:1446. doi:10.3390/ijms21041446.

Wang, Y., Mopper, S., and Hasentein, K.H. (2001). Effects of salinity on endogenous ABA, IAA, JA, and SA in *Iris hexagona*. *J. Chem. Ecol.* 27: 327–342.

Wang, J., Islam, F., Gill, A.R. et al. (2016). Salicylic acid mediates antioxidative defense system and ABA pathway related gene expression in *Orzya sativa* against quinclorac toxicity. *Ecotoxicol. Environ. Saf.* 133: 146–156.

Wani, S.H., Singh, N.B., Haribhushan, A., and Mir, J.I. (2013). Compatible solute engineering in plants for abiotic stress tolerance-role of glycine betaine. *Curr. Genomics* 14: 157–165.

Wasternack, C. and Hause, B. (2002). Jasmonates and octadecanoids: signals in plant stress responses and development. *Prog. Nucleic Acid Res. Mol. Biol.* 72: 165–221.

Werner, T., Motyka, V., Strnad, M., and Schmülling, T. (2001). Regulation of plant growth by cytokinin. *Proc. Natl. Acad. Sci. U. S. A.* 28: 10487–10492.

Wichern, F., Islam, M.R., Hemkemeyer, M. et al. (2020). Organic amendments alleviate salinity effects on soil microorganisms and mineralisation processes in aerobic and anaerobic paddy rice soils. *Front. Sustain. Food Syst.* 4: 30.

Wolbang, C.M., Chandler, P.M., Smith, J.J., and Ross, J.J. (2004). Auxin from the developing inflorescence is required for the biosynthesis of active gibberellins in barley stems. *Plant Physiol.* 134: 769–776.

Wu, X., He, J., Chen, J. et al. (2013). Alleviation of exogenous 6-benzyladenine on two genotypes of egg-plant (*Solanum melongena* Mill.) growth under salt stress. *Protoplasma* 251: 169–176.

Xiao, J., Jin, R., and Wagner, D. (2017). Developmental transitions: integrating environmental cues with hormonal signaling in the chromatin landscape in plants. *Genome Biol.* 18: 88.

Yamaguchi, S. and Kamiya, Y. (2000). Gibberellin biosynthesis, its regulation by endogenous and environmental signals. *Plant Cell Physiol.* 41: 251–257.

Yan, Z., Guo, S., Shu, S. et al. (2011). Effects of proline on photosynthesis, root reactive oxygen species (ROS) metabolism in two melon cultivars (*Cucumis melo* L.) under NaCl stress. *Afr. J. Biotechnol.* 10: 18381–18390.

Yang, L., Zhao, X., Zhu, H. et al. (2014). Exogenous trehalose largely alleviates ionic unbalance, ROS burst, and PCD occurrence induced by high salinity in *Arabidopsis* seedlings. *Front. Plant Sci.* 5: 570.

Yang, J., Duan, G., Li, C. et al. (2019a). The crosstalk between jasmonic acid and other plant hormone signaling highlight the involvement of jasmonic acid as a core component in plant response to biotic and abiotic stresses. *Front. Plant Sci.* 10: 1349.

Yang, Z., Wang, C., Xue, Y. et al. (2019b). Calcium-activated 14-3-3 proteins as a molecular switch in salt stress tolerance. *Nat. Commun.* 10: 1199.

Yeo, A.R., Flowers, S.A., Rao, G. et al. (1999). Silicon reduces sodium uptake in rice (*Oryza sativa* L.) in saline conditions and this is accounted for by a reduction in the transpirational bypass flow. *Plant Cell Environ.* 22: 559–565.

Yildirim, E., Ekinci, M., Turan, M. et al. (2015). Roles of glycinebetaine in mitigating deleterious effect of salt stress on lettuce (*Lactuca sativa* L.). *Arch. Agron. Soil Sci.* 61: 1673–1689.

Zahir, Z.A., Asghar, H.N., and Arshad, M. (2001). Cytokinin and its precursors for improving growth and yield of rice. *Soil Biol. Biochem.* 33: 405–408.

Zaki, N.M., Hassanein, M.S., Amal, G.A. et al. (2014). Foliar application of potassium to mitigate the adverse impact of salinity on some sugar beet varieties 2 effect on yield and quality. *Middle East J. Agric. Res.* 3: 448–460.

Zehr, J.P. (2011). Nitrogen fixation by marine cyanobacteria. *Trends Microbiol.* 19: 162–173.

Zeid, I.M. (2009). Trehalose as osmoprotectant for maize under salinityinduced stress. *Res. J. Agric. Biol. Sci.* 5: 613–622.

Zhu, Z., Wei, G., Li, J. et al. (2004). Silicon alleviates salt stress and increases antioxidant enzymes activity in leaves of salt-stressed cucumber (*Cucumis sativus* L.). *Int. J. Mol. Sci.* 167: 527–533.

Zhu, Y., Nomura, T., Xu, Y.O. et al. (2006). Elongated uppermost internode encodes a cytochrome P450 monooxygenase that epoxidizes gibberellins in a novel deactivation reaction in rice. *Plant Cell* 18: 442–456.

12

Crop Plants Develop Extracellular Signaling Products Against Salt Stress

Santwana Tiwari[1], Nidhi Verma[1], Shikha Singh[1], Shivam Gupta[2], Madhulika Singh[2], Pratibha Singh[1], Jitendra Pandey[2], and Sheo Mohan Prasad[1]

[1] *Ranjan Plant Physiology and Biochemistry Laboratory, Department of Botany, University of Allahabad, Allahabad, India*
[2] *Ganga River Ecology Research Laboratory, Environmental Sciences Division, Centre of Advanced study in Botany, Institute of Sciences, Banaras Hindu University, Varanasi, India*

12.1 Introduction

Plants nowadays regularly face abiotic and biotic stresses that make them sessile and harm their immune system but along with this they have intense defense strategies against various environmental stresses (Isah 2019). It involves wide varieties of extracellular products that govern signal cascade and serve as tools for combating stress constraints. These components are also called secondary metabolites. Secondary metabolites are valuable in prescription and are for the most part polyphenols, alkaloids, glycosides, terpenes, flavonoids, coumarins, tannins, and so forth (Uddin 2019; Bueno and Lopes 2020, Table 12.1). Plant signaling is a usual phenomenon that facilitates the transduction of external and internal signals into physiological reactions such as modification of activity of various enzymes, cytoskeleton structure, and gene expression (Pietrowska-Borek et al. 2020). In plants and animals, the extracellular signaling molecules are very important to govern the cellular metabolic processes, growth of tissues, protein synthesis, and their secretion and composition of intracellular and extracellular fluids. They regulate interaction and communication between unicellular organisms, and also they are mainly known as critical regulators of all the developmental and physiological processes in multicellular organisms (Table 12.1). Elaborating the putative structural and functional interactions of secondary metabolites contributes to the better understanding of their stress-protective role in plants as well as in cyanobacteria and can constitute a rationale for their extensive exploitation as pharmaceuticals, food additives, and cosmetic products. Salinity imposes limited plant productivity and cyanobacterial community by producing different toxic compounds. However, the

Physiology of Salt stress in Plants: Perception, Signalling, Omics and Tolerance Mechanism, First Edition.
Edited by Pratibha Singh, Madhulika Singh, Rajiv Kumar Singh, and Sheo Mohan Prasad.

Table 12.1 Detailed information about the role of several metabolites induced by salt/salinity stress condition in plant/cyanobacterial cell.

Sr. No.	Model organisms	Impact of salinity stress	Extracellular products	Functions	References
1.	*Salvinia auriculata*	Lower concentration of salt (NaCl) increased biomass and Chl *a* content while higher concentration decreased Chl b, Ca, K, Mg along with biomass and Chl *a* but proline content increased sharply at higher concentration of salt.	Proline	Act as an osmotic regulator and prevent the cell from dehydration.	**Gomes et al. (2017)**
2.	*Allium cepa* L.	The decrease in the amount of total phenolic content and flavonoids content noticed.	Flavonoids Phenolics	Act as ROS scavenger and reregulate photosynthesis.	**Amal and Mohamed (2008)**
3.	*Pistachio vera* L.	Decreased in growth was noticed due to depletion in photosynthetic performance.	Proline Sugar	Proline acts as an osmoprotectant and ROS scavenger, while sugar is considered as energy reservoir.	**Rahneshan et al. (2018)**
4.	*Vigna unguiculata* L.	Total lipid contents decreased in leaves and root regions, resulting in membrane damage.	Fatty acid	Maintained the cell membrane integrity and fluidity	**Guimarães et al. (2011)**
5.	*Carthamus tinctorius* L.	Decrease in fresh shoot weight, total lipid content and oil yield.	Fatty acid (palmitic, oleic and linoleic acid)	Maintained the membrane lipid unsaturation to maintain the membrane fluidity for proper functioning.	**Harrathi et al. (2012)**
6.	*Cucumis sativus* L.	Reduced growth of plant and membrane damage.	Total phenol Fatty acid	Alleviation of salt stress condition and retain the membrane function.	**Radhakrishnan and Lee (2013)**
7.	*Brasscia napus* var *Oleifera* Del.	Reduced germination at high salinity.	Total phenols	Acts as antioxidants	**Falcinelli et al. (2017)**

8.	*Zea mays*	Phytoalexins induction is root specific	Terpenoids	Acts as stress alleviator.	**Vaughan et al. (2015)**
9.	*Chlamydomonas mexicana, Scenedesmus obliquus*	Dry weight increased along with defensive metabolites	Fatty acid Lipids	Acts as an an osmoregulator.	**Salama et al. (2013)**
10.	*Microcoleus chthonoplastes*	Induction for several metabolites	MAA	MAA may play a role of an osmoprotectant.	**Gharib et al. (2020)**
11.	*Plectonema boryanum* *Lyngbya aestuarii* *Oscillatoria splendida* *Oscillatoria tenuis*	Varied growth and biomass concentration	Protein Carbohydrate Lipid	Enhanced tolerance capacity of cyanobacteria under salinity condition	**Shruthi and Rajashekhar (2014)**
12.	*Lessonia nigrescens* (brown algae)	None	Polysaccharides	Improved plant growth under salinity condition	**Zou et al. (2019)**
13.	*Synechococcus* sp. PCC 7942	Growth declined due to increase in ROS contents resulting in membrane damage.	Carbohydrate Fatty acid Lipid Proline Antioxidant enzymes (SOD, CAT, APX)	Acts as biofuel, natural resource of food and an element for cosmetic industries.	**Verma et al. (2019)**
14.	*Anabaena circinalis*	Growth declined in terms of absorbance and chlorophyll content.	Carbohydrate	Maintain the energy demand	**Hifney (2013)**

roles of these compounds are significantly nullified by the extracellular molecules produced by the plant/cyanobacteria itself and have different peculiar properties in different stress conditions (Table 12.1). During a stressful situation, the cellular level of reactive oxygen species (ROS) increased critically and antioxidants get misbalanced that cause various disorders in cell functioning and lead to ultimate cell death (Verma et al. 2020). To control this adverse situation, plants or other autotrophs secrete some defensive extracellular signaling products; which create signals to alert the cellular compartments to face this unexpected disaster. This type of cell communication involves six points: (i) synthesis of extracellular signals, (ii) release of these molecules by signaling cells, (iii) transport of these signals to the target cell, (iv) specific receptor proteins detected these signals, (v) cellular metabolic changes triggered by receptor-signal complex, and (vi) removal of those signals, which often terminates the cellular response. There are different types of signaling products, including proteins, which are membrane-anchored or secreted, peptides, small lipophilic molecules (e.g. steroid hormones, thyroxine), small hydrophilic molecules derived from amino acids (e.g. catecholamines), and gases. These signals can act at short range, long range, or both (Lodish et al. 2000). On the other and plant cell walls, dynamic and complex structures also secrete some defensive products such as wall polysaccharides, which come into three categories: pectins, hemicelluloses, and cellulose (Krzeslowska 2011). Liu et al. (2014) reported that the overexpression of the β-subunit of *the polygalacturonase*-1 enzyme leads to pectin degradation that ultimately makes rice plants sensitive to cold, drought, and salt stresses. Contradictory to this Xiong et al. (2015) revealed the involvement of H_2O_2 in increasing cell expansion and root diameter in rice plants by enhancing pectin accumulation. Induced expression of genes related to the synthesis of cell wall proteins, lignin, and cellulose was found in salt-stressed *Arabidopsis* and potato plants (Shafi et al. 2017, 2018). Moreover, polysaccharides have also very clear defensive strategies in which they protect from harmful abiotic factors (i.e. metal, salt, UV, heat, chilling stress, etc.) and prevent ROS accumulation inside the cell. In the same way, cyanobacteria and other algal organisms also secrete a defensive layer of exopolysaccharides (EPS), as Tiwari et al. (2019) noticed enhanced content of EPS in a paddy field cyanobacterium *Anabaena* PCC 7120 for the protection against aluminum stress. Singh (2014) noticed that field cyanobacteria secret several extracellular signaling products such as phytohormones, polysaccharides, peptides, vitamins, etc. and crop plants utilized these products to make themselves resistant against various abiotic and biotic stresses (Table 12.1). Salt stress is one of the major global problems that are massively responsible for a large number of barren lands. According to the report of Mustafa et al. (2019), about 831 million ha of land is affected by salt, which comprises ~7% of the total land in the whole world. Saline soil not only causes health hazard to crop plants by reducing their growth but also affects the soil micro flora, which finally results in the poor crop production. Under salinity stress, various extracellular signaling products were secreted by plants as well as soil micro flora (i.e. algae and cyanobacteria, etc.) that can make a possible way to cope with this situation. These extracellular signaling products can sense the environmental changes and work as primary defensive shield against adverse effects (Figures 12.1 and 12.2) (Franzoni et al. 2019).

So the prime focus of this chapter is to gather all the information about extracellular products secreted by crop plants as well as cyanobacteria with their working mechanisms under salt stress conditions and their role in economic values in agriculture and medicine.

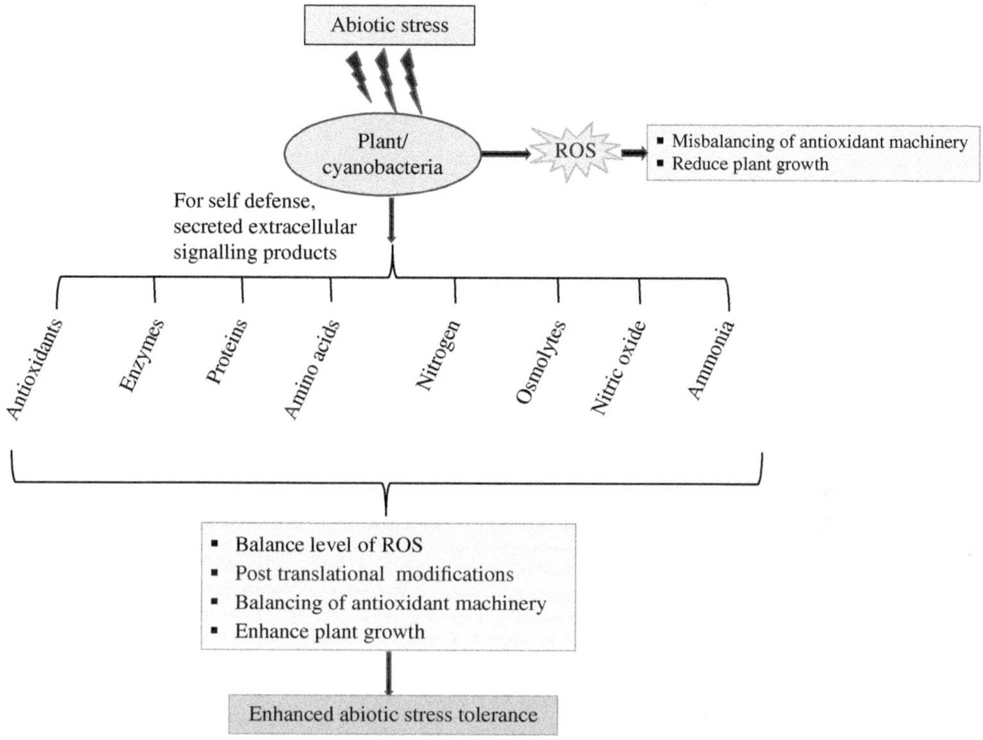

Figure 12.1 Schematic diagram representing the functional strategy of plant/cyanobacterial cell under abiotic (salinity) stress condition for enhancing the tolerance capacity.

12.2 Site of Synthesis of Extracellular Signaling Products

The biosynthesis of bioactive extracellular compounds in plants aimed to enhance crop tolerance to abiotic and biotic stresses and overcome stressful conditions. Environmental stress changes in the level of several extracellular products such as polysaccharides, amino acids, osmolytes, nitrogen compounds, proteins, phytols, sterols, terpenoid, fatty acids, phenolic compounds, phytohormones, etc (Figures 12.1 and 12.2). In these products polysaccharide is found as an active ingredient in plant cell wall preparation and represents up to 95% of cell wall mass (Krzeslowska 2011). There is a vast variety of polysaccharides that were secreted from different algal organisms, which are further utilized by plants to create stress resistance. These polysaccharides are laminarin (b-1,6-1,3 glucan) secreted from brown algae (*Laminaria digitata*), carrageenans (sulfated polysaccharides) secreted from red algae, ulvan secreted from green algae (*Ulva*) (Aziz et al. 2003; Jaulneau et al. 2010). Moreover, cyanobacterial polysaccharides are secreted as a mucilaginous external layer around the cell and it is organized either as a well-defined sheath or as a capsule or as a loosely bounded slime with the cell surface or sometimes as soluble polysaccharides, which are released into the external cell environment during cell growth (Pereira et al. 2009). These EPS have been noticed to play a significant role in providing cellular protection as a

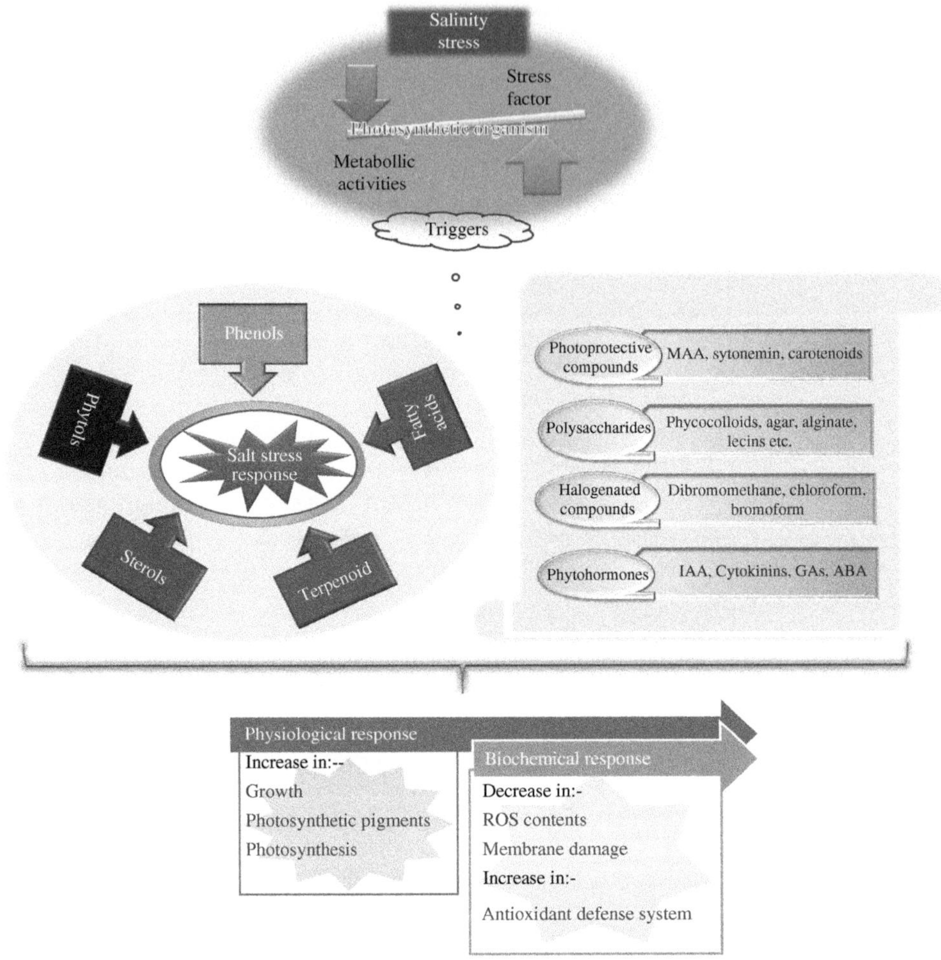

Figure 12.2 Diagrammatic representation of the role of several extracellular products secreted by plant/cyanobacterial cell under salinity stress condition.

defensive boundary layer (Tiwari et al. 2019). Galactose, fructose, xylose, and several unidentified sugars are the main constituents of these polysaccharides (Singh 2014).

Phenolic compounds such as lignin are the composition of three major types of monomers (monolignols), which are *p*-coumaryl, coniferyl, and sinapyl alcohols (Wang et al. 2013). These monolignols are first synthesized in cell cytoplasm, then translocated to the apoplast, and finally polymerized into lignin. Enzymes such as laccases and/or peroxidases are responsible for cross-linking of the monolignols in the wall (Wang et al. 2013). Biosynthesis of phenolic compounds in plants mainly occurs through pentose phosphate, shikimate, and phenylpropanoid pathways (Lattanzio 2013).

Amino acids are also known for their active triggering of plant defense responses. For the constitution of amino acids, a majority of peptides are supposed to be synthesized by NRPS (nonribosomal peptide synthetase) or NRPS/PKS (polyketide synthase) hybrid pathways, and to date, some 600 cyanobacterial peptides have been described (Welker and Döhren 2006; Kehr et al. 2011). Ribosomes mainly synthesized final protein structures by utilizing these units of amino acids. Proline further is an important flexible amino acid as well as a nonenzymatic antioxidant in determining for protein and membrane structures and scavenges ROS under and salinity stress (Singh et al. 2014). Proline not only acts as an osmotolerant, but also acts as a nutritional source.

Salinity stress also affects the nitrification process of the plants. Under this condition, other stable nitrogen forms in soil, such as ammonium, are also available to plants. Ammonia supplementation can reduce salinity stress more effectively than other nitrogen sources (Flores et al. 2001). In cyanobacteria, heterocyst cells are main fixers and producers of nitrogen compounds, which ultimately utilized plants (Chittora et al. 2020). One of the leading nitrogen derivatives, a signaling compound, nitric oxide (NO) is synthesized by oxidative and/or reductive pathways. Oxidative pathway nitric oxide synthase (NOS) enzyme or polyamines produce NO in the presence of L-arginine while in reductive pathway, NO is produced through nitrate reductase (NR), mitochondrial electron transport chain, or via nonenzymatic reactions. Moreover, the synthesis of several phytohormones such as salicylic acid, jasmonic acid, IAA, cytokinins, gibberellin, ethylene, and ABA is utilized at each developmental stage of the plant. These phytohormones also interact with other signaling molecules such as NO to make plants more tolerant against abiotic stresses (Verma et al. 2020).

12.3 Release of Extracellular Products by Cells of Cyanobacteria, Algae, and Crop Plants Under Salt Stress: Antioxidants, Enzymes, and Proteins

High salinity is a serious problem that can cause adverse effects on the growth of plants and the productivity of crops (Sytar et al. 2018; Singh et al. 2019). It can also affect the beneficial microflora present in crop fields. For example, some important nitrogen-fixing cyanobacteria such as *Anabaena* and *Nostoc* are found in rice crops that are also damaged due to excessive salinity (Singh and Dhar 2010). The level of antioxidants, enzymes, and proteins has fluctuated under salt stress (Figure 12.1).

12.3.1 Antioxidants

Antioxidants in any organism are considered as very important defense machinery, misbalance in this machinery can cause cell damage and death. Ascorbic acid (AsA) can actively quench the oxidative cascade and the consequent endogenously accumulated ROS and ultimately oxidized to monodehydroascorbic acid (MDHA) and dehydroascorbic acid (DHA) (Khan and Khan 2017). According to previous studies, it is considered that ROS plant stress signaling and AsA can maintain ROS homeostasis that shows the active participation of AsA in signaling process (Baxter et al. 2013; Martinez et al. 2018). Carotenoids are a large group of molecules that have the properties of both antioxidants and pigments; they are also considered as a precursor of an important plant hormone abscisic acid (Walter and Strack 2011; Nisar et al. 2015). β-Carotene is one of the main derivatives of carotenoid found accumulated due to ROS-induced oxidation. Similarly isoprene, a constituent unit of carotenoids, acts as signaling molecule (Havaux 2014). The previous study of Tiwari et al. (2019) noticed the enhanced level of nonenzymatic antioxidants such as proline, cysteine, and nonprotein thiols under exogenous application of SNP (sodium nitroprusside) to alleviate aluminum stress from cyanobacterium *Anabaena* PCC 7120.

12.3.2 Enzymes

The enzymatic antioxidants SOD (superoxide dismutase), POD (peroxidase), CAT (catalase), and GST (glutathione-*S*-transferase) are the key enzymes known for the removal of ROS mediated by abiotic stress plants as well as in algae and cyanobacteria. These enzymes can directly regulate oxidative stress or by interacting some other signaling molecules. A previous study of Hasanuzzaman et al. (2017) noticed that exogenously supplied H_2O_2 enhances the level of enzymatic antioxidants in *Brassica napus*. Likewise, Tiwari et al. (2019) found that NO enhanced the level of these enzymatic antioxidants to alleviate aluminum toxicity from the cells of *Anabaena* PCC 120. Similar behavior of NO was noticed in *Chlorella vulgaris* to alleviate herbicide stress (Qian et al. 2009).

12.3.3 Proteins

Shriram et al. (2016) noticed overexpression of SPL5, GA-Myb, and Glass III HD-Zip protein 4, which enhanced salt tolerance in *Saccharum* sp. On the other hand, Abbasi et al. (2007) have found about 98% reduction in total tocopherol accumulation under constitutively silenced activity of homogentisate phytyltransferase (HPT) and γ-tocopherol methyltransferase (γ-TMT) genes that ultimately decreased the capacity of salt tolerance in Tobacco plants. However, in the case of cyanobacteria, very few proteins are identified, which were secreted under abiotic stresses. However, some identified novel (extracellular) players involved in the mechanisms of plant resistance against various abiotic stresses are disclosed by previous studies. An outer-membrane protein (TolC) is a cytoplasmic-membrane ABC-type transporter and a periplasmic counterpart. It is a homotrimeric protein that can form a channel tunnel in the outer membrane to interact with different plasma-membrane protein complexes, which deliberate specificity with the substrate (Hahn et al. 2015; Oliveira et al. 2016).

12.4 Release of Extracellular Products by Cells of Cyanobacteria Algae and Crop Plants Under Salt Stress: Amino Acids, Osmolytes, Nitrogen Nitric Oxide Ammonia

Salt environment may be responsible for cellular dehydration, which causes osmotic stress by removing water from the cytoplasm resulting in reduction of the cytosolic and vacuolar volumes. Salt stress often generates both ionic and osmotic stress in plants, resulting in a greater disturbance in signaling and accumulation or decrease of specific metabolites in plants (Singh et al. 2017a). To acquire the salt stress tolerance, cellular release of bioactive molecules was noticed (Franzoni et al. 2019).

12.4.1 Amino Acids

Amino acids, the precursors, and constituents of proteins are the plant products that can act as regulatory and signaling molecules improving plant stress tolerance, and it also has an important role in plant nutrition, for example, glutamine synthetase is a key enzyme used for assimilation and transport of nitrogen across the cell membrane (Anjum et al. 2014; Planchet et al. 2015). In this queue, it is noticed that during abiotic stress, cytosolic pH generally decreases that correspondingly changes the level of amino acids. Serine, proline, and leucine are categorized as branched-chain amino acids (BCAA), which have been shown to have a direct activity as signaling molecules and some other amino acids are as the precursors for a variety of secondary metabolites or phytohormones with signaling function (Hildebrandt et al. 2015). Hausler et al. (2014) have noticed that some amino acids act as metabolic intermediates/end products, such as serine (both L-serine and D-serine) amino acids act as a metabolic signal in the communication between male gametophytes and pistil by regulating a glutamate receptor-like Ca^{2+} channel in pollen tubes. Similarly, very trending gamma-aminobutyric acid (GABA), a four-carbon nonprotein amino acid, also acts as a signal molecule, and its cellular accumulation was noticed under abiotic stress (Mekonnen et al. 2016). Studies of Karthikeyan et al. 2007, 2009 found that several amino acids, such as histidine, glycine, aspartic acid, arginine, cystine, and leucine, are secreted by some cyanobacterial strains such as *Anabaena azollae, Hapalosiphon sp., Calothrix sp., Westiellopsis prolifica,* and *Nostoc*; some of these amino acids are tested to enhance wheat seed germination.

12.4.2 Osmolytes

A range of organic solutes such as proline, glycine betaine, soluble sugars, polyamines, etc. are synthesized by plants, collectively known as osmolytes. Several studies reported the accumulation of osmolytes in plants including halophytes under salt stress by which it has been correlated that osmolytes can enhance the plant's capacity to tolerate salinity conditions (Mssedi et al. 2000; Slama et al. 2008). Proline accumulation was found to be much higher (about 300 μmol/g leaf dry matter) in *Sesuvium portulacastrum* in comparison to other plants (Slama et al. 2006). In addition to this, salinity stress also enhances the accumulation of glycine betaine (Lokhande et al. 2010), and this accumulation shows its positive role in the maintenance of membrane integrity and stability of other cellular structures under salt stress (Martinez et al. 2005).

12.4.3 Nitrogen and Its Derivatives

Nitrogen (N) is an essential macronutrient, required at every step of cellular metabolism, and its main role is synthesizing vital molecules, such as proteins, nucleic acids, and chlorophyll. It is well documented that uptake of nitrogen and its assimilation adversely affected under abiotic stresses. Previous studies also reported that the activity of NR enzyme was critically inhibited under salt stress (Goel and Singh 2015). Derivatives of nitrogen also play an important role in plant metabolisms such as nitric oxide (NO), which is a multitasking signaling molecule for plants as well as for algae and cyanobacteria. The previous study of Verma et al. (2020) revealed that under stressful environments, ROS mainly H_2O_2 mediated NO works for balancing of antioxidant defense system to protect the plant cells from damage. Tiwari et al. (2019) found that exogenous supplementation of sodium nitroprusside (SNP), as a donor of NO, alleviates aluminum stress from the cells of cyanobacterium *Anabaena* PCC 7120. Signaling mechanism of NO is not unidirectional; it can interact with other signaling molecules such as Ca^{2+}, H_2O_2, H_2S, and other phytohormones, and it is also necessary at every step of plant developmental processes (Verma et al. 2020). Another derivative of nitrogen, ammonia, has also a great potential to establish tolerance against abiotic stress. Hessini et al. (2009) noticed that the exogenous application of ammonia can activate the secretory glands in the leaf to expel out the excess salt from the cells of *Spartina alterniflora* and enhance the tolerance against salt.

12.5 Release of Extracellular Products by Cells of Cyanobacteria and Crop Plants Under Salt Stress: Phenols, Terpenoid, Phytols, Sterols, Fatty Acids

12.5.1 Phenols

Among several metabolites, polyphenols are group of several compounds such as phenolic acids, flavonoids, tannins, and lignins (Sun et al. 2020, Figure 12.2). As compared with non-stressed conditions, they are found profusely in plant tissues and are actively involved in defense mechanisms against biotic and abiotic stress (Viacava et al. 2020). Polyphenols act as effective antioxidants in vitro by free radical-scavenging activity as compared with ascorbate and tocopherols. Structurally, phenolic compounds possess one phenolic ring and show strong biological activities, when halogenated (Cabrita et al. 2010). Plants are subjected to produce these compounds but cyanobacteria are acknowledged for producing such metabolites under stress conditions for protecting the cells. In plant tissues, accumulation of phenolics is as an adaptive reaction for expanding evolutionary fitness resulted via activity of phenylalanine ammonia-lyase (PAL), chalcone synthase (CHS), and many more enzymes. Plant phenolics confer various physiological functions for survival and adaptation to environmental disturbances (Kabtni et al. 2020). Aras et al. (2020) reported the amelioration of salinity effect on onion plants by enhanced phenolic contents that could scavenge the oxygen free radicals and retain the normal metabolic function. A rise in the content of total phenols under salinity stress was observed by Falcinelli et al. (2017) in *Brassica napus* as an antioxidant. Several reports justified the occurrence of different phenolic compounds such as phlorotannins and phloroglucinols (act as antioxidants) (Gómez et al. 2016) in brown algae (Le Gall et al. 2015).

12.5.2 Terpenoid

Cyanobacteria and algae are considered as store house for terpenoids (Keeling and Bohlmann 2012; Pei et al. 2020, Figure 12.2). Terpenoids are not only essential for preliminary growth and development of plants but also facilitate pollination, and such features make them secondary biologically active compounds (Cai et al. 2020). Terpenoids have also been known for the supplementation of several products for their fragrances and flavors (Zhang and Hong 2020). They are also an essential tool for replacing biosynthetic diesel in the global market (Singh et al. 2017b). Very few studies have revealed the activation of terpenoids under salinity stress. In perennial evergreen sage (*Salvia officinalis*), leaf monoterpene contents were found to increase with increasing salinity; however, once the salinity reached a threshold level, monoterpene content began to decline (Taarit et al. 2009). Similar results were observed in marjoram (*Origanum majorana*) (Jelali et al. 2011). Monoterpenes have been suggested to act as anti-parasite (Bedoux et al. 2014).

12.5.3 Phytols

Phytols are vital constituents of chlorophyll molecules and photosynthesis as cofactors (Nowicka and Kruk 2010; Kiyota et al. 2014, Figure 12.2) and therefore, play a dynamic role in higher and lower plants. Along with phytols, carotenoids are also an essential part of pigment, quinone prenyl tails, hormones, and tocopherols. These components are known as antioxidants and play roles of ROS scavengers (Singh et al. 2017b; Tiwari et al. 2019; Nowruzi et al. 2020; Tiwari and Prasad 2020). In cyanobacteria, the inherent system of isoprenoids frequently contains phytol of photosynthetic pigments, and additionally, isoprenoids make plastoquinone and phylloquinone active via acting as cofactors for them (Kiyota et al. 2014; Singh et al. 2017a). Studies revealed that limonene (precursor of carotenoid and phytols) is derived by the activity of enzyme limonene synthase. Under abiotic stress, without any specific treatment, limonene may be secreted from the cell (Kiyota et al. 2014), which might help in enhancing its production for strength to photosynthetic organisms. Furthermore, it has been reported that *Nitzschia ovalis* and *Phaeodactylum tricornutum* synthesize phytol through the mevalonic acid pathway. In higher plants, isoprenoids are important component of photosynthetic machinery to cope with various abiotic stresses (Pulido et al. 2016).

12.5.4 Sterols

The significant role of sterols is in the maintenance of membrane fluidity (Scolaro et al. 2020, Figure 12.2). Sterols are produced during isoprenoid biosynthesis as a by-product. For the first time, Reitz and Hamilton (1968) reported the occurrence of sitosterol and cholesterol in *Anacystis nidulans* and *Fremyella diplosiphon*. It has been previously demonstrated that cyanobacteria produce sterols (saturated and unsaturated) (Kohlhase and Pohl 1988) and few of them are reported from *Anabaena cylindrical, Angraecum viguieri, Amanita solitaria, Nostoc carneum, Nodularia harveyana*, and *Microcystis aeruginosa* (Nishimura and Koyama 1977). In a recent study of Procházková et al. (2017), sterols with a concentration up to 2.25 mg/l in water have been reported in cyanobacterial blooms. Regarding abiotic stress condition, salt stress tolerance mechanism was developed in cashew seedlings by increased contents of free sterol in the root plasma membrane of CCPo6, which could favor the activation of plasma membrane H^{++}ATPase, as a regulator of cellular mechanism for ion omission from the shoot (Bybordi 2011). In the work of Chalbi et al. (2013), membrane fluidity in *Hordeum vulgare* was more increased than in *Hordeum maritimum* under salt stress condition.

12.5.5 Fatty Acids

Among different kinds of metabolites, fatty acids are also very much important due to their key role in metabolism (Figure 12.2). Some fatty acids (linolenic, linoleic, and arachidonic acids) are essential for growth of cyanobacteria and algae. Nowruzi et al. (2020) reported the diversity in fats, phospholipids, glycolipids, and waxes, which composite the lipids but in cyanobacteria, the structure of lipids may vary (Nowruzi et al. 2020). In plants, Guimarães et al. (2011) reported the increase in saturated and unsaturated fatty acid contents in leaves but only unsaturated fatty acid increment was noticed in roots under salinity stress reflecting its role in vital metabolic processes. Zhang et al. (2012) reported the role of fatty acid desaturase-6 (Fad6) (a chloroplast desaturase) in *Arabidopsis thaliana*, which developed the salt tolerance in early seedling development. An increase in palmitic, linolenic, and oleic acid content in the region of the root as well as the upper part of *Carthamus tinctorius* L. was noticed under saline condition for maintaining the membrane integrity (Harrathi et al. 2012). Similarly, Radhakrishnan and Lee (2013) also revealed the marked increment in fatty acid contents in *Cucumis sativus* L. to cope with salt stress. Unsaturated fatty acids protect the oxygen-evolving machinery in membranes against salt-induced impairment and speed up the retrieval of protein complex of the PSII from low-temperature photo inhibition, thus promoting photosynthetic machinery toward stress tolerance (Shu et al. 2015).

12.6 Release of Extracellular Products by Cells of Cyanobacteria and Crop Plants Under Salt Stress: Photoprotective Compounds, Polysaccharides, Halogenated Compounds, and Phytohormone

12.6.1 Photoprotective Compounds

Ultraviolet radiation (UVR) originates a wide variety of deleterious effects on living system. A number of biologically active compounds such as carotenoids, mycosporine-like amino acids (MAAs), and scytonemin have been isolated from cyanobacteria and algae (Figure 12.2) (Rastogi et al. 2015a, b; Joshi et al. 2018). The water-soluble, MAAs are thought to protect cyanobacteria from harmful UV radiation (Joshi et al. 2018). To date, more than 20 MAAs have been isolated and identified from different organisms. Significant and limited protection has been testified from various cyanobacteria with cytoplasmic MAAs. These compounds proficiently dissipate engrossed radiation in the form of heat without generating ROS (Singh et al. 2017b). In plants, flavonoids serve diverse functions in photo protection. They are known for scavenging the ROS species (SOR and H_2O_2) under stress conditions (Khanna-Chopra and Semwal 2020). Under UV stress condition, plants induce accumulation of flavonoids in leaf region and favor the growth by preventing the photosynthetic apparatus (Banerjee and Roychoudhury 2016). Under abiotic stress conditions, along with flavonoids, xanthophylls and glucosinolates are also reported as ROS scavenger and inhibitor of programmed cell death by inducing the antioxidant enzymatic activity and guiding phytohormone signaling (Banerjee and Roychoudhury 2016). Scytonemin is also a photo-protective compound that occurs in a dimer state (indolic and phenolic subunits).

For the first time, it was described as a yellowish-brown lipid-soluble pigment in some terrestrial cyanobacterial sp. and is positioned in the exopolysaccharides sheath (Rastogi et al. 2015a, b). Carotenoid pigments essentially occur in all living organisms. Two major forms of carotenoids, i.e. b-carotene and echinenone are of great significance. Additionally, astaxanthin, b-cryptoxanthin, etc. have a great consequence in cyanobacteria and algae and contribute to improving the antioxidant status in both plants and algae.

12.6.2 Polysaccharides and Halogenated Compounds

Plants and cyanobacteria possess some biochemical compounds attached to 40–50 different monosaccharides allied with some other constituents by glycosidic linkage and provide carbon and energy during ordinary/stressful conditions (Pereira et al. 2019, Figure 12.2). In cyanobacteria, extracellular polysaccharides (EPS) mainly act as storage particles and are found in cell envelopes. They exhibit the lowest diversity and are found as glycogen and starch. Lipopolysaccharides occur in the outer membrane and contribute 70–75% as part of the cyanobacterial cell envelope. Organisms' EPS collection determines the survival and growth rate of the organisms. Therefore, the cyanobacteria devote near about 70% of the total energy backup for assembly of EPS. Starch and glycogen are major storage products of photosynthesis in plants and bacteria, respectively (Kirsch et al. 2019; Pereira et al. 2019). To cope with stressful conditions, polysaccharides play a noteworthy role as antioxidants (Kılınç et al. 2013; Minicante et al. 2016). Few of them that are mainly secreted by cyanobacteria under normal as well as abnormal conditions are hydrocolloids, carrageenan, agar, alginate, lectins, etc. (Singh et al. 2017b). The extensive existence of halogenated complexes in cyanobacteria and algae may be categorized as acetogenins, phenols, terpenes, indoles, fatty acids, and unstable halogenated compounds (i.e. dibromomethane, chloroform, and bromoform) (Bondu et al. 2008; Singh et al. 2017b).

12.6.3 Phytohormone

It has been well known that different genera of cyanobacteria and algae expressively gather and release a varied collection of phytohormones, namely auxins, cytokinins (CKs), gibberellins (GA), and ethylene (ET) that contribute their role in growth and development of plants (Hussain and Hasnain 2011, Figure 12.2). Numerous plant growth regulators (PGRs) (amino acids, sugars, vitamins) have been isolated from different genera of cyanobacteria and algae that may up regulate the growth of the vascular plant (Singh et al. 2020). It is justified that auxins and CKs perform their role at very low concentrations among others phytohormones (Tiwari and Prasad 2020). In cyanobacteria, under salt stress conditions, ABA performs its role while in other group of algae, it acts as a stressor under varied conditions (Lu et al. 2014). ABA is well known as the signal molecules, which enable plants to survive under adverse environmental conditions such as salt and drought stresses (Raghavendra et al. 2010). A higher level of endogenous ABA for increasing salt stress tolerance in plants including *Oryza sativa* (Moons et al. 1995), *Brassica* (He and Cramer 1996), *Phaseolus vulgaris* (Cabot et al. 2009), and *Zea mays* (Cramer and Quarrie 2002) strongly correlated with growth defects, which might be due to rapid activation in ABA biosynthetic related genes (Chinnusamy et al. 2006). Several pieces of evidence indicated the involvement of auxin in counteraction of salinity stress in plants, but little information is available regarding mechanisms of salt stress regulation (Ribba et al. 2020; Tanveer et al. 2020). Higher auxin concentration by overexpression of auxin biosynthetic-related YUCCA3 caused hypersensitivity to salt stress. Similarly, the CKs are also known to lessen the opposing effects of salinity on plant growth (Tiwari et al. 2020). Seed priming with CKs was found to intensify the plant tolerance against salinity stress (Iqbal et al. 2006). Exogenous use of CKs enhanced salinity tolerance by augmented proline contents in egg plant. Another plant hormone gibberellic acid has been reported to relieve the negative effects of salinity strain on plant water relations and water use efficiency (Forghani et al. 2020; Moula et al. 2020). In a previous study on brassica, the use of GA in combination with nitrogen was supportive in improving salinity stress (Siddiqui et al. 2016). Another important function of GA was found to increase nitrogen and magnesium in leaves and roots under saline condition (Tuna et al. 2008). The positive influence of GA3 on salt-stressed mung bean seedlings was established due to its multiple functions (Mohammed 2007).

12.7 Uncovering Potential and Applications of Extracellular Signaling Products in Biology, Agriculture, and Medicine Current Status and Future Prospects

In view of the extensive application of plant and secondary metabolites of cyanobacteria, all of them are considered as a tool for medicines, agriculture, anti-irritant, antioxidants, and anti-inflammatory drugs, marketing components such as cosmetics, etc. such as photoprotective compounds, terpenoids, fatty acids, proteins, etc. are being used in several skincare products such as antiaging creams, regenerants, (Nowruzi et al. 2020). Some aquatic organisms such as *C. vulgaris* (green algae), *Dunaliella salina* (green algae), *Mastocarpus stellatus* (red algae), *Nannochloropsis oculata* (algae), and *Spirulina platensis* (blue-green algae) have highly gained a rank in cosmetic industries (Stolz and Obermayer 2005). With the dependency of people on safe and best sunscreens, cosmetic industries have exploited cyanobacterial MAAs and scytonemin, which can be used as effective natural UV protectors in these formulations. Moreover, it has been stated that keto carotenoid astaxanthin has a vibrant role in preventing pathological damages in human (Cardozo et al. 2007) and is recognized as an excellent and more influential antioxidant than any other such as vitamins C and E or other carotenoids, even though also conserving the important lipids and proteins found in human lymphocytes due to its superoxide dismutase and catalase enzyme activities (Vílchez et al. 2011). Variation in environmental conditions may cause heightening in ROS assembly, which may cause injury to the cells and cause oxidative stress. Concurrently, photosynthetic organisms have developed several strategies to avoid negative consequences of ROS. In this perspective, PUFAs (polyunsaturated fatty acids) play a crucial role in providing protection to the cell against oxidative damage (Kumar et al. 2012), and this is mostly due to a decline in PUFAs content by the action of antioxidants (i.e. catalase and superoxide dismutase). Detailed information on these phenomena has been reported by Mushtaq et al. (2020) under salt stress conditions.

In agriculture, cyanobacterial and plant's secondary metabolites are key tools for enhancing crop production. Cyanobacteria produce several cyanotoxins that prevent the plants against several microorganisms such as bacteria, fungi, few invertebrates, etc. (Elkomy 2020). Plant hormones produced by cyanobacteria directly benefitted the plant grown in their area and also protect them by inducing their defense system (Tiwari and Prasad 2020). Protein, amino acids, fatty acids, polysaccharides, and other metabolites also contribute their effort for the betterment of agriculture directly or indirectly; that is why they are also applied exogenously in the field under critical supervision.

Several metabolites are isolated from cyanobacteria as well as plants and are used for a medical purpose (Kumar et al. 2015; Mohotti et al. 2020). Many of the compounds possess the anti-inflammatory and anticancer properties and antibiotics isolated from cyanobacteria and algae (Nowruzi et al. 2020). Due to proficient medicinal properties, these organisms have a great demand in the market for developing essential compounds biotechnologically. Omega-3 fatty acids are PUFAs and crucial constituents for the growth of higher eukaryotes (Singh et al. 2017b), which are an essential component of human cell membranes. In the case of cardiovascular health, it is believed that regular consumption of omega-3 fatty acids reduces the risk of cardiac arrhythmia, hypertension, myocardial infarction, and thrombosis (Czumaj and Śledziński 2020). Carotenoids are very useful for eyes, and the two major carotenoids, i.e. lutein and zeaxanthin inhibit photo-oxidative damage to the human retina by accumulating in the macula of the retina (Leung et al. 2020). They also act as an anticancerous compound in the form of lycopene and therefore suggested for high dietary intake (Rajashri et al. 2020).

12.8 Current Status and Future Prospective

In this chapter, some of the most significant details regarding the production of extracellular signaling products, their role under normal as well as the saline condition, and their commercial use were reviewed. With respect to above-mentioned points, several studies have been pursued with highly appreciable and applicable findings; however, there is a skill gap between the knowledge whether these secondary metabolites may alter the stress-causing and defensive genes in both cyanobacteria and plants growing under the saline condition and if so, then what is the mechanism behind it. Regarding the values of extracellular signaling products produced mainly under stressful conditions, there is a huge scope for further research focused on their signaling pathways involved in the growth and developmental processes under saline conditions. Such information may result in the production of transgenic plants, which are more tolerant under salt stress condition.

References

Abbasi, A., Hajirezaei, M., Hofius, D. et al. (2007). Specific roles of α- and γ-tocopherol in abiotic stress responses of transgenic tobacco. *Plant Physiol.* 143: 1720–1738.

Amal, A. and Mohamed, A.A. (2008). Alterations of some secondary metabolites and enzymes activity by us in exogenous antioxidant compound in onion plants grown under sea water stress. *AEJSR* 3: 139–146.

Anjum, N.A., Gill, S.S., and Gill, R. (eds.) (2014). *Plant Adaptation to Environmental Change: Significance of Amino Acids and Their Derivatives*. Wallingford, UK: CABI.

Aras, S., Keles, H., and Eşitken, A. (2020). SNP mitigates malignant salt effects on apple plants. *Erwerbs-Obstbau* 62: 107–115.

Aziz, A., Poinssot, B., Daire, X. et al. (2003). Laminarin elicits defense responses in grapevine and induces protection against Botrytis cinerea and Plasmopara viticola. *Mol. Plant-Microbe Interact.* 16: 1118–1128.

Banerjee, A. and Roychoudhury, A. (2016). Plant responses to light stress: oxidative damages, photoprotection, and role of phytohormones. In: *Plant Hormones Under Challenging Environmental Factors* (eds. G.J. Ahammed and J.Q. Yu), 181–213. Dordrecht: Springer.

Baxter, A., Mittler, R., and Suzuki, N. (2013). ROS as key players in plant stress signalling. *J. Exp. Bot.* 65: 1229–1240.

Bedoux, G., Hardouin, K., Burlot, A.S., and Bourgougnon, N. (2014). Bioactive components from seaweeds: cosmetic applications and future development. In: *Advances in Botanical Research, Sea Plants Book*, vol. 71 (ed. N. Bourgougnon), 346–367. Amsterdam: Elsevier.

Bondu, S., Cocquempot, B., Deslandes, E., and Morin, P. (2008). Effects of salt and light stress on the release of volatile halogenated organic compounds by *Solieria chordalis*: a laboratory incubation study. *Bot. Mar.* 51: 485–492.

Bueno, P.C. and Lopes, N.P. (2020). Metabolomics to characterize adaptive and signaling responses in legume crops under abiotic stresses. *ACS omega* 5: 1752–1763.

Bybordi, A. (2011). Effects of NaCl salinity levels on lipids and proteins of canola (*Brassica napus* L.) cultivars. *Rom. Agric. Res.* 28: 197–206.

Cabot, C., Sibole, J.V., Barceló, J., and Poschenrieder, C. (2009). Abscisic acid decreases leaf Na^+ exclusion in salt-treated *Phaseolus vulgaris* L. *J. Plant Growth Regul.* 28: 187–192.

Cabrita, M.T., Vale, C., and Rauter, A.P. (2010). Halogenated compounds from marine algae. *Mar. Drugs* 8: 2301–2317.

Cai, Z., Ouyang, F., Su, J. et al. (2020). Attraction of adult *Harmonia axyridis* to volatiles of the insectary plant *Cnidium monnieri*. *Biol. Control* 143: 104189. https://doi.org/10.1016/j.biocontrol.2020.104189.

Cardozo, K.H., Guaratini, T., Barros, M.P. et al. (2007). Metabolites from algae with economical impact. *Comp. Biochem. Physiol. C* 146: 60–78.

Chalbi, N., Hessini, K., Gandour, M. et al. (2013). Are changes in membrane lipids and fatty acid composition related to salt stress resistance in wild and cultivated barley? *Soil Sci. Plant Nutr.* 176: 138–147.

Chinnusamy, V., Zhu, J. and Zhu, J.K. (2006). Salt stress signaling and mechanisms of plant salt tolerance. *Genet. Eng.* (NY) 27: 141–177.

Chittora, D., Meena, M., Barupal, T., and Swapnil, P. (2020). Cyanobacteria as a source of biofertilizers for sustainable agriculture. *Biochem. Biophys. Rep.* 22: 100737. https://doi.org/10.1016/j.bbrep.2020.100737.

Cramer, G.R. and Quarrie, S.A. (2002). Abscisic acid is correlated with the leaf growth inhibition of four genotypes of maize differing in their response to salinity. *Funct. Plant Biol.* 29: 111–115.

Czumaj, A. and Śledziński, T. (2020). Biological role of unsaturated fatty acid desaturases in health and disease. *Nutrients* 12: 356.

Elkomy, R.G. (2020). Screening for some marine cyanobacteria isolated from Red Sea Coast, Egypt producing antimicrobial activity. *Eur. Asia J. Bio. Sci.* 14: 11–19.

Falcinelli, B., Sileoni, V., Marconi, O. et al. (2017). Germination under moderate salinity increases phenolic content and antioxidant activity in rapeseed (Brassica napus var oleifera Del.) sprouts. *Molecules* 22: 1377.

Flores, P., Carvajal, M., Cerda, A., and Martines, V. (2001). Salinity and ammonium/nitrate interactions on tomato plant development, nutrition, and metabolites. *J. Plant Nutr.* 24: 1561–1573.

Forghani, A.H., Almodares, A., and Ehsanpour, A.A. (2020). The role of gibberellic acid and paclobutrazol on oxidative stress responses induced by in vitro salt stress in *Sweet Sorghum*. *Russ. J. Plant Physiol.* 67: 555–563.

Franzoni, G., Trivellini, A., Bulgari, R. et al. (2019). Bioactive molecules as regulatory signals in plant responses to abiotic stresses. In: *Plant Signaling Molecules: Role and Regulation Under Stressful Environment* (eds. M.I.R. Khan, P.S. Reddy, A. Ferrante and N.A. Khan), 169–182. Elsevier, Woodhead Publishing.

Gharib, R., Tabarzad, M., and Hosseinabadi, T. (2020). Effect of high salinity on mycosporine-like amino acid production in *Desmodesmus* sp. *Trends Peptide Protein Sci.* 5: 1–6.

Goel, P. and Singh, A.K. (2015). Abiotic stresses downregulate key genes involved in nitrogen uptake and assimilation in *Brassica juncea* L. *PLoS One* 10 (11): 0143645.

Gomes, M.A.C., Pestana, I.A., Santa-Catarina, C. et al. (2017). Salinity effects on photosynthetic pigments, proline, biomass and nitric oxide in *Salvinia auriculata* Aubl. *Acta Limnol. Brasilien.* 29: e9. https://doi.org/10.1590/S2179-975X4716.

Gómez, I., Español, S., Véliz, K., and Huovinen, P. (2016). Spatial distribution of phlorotannins and its relationship with photosynthetic UV tolerance and allocation of storage carbohydrates in blades of the kelp *Lessonia spicata*. *Mar. Biol.* 163: 110.

Guimarães, F.V.A., de Lacerda, C.F., Marques, E.C. et al. (2011). Calcium can moderate changes on membrane structure and lipid composition in cowpea plants under salt stress. *Plant Growth Regul.* 65: 55–63.

Hahn, A., Stevanovic, M., Brouwer, E. et al. (2015). Secretome analysis of *Anabaena* sp. PCC 7120 and the involvement of the TolC-homologue HgdD inprotein secretion. *Environ. Microbiol.* 17: 767–780.

Harrathi, J., Hosni, K., Karray-Bouraoui, N. et al. (2012). Effect of salt stress on growth, fatty acids and essential oils in safflower (*Carthamus tinctorius* L.). *Acta Physiol. Plant.* 34: 129–137.

Hasanuzzaman, M., Nahar, K., Gill, S.S. et al. (2017). Hydrogen peroxide pretreatment mitigates cadmium-induced oxidative stress in *Brassica napus* L.: an intrinsic study on antioxidant defense and glyoxalase systems. *Front. Plant Sci.* 8: 115.

Hausler, R.E., Ludewig, F., and Krueger, S. (2014). Amino acids a life between metabolism and signaling. *Plant Sci.* 229: 225–237.

Havaux, M. (2014). Carotenoid oxidation products as stress signals in plants. *Plant J.* 79: 597–606.

He, T. and Cramer, G.R. (1996). Abscisic acid concentrations are correlated with leaf area reductions in two salt-stressed rapid-cycling Brassica species. *Plant Soil* 179: 25–33.

Hessini, K., Gandour, M., Megdich, W. et al. (2009) How does ammonium nutrition influence salt tolerance in *Spartina alternifl ora* loisel? In: M. Ashraf, M. Ozturk, and H.R. Athar (Eds) Salinity and Water Stress, Springer Science, Business Media B.V. pp. 91–96.

Hifney, A.F. (2013). Improvement of growth and some metabolites of the salt affected Anabaena circinalis by calcium. *J. Biol. Earth Sci.* 3: 120–128.

Hildebrandt, T.M., Nunes, N.A., Araujo, W.L., and Braun, H.P. (2015). Amino acid catabolism in plants. *Mol. Plant* 8: 1563–1579.

Hussain, A. and Hasnain, S. (2011). Comparative assessment of the efficacy of bacterial and cyanobacterial phytohormones in plant tissue culture. *World J. Microbiol. Biotechnol.* 28: 1459–1466.

Iqbal, M., Ashraf, M., and Jamil, A. (2006). Seed enhancement with cytokinins: changes in growth and grain yield in salt stressed wheat plants. *Plant Growth Regul.* 50: 29–39.

Isah, T. (2019). Stress and defense responses in plant secondary metabolites production. *Biol. Res.* 52: 39.

Jaulneau, V., Lafitte, C., Jacquet, C. et al. (2010). Ulvan a sulfated polysaccharide from green algae, activates plant immunity through the jasmonic acid signaling pathway. *J. Biomed. Biotechnol.* 2010, 525291: 1–11.

Jelali, N., Dhifi, W., Chahed, T., and Marzouk, B. (2011). Salinity effects on growth, essential oil yield and composition and phenolic compounds content of marjoram (*Origanum majorana* l.) leaves. *J. Food Biochem.* 35: 1443–1450.

Joshi, D., Mohandass, C., and Dhale, M. (2018). Effect of UV-B radiation and desiccation stress on photoprotective compounds accumulation in marine *Leptolyngbya* sp. *Appl. Biochem. Biotechnol.* 184: 35–47.

Kabtni, S., Sdouga, D., Rebey, I.B. et al. (2020). Influence of climate variation on phenolic composition and antioxidant capacity of *Medicago minima* populations. *Sci. Rep.* 10: 1–15.

Karthikeyan, N., Prasanna, R., and Lata-Kaushik, B.D. (2007). Evaluating the potential of plant growth promoting cyanobacteria as inoculants for wheat. *Eur. J. Soil Biol.* 43: 23–30.

Karthikeyan, N., Prasanna, R., and Sood, A. (2009). Physiological characterization and electron microscopic investigations of cyanobacteria associated with wheat rhizosphere. *Folia Microbiol.* 54: 43–51.

Kehr, J., Picchi, D.G., and Dittmann, E. (2011). Natural product biosyntheses in cyanobacteria: a treasure trove of unique enzymes. *Beilstein J. Org. Chem.* 7: 1622–1635.

Khan, M.I.R. and Khan, N. (2017). Reactive oxygen species and antioxidant system in plants: role and regulation under abiotic stress. *Springer Nature* 10: 978–981.

Khanna-Chopra, R., and Semwal, V. K. (2020). Ecophysiology and response of plants under high temperature stress. In: Hasanuzzaman M (Ed) *Plant Ecophysiology and Adaptation under Climate Change: Mechanisms and Perspectives I*. Springer, Singapore. pp. 295–329.

Kirsch, F., Klähn, S., and Hagemann, M. (2019). Salt-regulated accumulation of the compatible solutes sucrose and glucosylglycerol in cyanobacteria and its biotechnological potential. *Front. Microbiol.* 10: 2139.

Kiyota, H., Okuda, Y., Ito, M. et al. (2014). Engineering of cyanobacteria for the photosynthetic production of limonene from CO_2. *J. Biotechnol.* 185: 1–7.

Kohlhase, M. and Pohl, P. (1988). Saturated and unsaturated sterols of nitrogen fixing blue-green algae (cyanobacteria). *Phytochemistry* 27: 1735–1740.

Krzeslowska, M. (2011). The wall cell in plant cell response to trace metals: polysaccharide remodeling and its role in defense strategy. *Acta Physiol. Plant.* 33: 35–51.

Kumar, S., Hahn, F.M., Baidoo, E. et al. (2012). Remodeling the isoprenoid pathway in tobacco by expressing the cytoplasmic mevalonate pathway in chloroplasts. *Metab. Eng.* 14: 19–28.

Kumar, J., Singh, V.P., and Prasad, S.M. (2015). NaCl-induced physiological and biochemical changes in two cyanobacteria *Nostoc muscorum* and *Phormidium foveolarum* acclimatized to different photosynthetically active radiation. *J. Photochem. Photobiol. Biol.* 151: 221–232.

Lattanzio, V. (2013). Phenolic compounds: introduction. In: *Natural Products* (eds. K.G. Ramawat and J.M. Merillon). Berlin, Heidelberg: Springer-Verlag https://doi.org/10.1007/978-3-642-22144-6-57.

Le Gall, H., Philippe, F., Domon, J.M. et al. (2015). Cell wall metabolism in response to abiotic stress. *Plants* 4: 112–166.

Leung, H.H., Galano, J.M., Crauste, C. et al. (2020). Combination of lutein and zeaxanthin, and DHA regulated polyunsaturated fatty acid oxidation in H 2 O 2-stressed retinal cells. *Neurochem. Res.* 22: 1–3.

Liu, H., Ma, Y., Chen, N. et al. (2014). Overexpression of stress-inducible OsBURP16, β subunit of polygalacturonase 1, decreases pectin content and increase abiotic stress sensitivity in rice. *Plant Cell Environ.* 37: 1144–1158.

Lodish, H., Berk, A., and Zipursky, S.L. (2000). *Molecular Cell Biology*, 4e. New York: W. H. Freeman; Section 20.1, Overview of Extracellular Signaling. Available from: https://www.ncbi.nlm.nih.gov/books/NBK21517.

Lokhande, V.H., Nikam, T.D., and Suprasanna, P. (2010). Biochemical, physiological and growth changes in response to salinity in callus cultures of *Sesuvium portulacastrum* L. *Plant Cell Tissue Organ Cult.* 102: 17–25.

Lu, Y., Tarkowská, D., Turečková, V. et al. (2014). Antagonistic roles of abscisic acid and cytokinin during response to nitrogen depletion in oleaginous microalga *N annochloropsis oceanica* expand the evolutionary breadth of phytohormone function. *Plant J.* 80: 52–68.

Martinez, J.P., Kinet, J.M., Bajji, M., and Lutts, S. (2005). NaCl alleviates polyethylene glycol-induced water stress in the halophyte species *Atriplex halimus* L. *J. Exp. Bot.* 56: 2421–2431.

Martinez, V., Nieves-Cordones, M., Lopez-Delacalle, M. et al. (2018). Tolerance to stress combination in tomato plants: new insights in the protective role of melatonin. *Molecules* 23: 535.

Mekonnen, D.W., Flugge, U.I., and Ludewig, F. (2016). Gamma amino butyric acid depletion affects stomata closure and drought tolerance of *Arabidopsis thaliana*. *Plant Sci.* 245: 25–34.

Minicante, S.A., Michelet, S., Bruno, F. et al. (2016). Bioactivity of phycocolloids against the Mediterranean protozoan *Leishmania infantum*: an inceptive study. *Sustainability* 8: 1131.

Mohammed, A.H.M.A. (2007). Physiological aspects of mung bean plant (Vigna radiata L. Wilczek) in response to salt stress and gibberellic acid treatment. *Res. J. Agric. Biol. Sci.* 3: 200–213.

Mohotti, S., Rajendran, S., Muhammad, T. et al. (2020). Screening for bioactive secondary metabolites in Sri Lankan medicinal plants by micro fractionation and targeted isolation of antimicrobial flavonoids from *Derris scandens*. *J. Ethnopharmacol.* 246: 112158.

Moons, A., Bauw, G., Prinsen, E. et al. (1995). Molecular and physiological responses to abscisic acid and salts in roots of salt-sensitive and salt-tolerant Indica rice varieties. *Plant Physiol.* 107: 177–186.

Moula, I., Boussadia, O., Koubouris, G. et al. (2020). Ecophysiological and biochemical aspects of olive tree (*Olea europaea* L.) in response to salt stress and gibberellic acid-induced alleviation. *South Afric. J. Bot.* 132: 38–44.

Mssedi, D., Sleimi, N., and Abdelly, C. (2000). Some plants: the origin of diacylglycerol moiety. Arch. Biophysiological and biochemical aspects of salt tolerchem. *Biophysics* 240: 851–858.

Mushtaq, Z., Faizan, S., and Gulzar, B. (2020). Salt stress, its impacts on plants and the strategies plants are employing against it: a review. *J. Appl. Biol. Biotechnol.* 8: 81–91.

Mustafa, G., Akhtar, M.S., and Abdullah, R. (2019). Global concern for salinity on various agro–ecosystems. In: *Salt Stress, Microbes, and Plant Interactions: Causes and Solution* (ed. M.S. Akhtar), 1–20. Singapore: Springer.

Nisar, N., Li, L., Lu, S. et al. (2015). Carotenoid metabolism in plants. *Mol. Plant* 8: 68–82.

Nishimura, M. and Koyama, T. (1977). The occurrence of stanols in various living organisms and the behavior of sterols in contemporary sediments. *Geochim. Cosmochim. Acta* 41: 379–385.

Nowicka, B. and Kruk, J. (2010). Occurrence, biosynthesis and function of isoprenoid quinones. *Biochim. Biophys. Acta* 1797: 1587–1605.

Nowruzi, B., Sarvari, G., and Blanco, S. (2020). The cosmetic application of cyanobacterial secondary metabolites. *Algal Res.* 49: 101959.

Oliveira, P., Martins, N.M., Santos, M. et al. (2016). The versatile TolC-like Slr1270 in the cyanobacterium *Synechocystis* sp. PCC6803. *Environ. Microbiol.* 18: 486–502.

Pei, Y., Xu, R., Hilt, S. and Chang, X. (2020). Effects of cyanobacterial secondary metabolites on phytoplankton community succession. In: Mérillon, Jean-Michel, Ramawat and Gopal K (Eds) *Co-Evolution of Secondary Metabolites.* Springer Nature Switzerland AG. pp. 323–344.

Pereira, S., Zille, A., Micheletti, E. et al. (2009). Complexity of cyanobacterial exopolysaccharides: composition, structures, inducing factors and putative genes involved in their biosynthesis and assembly. *FEMS Microbiol. Rev.* 33: 917–941.

Pereira, S.B., Sousa, A., Santos, M. et al. (2019). Strategies to obtain designer polymers based on cyanobacterial extracellular polymeric substances (EPS). *Int. J. Mol. Sci.* 20: 5693.

Pietrowska-Borek, M., Wojdyła-Mamon, A., Dobrogojski, J. et al. (2020). Purine and pyrimidine dinucleoside polyphosphates dierentially act the phenylpropanoid pathway in Vitis vinifera L. cv. Monastrell suspension cultured cells. *Plant Physiol. Biochem.* 147: 125–132.

Planchet, E., Limami, A.M. and De-Mello, J.P. (2015). Amino acid synthesis under abiotic stress. In: D′Mello. J. P. F. (Ed) *Amino Acids in Higher Plants.* CAB International, Wallingford, CT, 5. pp. 262–276.

Procházková, T., Sychrová, E., Javůrková, B. et al. (2017). Phytoestrogens and sterols in waters with cyanobacterial blooms - analytical methods and estrogenic potencies. *Chemosphere* 170: 104–112.

Pulido, P., Llamas, E., Llorente, B. et al. (2016). Specific Hsp100 chaperones determine the fate of the first enzyme of the plastidial isoprenoid pathway for either refolding or degradation by the stromal Clp protease in *Arabidopsis. PLoS Genet.* 12 (1): e1005824.

Qian, H., Chen, W., Li, J. et al. (2009). The effect of exogenous nitric oxide on alleviating herbicide damage in *Chlorella vulgaris. Aquat. Toxicol.* 92: 250–257.

Radhakrishnan, R. and Lee, I.J. (2013). Regulation of salicylic acid, jasmonic acid and fatty acids in cucumber (*Cucumis sativus* L.) by spermidine promotes plant growth against salt stress. *Acta Physiol. Plant.* 35: 3315–3322.

Raghavendra, A.S., Gonugunta, V.K., Christmann, A., and Grill, E. (2010). ABA perception and signalling. *Trends Plant Sci.* 15: 395–401.

Rahneshan, Z., Nasibi, F., and Moghadam, A.A. (2018). Effects of salinity stress on some growth, physiological, biochemical parameters and nutrients in two pistachio (*Pistacia vera* L.) rootstocks. *J. Plant Interact.* 13: 73–82.

Rajashri, R., Neralla, M., George, R. et al. (2020). Carica papaya leaf extracts: a novel herbal drug in cancer treatment. *Int. J. Res. Pharm. Sci.* 11: 1590–1594.

Rastogi, R.P., Madamwar, D., and Incharoensakdi, A. (2015a). Sun-screening bioactive compounds mycosporine-like amino acids in naturally occurring cyanobacterial biofilms: role in photoprotection. *J. Appl. Microbiol.* 119: 753–762.

Rastogi, R.P., Sonani, R.R., and Madamwar, D. (2015b). Cyanobacterial sunscreen scytonemin: role in photoprotection and biomedical research. *Appl. Biochem. Biotechnol.* 176: 1551–1563.

Reitz, R.C. and Hamilton, J.G. (1968). The isolation and identification of two sterols from two species of blue-green algae. *Comp. Biochem. Physiol.* 25: 401–416.

Ribba, T., Garrido, F., and Brien, J.A. (2020). Auxin-mediated responses under salt stress: from developmental regulation to biotechnological applications. *J. Exp. Bot.*

Salama, E.S., Kim, H.C., Abou-Shanab, R.A. et al. (2013). Biomass, lipid content, and fatty acid composition of freshwater *Chlamydomonas mexicana* and *Scenedesmus obliquus* grown under salt stress. *Bioprocess Biosyst. Eng.* 36: 827–833.

Scolaro, B., de Andrade, L.F. and Castro, I.A. (2020). Cardiovascular disease prevention: the earlier the better? A review of plant sterol metabolism and implications of childhood supplementation. *Int. J. Mol. Sci.* 21: 128.

Shafi, A., Pal, A.K., Sharma, V. et al. (2017). Transgenic potato plants overexpressing sod and apx exhibit enhanced lignification and starch biosynthesis with improved salt stress tolerance. *Plant Mol. Biol. Report.* 35: 504–518.

Shafi, A., Gill, T., Zahoor, I. et al. (2018). Ectopic expression of SOD and APX genes in Arabidopsis alters metabolic pools and genes related to secondary cell wall cellulose biosynthesis and improve salt tolerance. *Mol. Biol. Rep.* https://doi.org/10.1007/s11033-019-04648.

Shriram, V., Kumar, V., Devarumath, R.M. et al. (2016). MicroRNAs as potential targets for abiotic stress tolerance in plants. *Front. Plant Sci.* 7: 817.

Shruthi, M.S. and Rajashekhar, M. (2014). Effect of salinity and pH on the growth and biomass production in the four species of estuarine cyanobacteria. *J. Algal Biomass Utln* 5: 29–36.

Shu, S., Yuan, Y., Chen, J. et al. (2015). The role of putrescine in the regulation of proteins and fatty acids of thylakoid membranes under salt stress. *Sci. Rep.* 5: 14390.

Siddiqui, M.A., Shah, Z.H., Tunio, S., and Chacchar, Q. (2016). Effect of different nitrogen (N) phosphorus (P) fertilizer and plant growth regulators gibberellic acid (GA3) and indole-3-acetic acid (IAA) on qualitative traits of Canola (*Brassica napus* L.) genotypes. *Int. J. Pure Appl. Biosci.* 4: 238–244.

Singh, S. (2014). A review on possible elicitor molecules of cyanobacteria: their role in improving plant growth and providing tolerance against biotic or abiotic stress. *J. Appl. Microbiol.* 117: 1221–1244.

Singh, N.K. and Dhar, D.W. (2010). Cyanobacterial reclamation of salt-affected soil.In: Lichtfouse (Ed) *Genetic Engineering, Biofertilisation, Soil Quality and Organic Farming Sustainable Agriculture Reviews*. Springer, Netherlands. 4. pp. 243–275.

Singh, M., Singh, J.K., Singh, V.P., and Prasad, S.M. (2014). Proline and salinity tolerance in plants. *Biochem Phormacol.* 3: 6.

Singh, M., Singh, A., Singh, R.K., and Prasad, S.M. (2017a). Regulation of plants metabolism in response to salt stress: an omics approach. *Acta Physiol. Plant.* 39: 48.

Singh, R., Parihar, P., Singh, M. et al. (2017b). Uncovering potential applications of cyanobacteria and algal metabolites in biology, agriculture and medicine: current status and future prospects. *Front. Microbiol.* 8: 515.

Singh, M., Singh, V.P., and Prasad, S.M. (2019). Nitrogen alleviates salinity toxicity in *Solanum Lycopersicum* seedling by regulating ROS homeostasis. *Plant Physiol. Biochem.* 141: 466–476.

Singh, S.K., Kaur, R., Bansal, A., et al. (2020). Biotechnological exploitation of cyanobacteria and microalgae for bioactive compounds. In: Verma M., Chandel A. (Ed.) *Biotechnological Production of Bioactive Compounds*. Elsevier. pp. 221–259.

Slama, I., Messedi, D., Ghnaya, T. et al. (2006). Effects of water-deficit on growth and proline metabolism in *Sesuvium portulacastrum*. *Environ. Exp. Bot.* 56: 231–238.

Slama, I., Ghnaya, T., Savouŕe, A., and Abdelly, C. (2008). Combined effects of long-term salinity and soil drying on growth, water relations, nutrient status and proline accumulation of *Sesuvium portulacastrum*. *C. R. Biol.* 331: 442–451.

Stolz, P. and Obermayer, B. (2005). Manufacturing microalgae for skin care. *Cosmet. Toiletries* 120: 99–106.

Sun, C., Zhao, C., Guven, E.C. et al. (2020). Dietary polyphenols as antidiabetic agents: advances and opportunities. *Food Front.* 1: 18–44.

Sytar, O., Mbarki, S., Zivcak, M., and Brestic, M. (2018). The involvement of different secondary metabolites in salinity tolerance of crops. In: *Salinity Responses and Tolerance in Plants* (eds. V. Kumar, S.H. Wani, P. Suprasanna and L.S.P. Tran), 21–45. Switzerland: Springer.

Taarit, M.B., Msaada, K., Hosni, K., and Hammami, M. (2009). Plant growth, essential oil yield and composition of sage (*Salvia officinalis* L.) fruits cultivated under salt stress conditions. *Ind. Crop. Prod.* 30: 333–337.

Tanveer, M., Shaukat, R., Ali, M. and Pirdad, F. (2020). An overview of salinity tolerance mechanism in plants. In: Hasanuzzaman, M., Tanveer, M. (Eds.) *Salt and Drought Stress Tolerance in Plants*. Springer. Cham. pp. 1–16.

Tiwari, S. and Prasad, S.M. (2020). Regulation of insecticide toxicity by kinetin in two paddy field cyanobacteria: physiological and biochemical assessment. *Environ. Pollut.* 259: 113806.

Tiwari, S., Verma, N., Singh, V.P., and Prasad, S.M. (2019). Nitric oxide ameliorates aluminium toxicity in *Anabaena* PCC7120: regulation of aluminium accumulation, exopolysaccharides secretion, photosynthesis and oxidative stress markers. *Environ. Exp. Bot.* 161: 218–227.

Tiwari, S., Gupta, D., Fatima, A., et al. (2020). Phytohormonal metabolic engineering for abiotic stress in plants: New avenues and future prospects. In: Tripathi, D.K., Singh V.P., Chauhan D.,et al. (Eds) *Plant Life Under Changing Environment*. Academic Press. pp. 543–576.

Tuna, A.L., Kaya, C., Dikilitas, M., and Higgs, D. (2008). The combined effects of gibberellic acid and salinity on some antioxidant enzyme activities, plant growth parameters and nutritional status in maize plants. *Environ. Exp. Bot.* 62: 1–9.

Uddin, M. (2019). Environmental factors on secondary metabolism of medicinal plants. *Acta Scientific Pharm. Sci.* 3: 34–46.

Vaughan, M.M., Christensen, S., and Schmelz, E.A. (2015). Accumulation of terpenoid phytoalexins in maize roots is associated with drought tolerance. *Plant Cell Environ.* 38: 2195–2207.

Verma, E., Singh, S., and Mishra, A.K. (2019). Salinity-induced oxidative stress-mediated change in fatty acids composition of cyanobacterium *Synechococcus* sp. PCC7942. *Int. J. Environ. Sci. Technol.* 16: 875–886.

Verma, N., Tiwari, S., Singh, V.P., and Prasad, S.M. (2020). Nitric oxide in plants: an ancient molecule with new tasks. *Plant Growth Regul.* 90: 1–13.

Viacava, F., Santana-Gálvez, J., and Heredia-Olea, E. (2020). Sequential application of postharvest wounding stress and extrusion as an innovative tool to increase the concentration of free and bound phenolics in carrots. *Food Chem.* 307: 125551.

Vílchez, C., Forján, E., Cuaresma, M. et al. (2011). Marine carotenoids: biological functions and commercial applications. *Mar. Drugs* 9: 319–333.

Walter, M.H. and Strack, D. (2011). Carotenoids and their cleavage products: biosynthesis and functions. *Nat. Prod. Rep.* 28: 663–692.

Wang, Y., Chantreau, M., Sibout, R., and Hawkins, S. (2013). Plant cell wall lignification and monolignol metabolism. *Front. Plant Sci.* 4: 220.

Welker, M. and Döhren, H. (2006). Cyanobacterial peptides – nature's own combinatorial biosynthesis. *FEMS Microbiol. Rev.* 30: 530–563.

Xiong, J., Yang, Y., Fu, G., and Tao, L. (2015). Novel roles of hydrogen peroxide (H_2O_2) in regulating pectin synthesis and demethylesterification in the cell wall of rice (*Oryza sativa*) root tips. *New Phytol.* 206: 118–126.

Zhang, C. and Hong, K. (2020). Production of terpenoids by synthetic biology approaches. *Front. Bioeng. Biotechnol.* 8: 347.

Zhang, J., Liu, H., Sun, J. et al. (2012). Arabidopsis fattyacid desaturase FAD2 is required for salt tolerance during seed germination and early seedling growth. *PloS One* 7(1), e30355.

Zou, P., Lu, X., Zhao, H. et al. (2019). Polysaccharides derived from the brown algae *Lessonia nigrescens* enhance salt stress tolerance to wheat seedlings by enhancing the antioxidant system and modulating intracellular ion concentration. *Front. Plant Sci.* 10: 48.

Index

Physiology of Salt stress in Plants: Perception, Signalling, Omics and Tolerance Mechanism, First Edition.
Edited by Pratibha Singh, Madhulika Singh, Rajiv Kumar Singh, and Sheo Mohan Prasad.